LIFE SPAN:
EVOLUTIONARY, ECOLOGICAL,
AND DEMOGRAPHIC
PERSPECTIVES

LIFE SPAN: EVOLUTIONARY, ECOLOGICAL, AND DEMOGRAPHIC PERSPECTIVES

James R. Carey
Shripad Tuljapurkar
Editors

POPULATION AND DEVELOPMENT REVIEW

A Supplement to Volume 29, 2003

POPULATION COUNCIL
New York

Library of Congress Cataloging-in-Publication Data

Life span : evolutionary, ecological, and demographic perspectives / James R. Carey,
Shripad Tuljapurkar, editors.

    p. cm.

Includes bibliographical references.

ISBN 0-87834-111-0 (pbk. : alk. paper)

1. Life spans (Biology) 2. Longevity. 3. Population biology. 4. Demography. I. Carey,
James R. II. Tuljapurkar, Shripad, 1951-

QP85 .L583 2003

612.6'8—dc21   2003042903

ISSN 0098-7921
ISBN 0-87834-111-0

CONTENTS

This volume is based on a series of papers presented at the workshop "Life span: Evolutionary, ecological, and demographic perspectives," held 14–18 May 2001 at the Petros M. Nomikos Conference Centre on the Greek island of Santorini. The specific aims of the Santorini workshop were threefold: first, to bring together leading scientists from demography, evolutionary biology, and field ecology to integrate and synthesize concepts in gerontology, demography, ecology, and evolution as they pertain to aging and longevity at the level of the organism; second, to foster greater interaction between biologists and demographers, including the development of collaborative research on the biodemography of aging and longevity; third, to assemble a set of papers written to be accessible to a wide audience and designed to have lasting value to readers. To this end the chapters in this volume have been substantially developed beyond the conference presentations and rigorously reviewed.

The volume is concerned with the biodemography of life span, a subject that has recently emerged at the confluence of demography and biology. Over the twentieth century, human life span measured as period life expectancy has increased by more than 50 percent in most industrialized countries, driving concern about the benefits and costs of longer life and raising keen interest in whether we can maintain the past rate of increase in life span. Biodemography brings together methods, materials, theories, and analyses from both biology and demography with the aim of understanding the past and anticipating the future of life span.

Traditionally, demographers have focused on the age pattern of death within populations and on the consequences of mortality change for populations, societies, and economies. Biologists have studied aging in terms of its ultimate (i.e., evolutionary) and proximate (i.e., mechanistic and genetic) causes within species and in terms of variability within and between populations and species. To set the stage, the first chapter, by James Carey, presents a broad overview of these viewpoints and their contribution to understanding the nature and determinants of life span. Carey also discusses the role of social evolution as a potential accelerator of increase in life span.

The thirteen chapters that follow fall in four broad groups: those concerned primarily with biological mechanisms, observations, or theories of aging; those that develop new theories of aging that are not biologically based but use economic or statistical arguments; those that analyze old-age mortality among humans; and, finally, a synthesis.

The first group begins with a theoretical overview by Steven Orzack. He uses an explicitly evolutionary approach, based on allele frequencies and demographic theory, to examine the significance of environmental variability as a selective force that can shape life span in a number of significant ways. Jean-

Michel Gaillard and his coauthors present and analyze data on the age pattern of mortality in several species of large herbivorous mammals, providing evidence of senescence in populations in the wild and of sex differences in mortality patterns and life span. Marc Mangel describes progress toward a more proximate theory of mortality, showing how oxidative damage and repair can be used in an optimization model to predict mortality patterns in fish and potentially in other species.

These chapters challenge the classical evolutionary theory of senescence: that natural selection must decline with age and cease after the age of last reproduction. However, some empirical support for the classical theory is provided here by Robert Ricklefs and Alex Scheuerlein, who present data on phylogenetic variation in mortality and life span among wild and captive birds and mammals. Even so, they point out that the classical arguments need to be sharpened in terms of specific general or particular mechanisms if we are to properly understand and predict age patterns of mortality.

A different, mechanistic, dimension of biological work on aging is found in Lawrence Harshman's review of our rapidly expanding knowledge of the genetic determinants of life span in the fruit fly *Drosophila melanogaster*. In the final chapter of this first group, Shiro Horiuchi uses the general notion of differential selection to formulate demographic models that allow him to estimate biologically derived heterogeneity in mortality. He estimates and compares such variability in populations of humans and five other species.

The evolution of human life spans may require that we move away from strictly biological methods and paradigms to new theories about the determinants of aging and of adaptation in response to aging. Hillard Kaplan and his coauthors take a large step away from the classical view of biological evolution, building an economically based optimization model for the evolution of life span. In a related vein, Ronald Lee and Joshua Goldstein consider how the human life cycle may adapt to increases in life span. They use scaling ideas, similar to those employed in biology, to study the temporal extension of different parts of the life cycle in the perspective of life cycle consumption.

The third group of chapters considers an important question in biodemography: the age pattern of mortality among the oldest humans. This question became prominent when experiments on medflies revealed a fundamentally new pattern of mortality change: a slowdown in the rate of mortality increase with age. In search of the expected human parallel, Jean-Marie Robine and Yasuhiko Saito, and John Wilmoth and Robine, report here on the nature and implications of accurate data on human mortality at old ages, especially among centenarians. Their data add substantial support for the existence of a similar slowdown of mortality increase in humans. In a separate chapter, Robine explores the idea that late-age human mortality is shaped by plasticity, the response of similar genotypes to environmental differences, rather than by selection. A rather different explanation is offered by James Vaupel, who views postreproductive life as an epiphenomenon: natural selection cannot directly influence late-age mortality, which must therefore be shaped by a different force that he takes to be the pattern of reliability of a complex system.

The complexity of the phenomena and theories presented here points to a major task for biodemography: the integration of detailed work on proximate mechanisms with a broader general picture applicable across species. A start on a synthesis is made here by Kenneth Wachter. He focuses on the age pattern of mortality as a key variable, and shows how the various approaches described here contribute to our ability to explain and predict this age pattern. He argues that biodemographic understanding of mortality should inform us about the likely challenges of achieving continued progress in extending human life spans.

Shripad Tuljapurkar

James R. Carey

ACKNOWLEDGMENTS

We express our appreciation to staff at the Petros M. Nomikos Conference Centre for allowing us to use their beautiful facilities, for their efficiency throughout the workshop, and mostly for their warm hospitality.

We gratefully acknowledge financial support for the conference from the Center for the Economics and Demography of Aging at the University of California at Berkeley and the National Institute on Aging.

This publication was supported by major contributions from the Center for the Economics and Demography of Aging at the University of California at Berkeley, the College of Agricultural and Environmental Sciences at the University of California at Davis, and the Program on Population, Policy and Aging at the Terry Sanford Institute of Public Policy at Duke University, with additional funding provided by Stanford University. In particular we acknowledge support for the publication from National Institute on Aging grant P30-AG12839 awarded to Berkeley's Center for the Economics and Demography of Aging and from National Institute on Aging grant AG08761 awarded to Duke University; James W. Vaupel is the director of the program of research supported by the latter grant.

We thank Monique Verrier for technical assistance. We thank the editorial staff of *Population and Development Review* for their assistance in producing this volume.

J.R.C.

S.T.

Life Span:
A Conceptual Overview

JAMES R. CAREY

Life span is an evolved life-history characteristic of an organism that refers to the duration of its life course. Application of the concept is straightforward at both individual and cohort levels: it specifies the period between birth and death for the individual and the average length of life or life expectancy at birth for cohorts (including both real and synthetic cohorts). When life span is applied to a population or a species, however, it requires a modifier to avoid ambiguity (Goldwasser 2001). Maximum observed life span is the highest verified age at death, possibly limited to a particular population, cohort, or species in a specified time period. The overall highest verified age for a species is called its record life span. The theoretical highest attainable age is known as maximum potential life span.

Maximum observed life spans (i.e., longevity records) are not synonymous with theoretical maximums for at least two reasons. First, maximum longevity is an inappropriate general concept because an animal dies before the age of infinity. It does so not because it cannot pass some boundary age but because the probability of its avoiding the ever-present risk of death for that long is infinitesimally small (Gavrilov and Gavrilova 1991). In other words, there is no specific identifiable age for each species to which some select individuals can survive but beyond which none can live. Second, the record age of a species is heavily influenced by the number of individuals observed. That is, the longevity records for a species in which the life spans of large numbers of individuals have been observed will be significantly greater than the corresponding figure for a species with the same longevity but represented by a few dozen individuals. For the vast majority of longevity records by species, the number of individuals at risk is unknown.

Abstract perspectives on life span

Life span should be considered within a larger biological framework concerned with life duration—a model where the end points of life for an individual do not necessarily correspond to the conventional events of birth

1

and death. The concept of life span does not apply to bacteria that reproduce by binary fission, to plant species that reproduce by cloning, or to modular organisms with iterated growth such as coral or honeybee colonies. The state of existence for an individual refers to one of several possible states of living such as "normal state," metabolic reduction (e.g., hibernation), or arrest (e.g., frozen embryo; dormancy); its time of existence refers to the sum total of all stages of the individual's life up to its extinction. What constitutes an individual is conditional on the level of individualism. For example, the individual cell within a multicellular organism and the individual worker bee in a colony of honeybees are individuals on one level but are part of a greater whole on another. Death of a component part (e.g., through apoptosis) may or may not determine the death of the whole, although in each case the components (cells; worker bees) relinquish both their evolutionary heritage and their autonomy in favor of the higher organisms (or in the case of honey bees, the colony as "super-organism") of which they are a part (Michod 1999).

The distinction between mortal macroorganisms and immortal microorganisms is unclear with certain organisms. For example, slime molds such as *Dictostelium discoideum* feed most of the time as millions of individual, potentially immortal single-celled amoeba in the forest soil and litter. Under certain conditions, however, these millions form aggregate "swarms" in dense, coordinated, visible, mortal masses that look slug-like. These aggregates then sprout into a "flower" to create and disseminate spores (Bonner 1967; Murchie 1967). A different life span issue concerns the concept of replacement of constituent parts (i.e., cells). The hydra, *Hydra littoralis*, replaces all of its cells over the course of a few weeks and thus can potentially live forever (Campbell 1967).

When a single reproductive event that occurs at the end of the life course results in the death of the individual, then life span is linked deterministically to the species' natural history. This occurs with the seed set of annual plants (grasses), in drone (male) honeybees as a consequence of the mechanical damage caused by mating, in many mayfly species when a female's abdomen ruptures to release her eggs after she drops into a lake or stream, and in anadromous salmon that die shortly after spawning. Life span can be considered indeterminate for species (including humans) that are capable of repeated (iteroparous) reproduction.[1] This concept is consistent with everything that is known about the lack of physiological cutoff points in biology: all evidence suggests that species do not have an internal clock for terminating life.

Death and extinction

Death is the unique and singular event that ends life. Dying is a graded process in virtually all organisms starting with substages of infirmity (weakness) and morbidity (sickliness) and progressing to coma, breath stoppage, heart stoppage, brain death, cessation of metabolism, tissue degradation, and loss of DNA integrity. Because of this progression of death by stages, it is useful to distinguish between the death of the organism as a whole, which refers to somatic or organismic death (the organism as an integrated functional unit), and the death of the whole organism, which refers to death of the various tissues and cells (Kass 1983). Whereas the death of the organism as a whole is usually an event—an abrupt crossing-over from the live to the dead state—the death of different parts of the whole organism is a protracted process of degradation.

In broader biological contexts, the ways in which individuals can become extinct (as opposed to extinction by dying) as singular entities are through fission in bacteria where one individual becomes two or through fragmentation as in fungi where the filaments of the original fungal growth break into many pieces, each of which is essentially a seed for new growth. A remarkable example is splitting in flatworms where the original individual can be divided into parts, each being capable of growing into a new worm. In each of these cases, the original individual no longer exists; no part or division has precedence over the other.

End points and perpetuity

Mortal refers to individuals destined to die, immortal refers to individuals living forever (however, it also is used in the context of immortal germ lines), finitude (or finite) refers to limits, infinitude (or infinity) to anything without limits, determinate refers to a fixed process or program, and indeterminate refers to an open-ended process or rate. The term immortalization describes cell cultures that can be propagated generation after generation with no loss of viability (Blackburn 2000) such as the HeLa cell line (the first human cells to live indefinitely outside the body), which was derived from and named after cancer victim Henrietta Lack's cells in the 1940s (Murchie 1967; Skloot 2001), and immortalized B lymphocytes, which are used to produce virtually unlimited quantities of identical antibody molecules (Köhler and Milstein 1975).[2] The term preservation refers to maintenance in the existing state. The bacteria trapped in 25–30-million-year-old amber (Cano and Borucki 1995) or in a 250-million-year-old primary salt crystal (Vreeland et al. 2000) and the mitochondrial DNA isolated from a 29,000-year-old Neanderthal (Ovchinnkov et al. 2000) are all examples of preservation. Cryopreservation is a special type of preservation using ultra-low temperatures (–196° C) to preserve human sperm and embryos (Silver 1998; Trounson 1986). Dormancy describes any one of a number of evolved types of arrested growth ranging from torpor to hibernation (Carson and Stalker 1947; Hairston and Caceres 1996; Mansingh 1971; Masaki 1980).

Evolution

Evolutionary origins of senescence

Bell (1988) established the deep connection between the two invariants of life—birth and death—by demonstrating that protozoan lineages senesce as the result of an accumulated load of mutations. The senescence can be arrested by recombination of micronuclear DNA with that of another protozoan through conjugation. Conjugation (sexual reproduction) results in new DNA and in the apoptotic-like destruction of old operational DNA in the macronucleus. Thus, rejuvenation in the replicative DNA and senescence of operational DNA are promoted by sexual reproduction.

When this discussion is extended to multicellular organisms, sexual reproduction and somatic senescence are inextricably linked (Clark 1996). In multicellular sexually reproducing organisms, the function of somatic cells (i.e., all cells constituting the individual besides the germ cells) is survival, and the function of the replicative DNA—the germ cells—is reproduction. Before the arrival of bacteria, the somatic DNA was the germ line DNA; prior to multicellular animals, the somatic cell and the germ cell were undifferentiated. Like the macronuclei in the paramecia, the somatic cells senesce and die as a function of their mitotic task of ensuring the survival and development of the germ cells. The advent of sexual reproduction allowed internal repair of replicative DNA (Clark 1996).

Life span as an evolutionary adaptation

In evolutionary biology an "adaptation" is a characteristic of organisms whose properties are the result of selection in a particular functional context. Just as different bird beaks are adaptations for exploiting different niches that must be balanced with the other traits such as body size and flight propensity, the longevity of an animal is also an adaptation that must be balanced with other traits, particularly with reproduction. The variations in the relationship between reproduction and longevity only make sense when placed within the context of such factors as demographic circumstances, duration of the infantile period, number of young, and the species' ecological niche—the organism's overall life-history strategy. Inasmuch as life spans differ by 5,000-fold in insects (2-day mayflies to 30-year termite queens), by 100-fold in mammals (2-year mice to 200-year whales), and by 15-fold in birds (4-year songbirds to 60-year albatrosses), it is clear that life span is a life-history adaptation that is part of the larger life history of each species.

Evolutionary ecology of life span

The literature on aging and longevity describes only a small number of broad life span correlates,[3] including the well-known relationship between longevity and both body mass and relative brain size, and the observation that animals that possess armor (e.g., beetles; turtles) or capability of flight (e.g., birds; bats) are often long-lived. But major inconsistencies exist within even this small set of correlates (Carey and Judge 2000, 2001). For example, there are several exceptions to the correlation between extended longevity and large body size. This positive relationship may be either absent or reversed within certain orders, including a negative correlation within the Pinnipeds (seals and walruses) and no correlation within the Chiroptera (bats). Likewise, the observation that flight ability and extended longevity are correlated does not provide any insight into why within-group differences in life span exist (e.g., among birds), nor does it account for the variation in longevity in insects where adults of the majority of species can fly.

A classification system regarding the life span correlates of species with extended longevity has been developed and applied to a wide range of invertebrate and vertebrate species (Carey and Judge 2001). It proposes two categories. The first, environmentally selected life spans, includes animals whose life histories evolved under conditions in which food was scarce and where resource availability was uncertain or environmental conditions were predictably adverse at times.[4] Some of the longest-lived small and medium-sized mammals (gerbils, rock hyrax, and feral asses) live in deserts where rainfall, and thus reproduction, are episodic and unpredictable.

The second category, socially selected life spans, includes species that exhibit extensive parental investment, extensive parental care, and eusociality (social strategy of, e.g., ants, bees, wasps, and termites, which have overlapping generations, cooperative care of young, and a reproductive division of labor). The extended longevity of animals in this category results from natural, sexual, and kin selection and involves all of the social primates, including humans.

Roles of the elderly

The long life span of many nonsocial species is an adaptation designed to increase the likelihood of population replacement by enabling the species to survive through extended periods that are unfavorable to reproduction; to acquire scarce resources over a long period; or to repeatedly produce and invest heavily in singleton offspring. The elderly in social species play important roles in the cohesion and dynamics of groups, populations, and communities. In species such as elephants, whales, and primates, the elderly serve as caregivers, guardians, leaders, teachers, sexual consorts, and midwives (Carey and Gruenfelder 1997). Older individuals in human populations may provide economic and emotional support; assist with child care, nursing, counseling, and homemaking; teach skills and crafts; conserve and transmit the traditions and values of culture groups; serve as valued voices in communal council; and provide historical continuity for younger members (Neugarten 1996).

Minimal life spans

Understanding life span evolution requires an understanding of the conditions in which natural selection favors a particular life span, especially one that is extremely long or extremely short. Thus studies of short-lived species are nearly as important as studies of long-lived species; both models can be used to test theories of aging and they provide comparative context. Studies of short-lived species also bring into focus the evolutionary bottom line of reproduction—like any trait, a particular life span coevolved with other life-history factors and is thus only part of an organism's mosaic of adapted characteristics. One example of a "longevity minimalist" is the mayfly, whose life history has several generic properties that favor the evolution of a short life span, including (Carey 2002): (1) synchronous emergence to produce a concentration of adults so that the likelihood of finding a mate is maximized (this can be generalized as minimal mating requirements, including parthenogenesis); (2) inability to feed owing to vestigial mouthparts, which obviates the need to forage for food (this trait can be generalized as minimal food requirement); (3) general nymphal habitat, which requires females to travel only a short distance (i.e., minimal energy requirements) to deposit their eggs; and (4) a single, large reproductive bout that reduces the risk that some eggs are not deposited before death.

Absence of evidence for life span limits

Although there has been considerable debate whether the human life span (and life span in general) is fixed, the evidence now appears to support the notion that no single number can be assigned as the life span limit of a species, particularly a species with iteroparous reproduction (as distinct from semelparous species like salmon where individuals die soon after reproducing). The reliability theory of aging predicts a deceleration in mortality late in life, with subsequent leveling off as well as mortality plateaus late in life, as inevitable consequences of the loss of redundancy of components necessary for life (Gavrilov and Gavrilova 2001b). A deceleration of mortality at advanced ages is inconsistent with the concept of a specific limit to life span. The evidence can be outlined as follows. The results of large-scale life table studies in virtually all species that have been examined reveal a deceleration in mortality at older ages (Clark and Guadalupe 1995; Vaupel et al. 1998). If a single life span limit existed for all species, a pattern of records of the long-lived should begin to appear for a least some of the species for which longevity information is available. However, this is not the case: there is no clustering of longevity records from independent sources for thousands of species for which life span data are available (Carey and Judge 2000). Moreover, the concept of a fixed species-specific life span is itself misleading because of issues such as species strains, biotypes, sex, hybrids, and seasonal variants. For example, no single life span exists for honeybees (*Apis melifera*) even in theory since queens live 4 to 6 years, workers 6 weeks, and drones from a few days to several weeks (Wilson 1971). A similar problem exists for the domestic dog (*Canis familiaris*) with its scores of breeds each with different life spans (Patronek et al. 1997). Finally, the human life span has been increasing for over a century as revealed by the records of extreme ages at death in Sweden for the last 130 years (Wilmoth et al. 2000).

Humans

Life span patterns: Humans as primates

Estimates based on regressions of anthropoid primate subfamilies or limited to extant apes indicate that a major increase in longevity between *Homo habilis* (52–56 years) and *H. erectus* (60–63 years) occurred 1.7 to 2 million years ago (Judge and Carey 2000). The predicted life span for small-bodied *H. sapiens* is 66–72 years. From a catarrhine (Old World monkeys and apes) comparison group, a life span of 91 years is predicted when contemporary human data are excluded from the predictive equation. For early hominids to live as long as or longer than predicted was probably extremely rare; the important point is that the basic Old World primate design resulted in an organism with the potential to survive long beyond a contemporary mother's ability to give birth. This suggests that postmenopausal survival is not an artifact of modern life style but may have originated between 1 and 2 million years ago coincident with the radiation of hominids out of Africa.[5]

Sex differentials in life span

The general biodemographic principle to emerge from recent studies on male–female differences in life span is not that the female life span advantage is a universal "law" of nature.[6] Rather the deeper principle is that the mortality response and, in turn, the life spans of the two sexes will always be different in similar environments. This is because the physiology, morphology, and behavior of males and females are different (Glucksmann 1974). Men have heavier bones, muscles, hearts, lungs, salivary glands, kidneys, and gonads in proportion to body weight, while women have proportionately heavier brains, livers, spleens, adrenals, thymus, stomach, and fat deposits. The basal metabolic rate of women is lower than that of men. Men have more red corpuscles and hemoglobin per unit volume of blood than women. Sex differences are reflected in the rates of growth and maturation, metabolism, endocrine activity, blood formation, immune responses, disease patterns, behavior, and psychology. Males are generally more susceptible to reduction in food (Clutton-Brock 1994). Sexual dimorphism was highly pronounced in our *Australopithicus* ancestors, suggesting that sexual selection via competition between males had a strong influence on male–female differences in our progenitors. The height differences persist as a residual of our ancestry. In short, the mortality response of males and females in humans and virtually all other animals differs because of differences in their physiology, reproductive biology, and behavior (Carey et al. 1995).

Age classification

Age is as much a constituent of any living organism as is its physical self; its existence is identical with its duration. The time an organism lives is measured, on the one hand by hours, days, years, and decades according to chronological (solar) time that flows evenly, and, on the other hand, by rhythms and irreversible changes according to a physiological time scale that flows unevenly (Carrel 1931). Just as drying a rotifer can stop the stream of its duration or raising or lowering the temperature can increase or decrease the rate of living of *Drosophila*, freezing a human embryo arrests its development. In all cases the change in metabolism changes the rate of flow of physiological time and thus changes the chronological duration.

Age and time can be considered in a number of dimensions for humans (Neugarten and Datan 1973) including historical time, which shapes the social system and, in turn, creates changing sets of age norms and a changing age-grade system that shapes the individual life cycle; social time, which refers to the dimension that underlies the age-grade system of a society (e.g., rites of passage); psychological time, which refers to the perceived increase in speed at which time progresses as a person grows older because each year that passes becomes a progressively smaller fraction of the accumulated years (Carrel 1931); and life time, which refers to the chronological age as a series of orderly events from infancy through childhood, adolescence, maturity, and old age. Demographers and sociologists have created new age categories and terms for the life times of seniors and elderly (Neugarten 1974). Thus persons classified as young-old, old, and oldest-old are considered in the age categories 55–65, 66–85, and over 85 years, respectively. The terms centenarian and supercentenarian refer to persons in their 100s, and 110s, respectively, although decacentenarian and dodecacentenarian have recently been suggested for persons in their 110s and 120s (Allard et al. 1998). In general, distinctions among older people by age classifications are useful for understanding the human life cycle, as a framework for inter- and intragroup comparisons, and for health planners concerned with social and health characteristics of populations.

The contemporary human life span of over 120 years based on the maximal age recorded may be considered as consisting of two segments (Judge and Carey 2000): (1) the Darwinian or "evolved" segment of 72–90 years; and (2) the post-Darwinian segment, which is the artifactual component that exceeds expectations from a primate of human body and brain size. The argument that human life span has not changed in 100,000 years can be considered substantially correct when the "evolved" life span is considered. It is clear, however, that this is not correct when the nonevolved segment of human life span is considered. There is evidence from Swedish death registers that the record age in humans has been increasing for well over a century (Wilmoth et al. 2000).

Jeune (1995) suggested that most of the literature on centenarians is based on the hypothesis of a secret of longevity, which is summarized in four assertions, each of which is debatable: maximum longevity is fixed, longevity is genetically determined, centenarians always existed, and centenarians are qualitatively different. The first assertion is inconsistent with mortality patterns in both humans and nonhuman species. The second assertion is inconsistent with estimates of the degree of heritability of longevity, which are between 20 percent and 30 percent.[7] The third assertion is improbable if one examines the estimated mortality levels that persisted before 1800. The last assertion is tautological and thus unverifiable since the criteria for being different (i.e., being a centenarian) cannot be tested any other way than actually living to at least 100 years. Vaupel and Jeune (1995) claimed that supercentenarians did not exist before 1950 and centenarians did not exist before 1800 in any population.

The oldest verified age to which a human being has ever lived is 122 years and 167 days, a record held by the Frenchwoman Jeanne Calment, who was born on 21 February 1875 and died on 4 August 1997 (Allard et al. 1998). Other verified long-livers include Sarah Knauss, Mare Meiller, Chris Mortinson, and Charlotte Hughs, who lived 118.1, 117.6, 115.8, and 115.6 years (Robine and Vaupel 2001). Madame Calment and the four other long-livers all died between 1993 and 1999. Thus the five oldest persons whose ages of death have been verified died within the last decade.

Theory of longevity extension in social species: A self-reinforcing process

Improved health and increased longevity in societies may set in motion a self-perpetuating system of longevity extension (Carey and Judge 2001). This positive feedback relationship may be one reason why human life span is continuing to increase. This hypothesis is based on the demographic tenet that increased survival from birth to sexual maturity reduces the number of children desired by parents, ceteris paribus. Because of the reduced drain of childbearing and rearing, parents with fewer children remain healthier longer and raise healthier children with higher survival rates; this, in turn, fosters further reductions in fertility. Greater longevity of parents also increases the likelihood that they can contribute as grandparents to the fitness of their children and grandchildren. Thus does the self-reinforcing cycle continue. Whereas the positive correlation between health and income per capita is very well known in international development, the health and income correlation is partly explained by a causal link running the other way—from health to income (Bloom and Canning 2000). In other words, productivity, education, investment in physical capital, and the "demographic dividend" (positive changes in birth and death rates) are all self-reinforcing: these factors can contribute to health, and better health (and greater longevity) contributes to their improvements.

The future of human life span

Various approaches such as extrapolation, relational, cause-of-death, and stochastic models have been used to make near- and medium-term mortality forecasts (Ahlburg and Land 1992; Keyfitz 1982; Olshansky 1988; Tuljapurkar et al. 2000).). For example, Tuljapurkar and colleagues (2000) used stochastic models to predict that life expectancies in 2050 will range from 80 to 83 years in the United States and from 83 to 91 years in Japan. Oeppen and Vaupel (2002) used "best-practice" (i.e., highest worldwide life expectancy) models to forecast that life expectancy of American females in 2070 will range from 92.5 to 101.5 years and used a rapidity of mortality improvement model to predict that life expectancy of Japanese females in 2070 will range between 87.6 and 94.2 years. These forecasts are to be distinguished from those based on expected medical discoveries or risk of mass starvation.

The future of human life span should also be considered in the context of possibility—the constructive, scenario-building aspects of science akin to the conceptual arts (Gill 1986). Whereas predictions of changes in life expectancy are usually considered to point to precise results, the recognition of life span possibilities includes the effects on longevity of scientific and medical breakthroughs. These multiple demographic alternatives are not based on rates of change in life expectancy per se, but rather on elements that will alter the rate of change itself.

Integrating concepts related to different life span possibilities into research and policy planning is important for several reasons. First, integration will establish closer connections between the biological discoveries on the nature of aging and will indicate how these discoveries might affect individuals and society. Biologists and demographers seldom exchange ideas in this context. Second, considering possibilities may suggest different analytical models in which different types of scientific breakthroughs could be included in the predictions. For example, a breakthrough in therapeutic cloning would have sweeping implications for organ replacement, potentially saving many lives. Third, considering future possibilities for human life span will provide a foundation for policymakers to consider medical, economic, and political contingencies in which life span increases are much greater than predicted by standard methods. Exploring possibilities based on potential scientific breakthroughs will provide policymakers with a wide variety of scenarios for the demographic future of both developed and developing countries.

Scientific and biomedical determinants

Developments in four areas of science and public health will likely determine the biological future of human life span. First, healthful living—elimination of vices (smoking; drugs; alcohol abuse), proper diet, exercise, mental stimulation, and social support (Fries 1980; Verbrugge 1990; Willett 1994). Healthful living is framed around the concept of "successful aging," which consists of three components: avoiding disease and disability, maintaining mental and physical function, and continuing engagement with life (Rowe and Kahn 1998). Adherence to healthful living probably accounts for the substantial reductions in mortality over the last several decades and will serve as the foundation for further reductions in the future. Second, disease prevention and cure—developing treatments and cures for debilitating diseases such as arthritis (Persidis 2000a), cancer (Greaves 2000; Pardoll 1998), cardiovascular diseases, stroke, and autoimmune diseases (Heilman and Baltimore 1998). New fields that will revolutionize medicine include molecular and systems biology (Idelker et al. 2001); molecular cardiovascular medicine; gene therapy (Pfeifer and Verma 2001); pharmacogenomics, where the drugmaker tailors a therapy to the individual's genetic needs; and nanomedicine, which involves the monitoring, repair, construction, and control of human biological systems at the molecular levels. Third, organ replacement and repair—tissue engineering (Colton 1999; Persidis 2000b), including skin equivalents and tissue-engineered bone, blood vessels, liver, muscle, and nerve conduits. Significant research and development are underway into xenotransplantation (organs from other animal species into humans) (Cooper and Lanza 2000; Persidis 2000c; Steele and Auchincloss 1995), stem cell research (Blau et al. 2001), and therapeutic cloning (Colman 1999; Yang and Tian 2000). Fourth, aging arrest and rejuvenation—identifying aging processes in humans that can be studied in model organisms (Guarente and Kenyon 2000) and developing therapeutic interventions (Hadley et al. 2001). Research on caloric restriction (CR) is designed to identify a CR-mimetic (molecular mimic) with a known mechanism of action that produces effects on life span and aging similar to those of CR (Finch and Ruvkun 2001). This would implicate this mechanism as a likely mediator of CR's effects (Lee et al. 1999; Masoro and Austad 1996; Sohal and Weindruch 1996; Weindruch 1996). Recent studies on ovarian transplants in mice (Cargill et al. 2003) reveal a second method by which the rate of aging in mammals can be reduced. Reducing the rate of actuarial aging would increase longevity and life span far more than would improvements from any other single mechanism.

Demographic ontogeny

The concept of adaptive demography that Wilson (1977) developed in the context of social insect evolution at both individual and societal levels applies equally to humans. Future demographic changes that will likely occur at the individual level in humans include the following. (1) Appearance of new extreme ages. Changes in record ages are important, not because the presence of these record-holders changes society in any substantive way, but because they are the harbingers of the future and the extreme manifestations of improved health. Just as the first appearance of nonagenarians and centenarians was probably in the nineteenth century and of super-centenarians in the twentieth century, someone may break the current age record of 122 years in the twenty-first century. (2) Emergence of new functional and healthy age classes. Using equivalencies of age-specific mortality as proxies for similarities in health and functionality, the 60- and 70-year-olds of today have the same mortality risk as the 45- and 61-year-olds in 1900 (Bell et al. 1992). In other words, the frail elderly of yesterday are the more robust elderly of today. Large numbers of individuals are staying healthier longer and therefore are functional at more advanced ages. (3) Life cycle and event-history modification. As life span and life expectancy increase, people plan their lives differently and thus change the timing, sequence, duration, and spacing of key events including marriage and child-bearing, education, working life, and retirement (Elder 1985; O'Rand and Krecker 1990; Settersten and Mayer 1997).

     Changes at the societal level will occur as the result of changes in event histories at the level of the individual. The family life cycle will probably continue to quicken as marriage, parenthood, empty nest, and grandparent-hood all occur earlier, partly because of a decreasing number of children. A new rhythm of social maturity would impinge upon other aspects of family life in subtle ways including the age of parents, which affects the degree of authoritarianism; and the relative youth of parents and grandparents, which in turn affects the patterns of health assistance, flows of money in both directions, and child care (Neugarten and Datan 1973). The life cycle may become more fluid with an ever-increasing number of role transitions, the disappearance of traditional timetables, and the continued lack of synchrony among age-related roles. New generational constellations may emerge including four- and even five-generation families and new contexts for sorting out needs and resources of the family. The age at which full retirement benefits start is likely to continue to rise in developed countries, to the mid-70s in the near term (Drucker 2001) and to the late 70s, early 80s, and beyond as life span increases. Manton and Gu (2001) note that to maintain the stock of human capital available to the US economy, strategies will have to be adopted that preserve human capital and keep it in the labor force for longer periods of time. Highly skilled and educated older people, instead of being retired, may be offered continuing professional relationships that preserve their skill and knowledge for the enterprise while giving them flexibility. A model for this already in use comes from academia: the professor emeritus (Drucker 2001). Education may become more of a seamless web in which primary and secondary education, undergraduate and graduate training, professional schooling and apprenticeships, internships and on-the-job training, and continuing education and lifelong learning and enrichment become a continuum (Duderstadt 1999, 2000). Just as the presence of stages and their duration in the life cycle are considered adaptations and thus of evolutionary value in ecological contexts (Bogin 2000; Bonner 1965), the evolution and emergence of new stages at the end of the human life course can also be considered adaptations in a biocultural context; the life-history properties of every age class in the human life course shape and, in turn, are shaped by every other age class.

In short, adding new age-class extremes, expanding the duration of healthy living, and reconfiguring the event-history schedule create new and unpredictable societal dynamics. Although the fate of the human species is impossible to predict far into the future, molecular biologist Lee Silver's (1998) perspective on scientific and technological possibilities is probably the most insightful: look to the 4-billion-year creative history of life on Earth for ideas, most of which are locked within the genetic code of all living creatures. We have just begun the "century of biology" (Idelker et al. 2001), and sooner or later some of the deepest and most profound secrets of life will be unlocked. It is likely that some of these secrets will involve the keys to extending life span.

Postscript

Life span has a special character quite apart from mortality and longevity in that the concept extends beyond measures of biological efficiency and statistical expectations. It has multiple characteristics including many etymological, semantic, and scientific nuances and also signals ways of thinking about the evolution, ecology, and life course of humans. On one level the concept of life span is decoupled from death because it is explicitly a longevity rather than a mortality concept; it implies living rather than dying. However, on another level the concept is inextricably linked to death because it delimits an age that extends beyond what is attainable by virtually the entire population: it implies life's ultimate age of closure.

Lovejoy (1981) noted that the social bonds, intelligence, learning ability, and intensive parenting that contributed to reducing mortality levels in primates both required and contributed to a long life span in higher primates. Thus the most distinctive qualities of the human species, including our language, ability to teach and learn, culture, and, most importantly, our humanity could not have emerged were it not for the tribal elders, the shamans, the seers, and the grandmothers. We impede evolution and the development of the human species if we neglect the aged because, as Hillman (1999) notes, it is the elderly who sheltered and promoted our civilization in the past and it is they who will continue to do so in the future.

Notes

I thank Debra Judge for comments on earlier drafts. The research on which this chapter is based was supported by the National Institute on Aging (P01-08761).

1 Indeterminate life span means a life span that is not precisely fixed. It does not imply immortality.

2 Although certain "transformed" or "immortalized" cell cultures do exhibit indefinite replicative potential (a more accurate term than the oft-used "immortalization"), it is the population of cells that appears to have indefinite replicative potential; individual cells often undergo cell death or exit the mitotic cell cycle via terminal differentiation.

3 There are, however, a number of detailed life-history correlates such as age at first breeding, age at weaning, number of litters, and metabolic rate (See Tables A and B in Gavrilov and Gavrilova 1991).

4 Several authors have argued that predation is one of the primary determinants of life span (Austad 1993; Reznick 2001). Multiple confounding factors in each investigation, however, allow for alternative arguments to the predation hypothesis (e.g., differences in physical environments). Other reasons why predation should not be considered a stand-alone determinant of longevity are: (1) a more likely evolutionary response to intense predation involves selection for adaptations that increase predator avoidance (e.g., cryptic coloration; Futuyma 1998); (2) predators constitute only one of three categories of biological risk factors that organisms face in nature, the other two being parasites and disease. Thus a theory of longevity change based on intensity of predation is incomplete both biologically and actuarially.

5 There is an ongoing debate about whether menopause is an evolved trait in both nonhuman species (Packer et al. 1998) and humans (Diamond 1996; Hawkes et al. 1998).

6 The two most common explanations supporting the notion that males (putatively) have shorter life spans are: (1) behavioral aspects where males of many species are at higher risk than females owing to different life-history requirements such as mate finding and territory defense (Zuk et al. 1990); and (2) the chromosomal hypothesis where, it is suggested, females have an advantage because in most species they are the homogametic sex (XX) whereas males are the heterogametic sex (XY). It is argued that having two X chromosomes is advantageous because the X chromosome is three times the size of the Y chromosome and contains far more expressed genetic information, most of which is for functions and molecules unrelated to the female geneotype (Greenwood and Adams 1987; Smith and Warner 1989).

7 This statement requires the caveat that, although the heritability estimates for total life span studied for the entire range of ages may indeed be small, an exceptional longevity may have a much higher genetic component (Gavrilov and Gavrilova 2001a).

References

Ahlburg, D. A. and K. C. Land. 1992. "Population forecasting: Guest editors' introduction," *International Journal of Forecasting* 8: 289–299.

Allard, M., V. Lebre, and J.-M. Robine. 1998. *Jeanne Calment: From van Gogh's Time to Ours. 122 Extraordinary Years.* New York: W. H. Freeman and Company.

Austad S. N. 1993. "Retarded senescence in an insular population of Virginia opossums (*Didelphis virginiana*)," *Journal of Zoology* 229: 695–708.

Bell, F. C., A. H. Wade, and S. C. Goss. 1992. *Life Tables for the United States Social Security Area 1900–2080.* US Department of Health and Human Services, Social Security Administration, Office of the Actuary, Washington, DC.

Bell, G. 1988. *Sex and Death in Protozoa.* Cambridge: Cambridge University Press.

Blackburn, E. H. 2000. "Telomere states and cell fates," *Nature* 408: 53–56.

Blau, H. M., T. R. Brazelton, and J. M. Weimann. 2001. "The evolving concept of stem cell: Entity or function?," *Cell* 105: 829–841.

Bloom, D. E. and D. Canning. 2000. "The health and wealth of nations," *Science* 287: 1207–1209.

Bogin, B. 2000. "Evolution of the human life cycle," in S. Stinson, B. Bogin, R. Huss-Ashmore, and D. O'Rourke (eds.), *Human Biology: An Evolutionary and Biocultural Perspective.* New York: Wiley-LISS, pp. 377–424.

Bonner, J. T. 1965. *Size and Cycle.* Princeton: Princeton University Press.

———. 1967. *The Cellular Slime Molds.* Princeton: Princeton University Press.

Campbell, R. D. 1967. "Tissue dynamics of steady state growth in *Hydra littoralis*," *Journal of Morphology* 121: 19–28.

Cano, R. J. and M. Borucki. 1995. "Revival and identification of bacterial spores in 25 to 40 million year old Dominican amber," *Science* 268: 1060–1064.

Carey, J. R. 2002. "Longevity minimalists: Life table studies of two northern Michigan adult mayflies," *Experimental Gerontology* 37: 567–570.

Carey, J. R. and C. A. Gruenfelder. 1997. "Population biology of the elderly," in K. Wachter and C. Finch (eds.), *Biodemography of Longevity.* Washington, DC: National Academy Press, pp. 127–160.

Carey, J. R. and D. S. Judge. 2000. *Longevity Records: Life Spans of Mammals, Birds, Reptiles, Amphibians and Fishes.* Vol. 8, Odense, Denmark: Odense University Press.

———. 2001. "Life span extension in humans is self-reinforcing: A general theory of longevity," *Population and Development Review* 27: 411–436.

Carey, J. R., P. Liedo, D. Orozco, M. Tatar. and J. W. Vaupel. 1995. "A male-female longevity paradox in medfly cohorts," *Journal of Animal Ecology* 64: 107–116.

Cargill, S., J. Carey, G. Anderson and H.-G. Müller. 2003. "Young ovaries extend life expectancy in old ovariectomized mice," submitted.

Carrel, A. 1931. "Physiological time," *Science* 74: 618–621.

Carson, H. L. and H. D. Stalker. 1947. "Reproductive diapause in *Drosophila robusta*," *Proceedings of the National Academy of Sciences* 34: 124–129.

Clark, A. G. and R. N. Guadalupe. 1995. "Probing the evolution of senescence in *Drosophila melanogaster* with P-element tagging," *Genetica* 96: 225–234.

Clark, W. R. 1996. *Sex and the Origins of Death.* New York: Oxford University Press.

Clutton-Brock, T. H. 1994. "The costs of sex," in R. V. Short and E. Balaban (eds.), *The Differences Between the Sexes.* Cambridge: Cambridge University Press, pp. 347–362.

Colman, A. 1999. "Somatic cell nuclear transfer in mammals: Progress and application," *Cloning* 1: 185–200.

Colton, C. K. 1999. "Engineering a bioartificial kidney," *Nature Biotechnology* 17: 421–422.

Cooper, D. K. C. and R. P. Lanza. 2000. *Xeno: The Promise of Transplanting Animal Organs into Humans.* New York: Oxford University Press.

Diamond, J. 1996. "Why women change," *Discover Magazine* 17: 129–137.

Drucker, P. 2001. "The next society," *The Economist* 361: 3–20.

Duderstadt, J. J. 1999. "The twenty-first century university: A tale of two futures," in W. Z. Hirsch and L. E. Weber (eds.), *Challenges Facing Higher Education at the Millennium.* Phoenix: The Oryx Press, pp. 37–55.

———. 2000. Fire, ready, aim! University decision making during an era of rapid change. The Blion Colloquium II, La Jolla, California.

Elder, G. H. J. 1985. "Perspectives on the life course," in G. H. F. Elder (ed.), *Life Course Dynamics.* Ithaca: Cornell University Press, pp. 23–49.

Finch, C. E. and G. Ruvkun. 2001. "The genetics of aging," *Annual Review of Genomics and Human Genetics* 2: 435–462.

Fries, J. F. 1980. "Aging, natural death, and the compression of morbidity," *The New England Journal of Medicine* 303: 130–135.

Futuyma, D. 1998. *Evolutionary Biology.* 3rd Ed. Sunderland, Massachusetts: Sinauer Associates, Inc.

Gavrilov, L. and N. Gavrilova. 1991. *The Biology of Life Span.* Chur, Switzerland: Harwood Academic Publishers.

———. 2001a. "Biodemographic study of familial determinants of human longevity," *Population: An English Selection* 13: 197–222.

———. 2001b. "The reliability theory of aging and longevity," *Journal of Theoretical Biology* 213: 527–545.

Gill, S. P. 1986. "The paradox of prediction," *Daedalus* 115: 17–48.

Glucksmann, A. 1974. "Sexual dimorphism in mammals," *Biological Reviews* 49: 423–475.

Goldwasser, L. 2001. "The biodemography of life span: Resources, allocation and metabolism," *Trends in Ecology and Evolution* 16: 536–538.

Greaves, M. 2000. *Cancer: The Evolutionary Legacy.* London: Oxford University Press.

Greenwood, P. J. and J. Adams. 1987. *The Ecology of Sex.* London: Edward Arnold.

Guarente, L. and C. Kenyon. 2000. "Genetic pathways that regulate ageing in model organisms," *Nature* 408: 255–262.

Hadley, E. C., C. Dutta, J. Finkelstein, T. B. Harris, M. A. Lane, G. S. Roth, S. S. Sherman, and P. E. Starke-Reed. 2001. "Human implications of caloric restriction's effects on aging in laboratory animals: An overview of opportunities for research," *Journal of Gerontology: Series A* 56A (Special Issue): 5–6.

Hairston, N. G. J. and C. E. Caceres. 1996. "Distribution of crustacean diapause: Micro- and macroevolutionary pattern and process," *Hydrobiologia* 320: 27–44.

Hawkes, K., J. O'Connell, N. G. B. Jones, H. Alvarez, and E. L. Charnov. 1998. "Grandmothering, menopause, and the evolution of human life histories," *Proceedings of the National Academy of Sciences, USA* 95: 1336–1339.

Heilman, C. A. and D. Baltimore. 1998. "HIV vaccines-where are we going?," *Nature Medicine* 4: 532–534.

Hillman, J. 1999. *The Force of Character.* New York: Ballantine Books.

Idelker, T., T. Galitski, and L. Hood. 2001. "A new approach to decoding life: Systems biology," *Annual Review of Genomics and Human Genetics* 2: 343–372.

Jeune, B. 1995. "In search of the first centenarians," in B. Jeune and J. W. Vaupel (eds.), *Exceptional Longevity: From Prehistory to the Present.* Odense: Odense University Press, pp. 11–24.

Judge, D. S. and J. R. Carey. 2000. "Post-reproductive life predicted by primate patterns," *Journal of Gerontology: Biological Sciences* 55A: B201–B209.

Kass, L. R. 1983. "The case for mortality," *American Scholar* 52: 173–191.

Keyfitz, N. 1982. "Choice of function for mortality analysis: Effective forecasting depends on a minimum parameter representation," *Theoretical Population Biology* 21: 329–352.

Köhler, G. and C. Milstein. 1975. "Continuous cultures of fused cells secreting antibody of predefined specificity," *Nature* 256: 495–497.

Lee, C.-K., R. G. Klopp, R. Weindruch, and T. A. Prolla. 1999. "Gene expression profile of aging and its retardation by caloric restriction," *Science* 285: 1390–1393.

Lovejoy, O. 1981. "The origin of man," *Science* 211: 341–350.

Mansingh, A. 1971. "Physiological classification of dormancies in insects," *The Canadian Entomologist* 103: 983–1009.

Manton, K. G. and X. Gu. 2001. "Changes in the prevalence of chronic disability in the United States black and nonblack population above age 65 from 1982 to 1999," *Proceedings of the National Academy of Sciences* 98: 6354–6359.

Masaki, S. 1980. "Summer diapause," *Annual Review of Entomology* 25: 1–25.

Masoro, E. J. and S. N. Austad. 1996. "The evolution of the antiaging action of dietary restriction: A hypothesis," *Journal of Gerontology: Biological Sciences* 51A: B387–B391.

Michod, R. E. 1999. "Individuality, immortality, and sex," in L. Keller (ed.), *Levels of Selection in Evolution.* Princeton: Princeton University Press, pp. 53–74.

Miller, J. D. 1997. "Reproduction in sea turtles," in P. L. Lutz and J. A. Musick (eds.), *The Biology of Sea Turtles.* Boca Raton: CRC Press, pp 51–81.

Moyle, P. B. and B. Herbold. 1987. "Life-history patterns and community structure in stream fishes of Western North America: Comparisons with Eastern North America and Europe," in W. J. Matthews and D. C. Heins (eds.), *Community and Evolutionary Ecology of North American Stream Fishes.* Norman, Oklahoma: University of Oklahoma Press, pp. 25–34.

Murchie, R. 1967. *The Seven Mysteries of Life.* New York: Houghton Mifflin Company.

Neugarten, B. 1974. "Age groups in American society and the rise of the young-old," *Annals of the American Academy of Political and Social Science* 415: 187–198.

Neugarten, B. 1996. "Family and community support systems," paper presented to Committee #7, White House Conference on Aging, Washington, DC, November 1981, in D. Neugarten (ed.), *Edited Papers of Bernice Neugarten,* Chicago: University of Chicago Press, pp. 355–376.

Neugarten, B. and N. Datan. 1973. "Sociological perspectives on the life cycle," in P. B. Baltes and K. W. Schaie (eds.), *Life-Span Development Psychology: Personality and Socialization.* New York: Academic Press, pp. 53–79.

Oeppen, J. and J. W. Vaupel. 2002. "Broken limits to life expectancy," *Science* 296: 1029–1030.

Olshansky, S. J. 1988. "On forecasting mortality," *The Milbank Quarterly* 66: 482–530.

O'Rand, A. M. and M. L. Krecker. 1990. "Concepts of the life cycle: Their history, meanings, and uses in the social sciences," *Annual Review of Sociology* 16: 241–262.

Ovchinnkov, I. V., A. Gotherstrom, G. P. Romanova, V. M. Kharitonov, K. Liden, and W. Goodwin. 2000. "Molecular analysis of Neanderthal DNA from the northern Caucasus," *Nature* 404: 490–493.

Packer, C., M. Tatar, and A. Collins. 1998. "Reproductive cessation in female mammals," *Nature* 392: 807–811.

Pardoll, D. M. 1998. "Cancer vaccines," *Nature Medicine* 4: 525–531.

Patronek, G. J., D. J. Waters, and L. T. Glickman. 1997. "Comparative longevity of pet dogs and humans: Implications for gerontology research," *Journal of Gerontology* 52A: B171–B178.

Persidis, A. 2000a. "Cancer multidrug resistance," *Nature Biotechnology* 18: IT18–IT20.

———. 2000b. "Tissue engineering," *Nature Biotechnology* 18: IT56–IT58.

———. 2000c. "Xenotransplantation," *Nature Biotechnology* 18: IT53–IT55.

Pfeifer, A. and I. M. Verma. 2001. "Gene therapy: Promises and problems," *Annual Review of Genomics and Human Genetics* 2: 177–211.

Reznick, D., M. J. Butler IV, and H. Rodd. 2001. "Life-history evolution in guppies. VII. The comparative ecology of high- and low-predation environments," *The American Naturalist* 157: 126–140.

Riedman, M. 1990. *The Pinnipeds.* Berkeley: University of California Press.

Robine, J.-M. and J. W. Vaupel. 2001. "Supercentenarians: Slower ageing individuals or senile elderly?," *Experimental Gerontology* 36: 915–930.

Rowe, J. W. and R. L. Kahn. 1998. *Successful Aging.* New York: Pantheon Books.

Settersten, R. A. J. and K. U. Mayer. 1997. "The measurement of age, age structuring, and the life course," *Annual Review of Sociology* 23: 233–261.

Silver, L. 1998. *Remaking Eden.* New York: Avon Books.

Skloot, R. 2001. "Cells that save lives are a mother's legacy," *New York Times*, pp. A21, A23, New York.

Smith, D. W. E. and H. R. Warner. 1989. "Does genotypic sex have a direct effect on longevity?," *Experimental Gerontology* 24: 277–288.

Sohal, R. S. and R. Weindruch. 1996. "Oxidative stress, caloric restriction, and aging," *Science* 273: 59–63.

Steele, D. J. R. and H. Auchincloss. 1995. "Xenotransplantation," *Annual Review of Medicine* 46: 345–360.

Trounson, A. 1986. "Preservation of human eggs and embryos," *Fertility and Sterility* 46: 1–12.

Tuljapurkar, S., N. Li, and C. Boe. 2000. "A universal pattern of mortality decline in the G7 countries," *Nature* 405: 789–792.

Vaupel, J. W., J. R. Carey, K. Christensen, T. E. Johnson, A. I. Yashin, N. V. Holm, I. A. Iachine, V. Kannisto, A. A. Khazaeli, P. Liedo, V. D. Longo, Y. Zeng, K. G. Manton, and J. W. Curtsinger. 1998. "Biodemographic trajectories of longevity," *Science* 280: 855–860.

Vaupel, J. W. and B. Jeune. 1995. "The emergence and proliferation of centenarians," in B. Jeune and J. W. Vaupel (eds.), *Exceptional Longevity: From Prehistory to the Present.* Odense: Odense University Press, pp. 109–116.

Verbrugge, L. M. 1990. "Pathways of health and death," in R. D. Apple (ed.), *Women, Health, and Medicine in America: A Historical Handbook,* New York: Garland.

Vreeland, R. H., W. D. Rosenzweig, and D. W. Powers. 2000. "Isolation of a 250 million-year-old halotolerant bacterium from a primary salt crystal," *Nature* 407: 897–900.

Weindruch, R. 1996. "Caloric restriction and aging," *Scientific American* 274: 46–52.

Willett, W. C. 1994. "Diet and health: What should we eat?," *Science* 264: 532–537.

Wilmoth, J. R., L. J. Deegan, H. Lundstrom, and S. Horiuchi. 2000. "Increase of maximum life-span in Sweden, 1861–1999," *Science* 289: 2366–2368.

Wilson, E. O. 1971. *The Insect Societies.* Cambridge, MA: Harvard University Press.

———. 1977. *Sociobiology: The New Synthesis.* Cambridge, MA: Harvard University Press.

Yang, X. and X. C. Tian. 2000. "Cloning adult animals—What is the genetic age of the clones?," *Cloning* 2: 123–128.

Zuk, M., R. Thornhill, J. D. Ligon, K. Johnson, S. Austad, S. H. Ligon, N. W. Thornhill, and C. Costin. 1990. "The role of male ornaments and courtship behavior in female mate choice of red jungle fowl," *The American Naturalist* 136: 459–473.

How and Why Do Aging
and Life Span Evolve?

STEVEN HECHT ORZACK

"Why do we get old?" and "How long will we live?" are questions that have been posed and answered in a myriad of ways since ancient times. Despite this legacy of attention, there has been little agreement as to what are the most meaningful answers to these questions. In the scientific realm, this lack of resolution has spurred on many to work ever harder, and in the last decade or so there has been a veritable explosion of new claims and studies about aging and life span. These dwarf past efforts, so much so as almost to obscure them.

What sense is one to make of this explosion of new information? In this chapter, I provide an overview of recent claims about aging and life span so as to order the conceptual landscape and to provide a context for understanding what we have and have not learned. My primary aim is to encourage new directions for research.

Aging was recently defined by Kirkwood and Austad (2000: 233) as "the progressive loss of function accompanied by decreasing fertility and increasing mortality with advancing age." This kind of definition, with its focus on ever-increasing debility and ever-decreasing fertility and survival rates, is controversial, especially since new data on the oldest age classes in humans and several other species indicate that survival rates may stop decreasing at advanced ages (Vaupel et al. 1998). Nonetheless, Kirkwood and Austad's definition, with its description of organismic decline, captures the essence of much of the biology that generates interest and motivates attempts to provide scientific explanation.

Naively, aging is hard to explain given a Darwinian view of the world. After all, why should deleterious characteristics appear in the lifetime of an organism if evolution is concerned with producing what is best? This apparent Darwinian paradox has been resolved in a number of ways by biologists.

One of the proposed solutions is the "rate of living hypothesis": animals with higher metabolic rates have shorter life spans as a result (Rubner 1908; Pearl 1928). This claim is supported by the overall positive association between metabolic rate (as measured via the proxy of body size) and average life span for many organisms, but especially for vertebrates. However, groups of vertebrates exhibit substantial enough differences to cast doubt on this hypothesis as being the sole explanation of aging and life span differences. For example, it does not explain why many species of birds have substantially higher metabolic rates *and* longer life spans than similarly sized mammals (Holmes and Austad 1995). Nonetheless, the overall positive association between metabolic rate and life span is substantial enough to suggest that metabolic differences are partially responsible for the life span differences observed among major groups of taxa.

One possible mechanistic basis for the association between metabolic rate and life span is that a higher metabolic rate generates a higher level of chemical byproducts that have deleterious consequences on the somatic cells of the body. In particular, free radicals and other highly reactive compounds such as hydrogen peroxide are generated by enzymatic reactions especially in the mitochondria; these compounds can cause damage to DNA as well as damage to proteins. An extensive literature exists on this phenomenon (see Stadtman 1992 and Finkel and Holbrook 2000 for further discussion).

Another proposed solution to the Darwinian paradox presented by aging is the "mutation accumulation hypothesis": the force of natural selection is too weak to oppose the accumulation of mutations with late deleterious effects (Medawar 1952). The basis for this claim is that random mortality unrelated to age has usually reduced later age groups to such small sizes that the power of natural selection to eliminate deleterious mutations is dramatically weakened. In effect, natural selection cannot distinguish later in life between the signal of deleterious mutations and the signal of beneficial mutations. This explanation posits a limitation on the power of natural selection to alter phenotypes.

A third proposed solution for the evolutionary paradox does not posit such a limitation. Instead, the "antagonistic pleiotropy hypothesis" posits the presence of pleiotropic mutations (those with multiple effects) that can increase, say, fertility early in life and decrease survival or fertility later in life. According to this hypothesis, it is the expression of the later deleterious effects of pleiotropic mutations that we term aging. This idea was first fully elaborated by Williams (1957). The beneficial effect of a mutation can outweigh the later deleterious effect, all other things being equal (Hamilton 1966). The reason is that it is almost always evolutionarily beneficial to increase one's contribution of offspring to the population earlier rather than later; the reason is essentially the principle of compound interest, which makes it more beneficial to put money in the bank today as opposed to tomorrow. For the antagonistic pleiotropy hypothesis, instead of a limitation on selection there is a limitation on mutation and selection in that the positive and negative consequences of mutations are regarded as inseparable.

A later refinement of Williams's idea, one that makes explicit the possible physiological mechanisms underlying antagonistic pleiotropy, is the "disposable soma" hypothesis advanced by Kirkwood and Holliday (1979). Their claim is that it is selectively advantageous for an organism to save energy by reducing "proofreading" in somatic cells as it gets older. The positive consequence is that reproduction is enhanced since more energy is allocated toward reproduction; the negative consequence is the deterioration of the soma because of a lack of scrupulous biochemical maintenance. Kirkwood and Holliday's mechanistic claim is consistent with the inferences and observations that cells are in need of proofreading because proteins and DNA are subject to more-or-less constant "challenges" owing to the presence of free radicals within cells (see above). They go further by claiming that the pattern of aging we observe in most species is the result of an optimal balance between the energy allocated to reproduction and the energy allocated to somatic maintenance. This is a strong claim, as it implies that the pattern of aging observed in a given species is the consequence of a perfect balance between reproduction and somatic maintenance. Such a balance cannot be improved upon by natural selection, given the basic constraints on the organism engendered by such "fixed" aspects of the species as body size, mating biology, and feeding habits; these fixed aspects create the overall context for evolutionary change. Although there is evidence to support the claim that aging is at least partially the result of antagonistic pleiotropy (see below), there is no evidence that aging is associated with an optimal balance of energy. In this respect, the disposable soma hypothesis is not supported; however, the data supporting antagonistic pleiotropy may be consistent with a weak form of the disposable soma hypothesis, one in which natural selection is the main determinant of an organism's energy balance. Such a claim for the role of natural selection is quite distinct from a claim about optimality (Orzack and Sober 1994).

Additional hypotheses have been proposed as evolutionary explanations for aging, but those I have outlined are the most widely recognized at present.

What is the truth? Do experimental data indicate an accumulation of mutations that affect older age groups, as the mutation accumulation theory predicts? Do experimental data provide clear associations between positive effects on reproduction and survival early in life and negative effects later in life, as the antagonistic pleiotropy theory predicts? Most recent experimental studies on aging and life span have involved model organisms, especially the nematode worm *Caenorhabditis elegans*, the medfly *Ceratitis capitata*, the fruitfly *Drosophila melanogaster*, and the mouse *Mus musculus*. Each of these species has virtues and limitations as a model system. For the most part, experiments focus primarily on either demography or genetics. The former type of experiment now often involves extensive manipulation of large cohorts of individuals so as to exactly characterize age- and sex-specific changes in mortality and fertility rates (e.g., Carey et al. 1992). Genetic analyses have largely involved either an examination of the effects of new mutations on the life history of the organism (e.g., Hughes and Charlesworth 1994) or an examination of the effects on one part of the life history of genetically changing another part of the life history. So, in the latter case an investigator might carry out an experiment in which only longer-lived flies are allowed to contribute offspring to the next generation, the intent being to determine whether such selection results in longer-lived flies that have reduced reproduction early in life (Luckinbill et al. 1984). The motivation is the prediction of the antagonistic pleiotropy theory that mutations that cause longer life also unavoidably cause reduced early reproduction because energy allocated to survival must reduce energy allocated to reproduction (and vice versa). Alternatively, if the mutation accumulation theory is correct, selection of mutations for longer life would not result in decreased early reproduction, since such mutations would not tend to have such a pleiotropic effect.

Data in some experiments support the mutation accumulation hypothesis (e.g., Partridge and Fowler 1992; Charlesworth and Hughes 1996). Other experiments support the antagonistic pleiotropy hypothesis in that extending life span does reduce early fertility (e.g., Luckinbill et al. 1984; Zwaan, Bijlsma, and Hoekstra 1995), although it appears that such negative associations can be evolutionarily modifiable, implying that they are not a result of inevitable energetic tradeoffs (Leroi, Chippindale, and Rose 1994). For the most part, there appears to be more support for the antagonistic pleiotropy hypothesis since many experiments provide at least partial evidence for a negative association between survival and fertility rates. Such an association is not universally observed, however (see Le Bourg 2001 for discussion). In addition, a negative association between survival and fertility does not rule out mutation accumulation as also contributing to aging. Direct tests of the latter hypothesis have not yielded clear evidence that it is broadly applicable. Nonetheless, in this context a negative result, one in which there is no accumulation of mutations with age, is often thought to be less compelling than a positive result, one in which there is such an accumulation. Since mutation rates are small, it is quite possible that a typical experiment, say, one lasting at most a few dozen generations and involving a few thousand individuals, might fail to detect an increase in genetic variation resulting from the accumulation of deleterious mutations.

Rightly or wrongly, the ensemble of results just described causes many researchers to believe that the evolution of aging often involves antagonistic pleiotropy. Part of this support stems from the suggestive (but not universal) evidence for such effects. An additional reason is a general commitment to energetic tradeoffs as being central to life-history evolution. For many investigators (e.g., Roff 1992), a negative relationship between, say, early reproduction and late reproduction or survival *must* underlie evolutionary change of life histories. Otherwise, the reasoning goes, all individuals would reproduce indefinitely; in other words, these investigators believe there must be a "cost of reproduction." A corollary of this attitude is that tradeoffs are "there to be found" and that a lack of evidence for them is simply a result of a lack of experimental power. While this logic is undoubtedly correct in some instances, it is not clear that it can or should serve as an explanatory principle for all facets of life-history evolution. The reasons are simple. If one views the data at face value and ignores expectations of what "must be," one must take seriously the fact that empirical support for the existence of tradeoffs is mixed (cf. Roff 1992). In addition, one can formulate evolutionary models of life-history phenomena such as aging and life span in which important aspects of the dynamics are not determined by tradeoffs (see below).

It is essential to understand what the present body of theoretical and experimental work on aging and life span does and does not include in order to fully appreciate its meaning as well as its limitations. I first discuss the theoretical work.

Theoretical work on aging and life span

Present theoretical work invokes natural selection as the primary (or only) force affecting aging and life span; consequently, little or no role is posited for nonadaptive forces such as genetic drift (evolutionary change engendered by the finite size of populations, which causes random sampling of gametes and individuals; see Wright 1931) or phylogenetic inertia (the presence of traits that have evolved in an ancestral environment or species; see Orzack and Sober 2001 and references therein). More specifically, observed aging and life span patterns are assumed to be the result of natural selection only *within* populations; it is often explicitly claimed that evolutionary processes operating at the level of the population or species could not play a substantial role, if any (e.g., see Williams 1957 and Charlesworth 1994).

Second, all of these theories are deterministic in that the underlying demographic analysis does not explicitly account for temporal variation in vital rates within the lifetime of an individual. Changes in, say, the survival rate of newborn individuals from one time unit or from one census to the next can have a dramatic effect on life-history evolution and population dynamics (see below and Tuljapurkar 1989). Changes of this kind are to be distinguished from changes in the vital rates observed over the longer term that are due to the waxing and waning of different life histories as evolution proceeds. In and of itself, the assumption that the environment is deterministic in the sense described above is not necessarily problematic; such an assumption is arguably an appropriate starting point for many evolutionary analyses. What *is* potentially quite problematic, however, are claims that characterize evolutionary responses to environmental variation even though the underlying analysis does not mathematically account for such variation. For example, Kirkwood and Austad (2000: 235) write, "Many organisms live their lives in highly variable environments. In such circumstances we can expect…the co-adapted set of traits influencing survival and fecundity to possess a degree of evolved plasticity that permits a range of optimal responses suited to different circumstances." While this claim about evolution in a variable environment is ambiguous enough to be plausible, it lacks an explicit theoretical underpinning and should not be viewed as meaningful until it is based on analyses that incorporate stochasticity. Why would an omission of such analyses be a concern? Apart from specific reasons apparent from a stochastic analysis (see below), there is no general reason why the average behavior of a stochastic model should even crudely match the "average" behavior of the associated deterministic model (see Taylor 1992 for examples of agreement and disagreement).

Third, present theories implicitly assume that aging and life span evolve independently of one another. So, for example, the antagonistic pleiotropy theory predicts how aging evolves but it does so only given an arbitrarily fixed life span. While this independent evolution is plausible in some circumstances, other scenarios are also plausible. For example, a larger ensemble of life-history traits (such as vital rates *and* life span) might coevolve with one another, with none preceding the evolution of the other. It is also possible that life span evolution is the primary determinant of the evolution of aging. The overall point is that there is a need for more exploration of the overall demographic context for the joint evolution of aging and life span.

The particularity of present theories for the evolution of aging and life span is neither unusual as compared to many other evolutionary theories nor a necessary indication of their weakness. After all, particularity of assumptions does not necessarily generate particularity of predictions; predictions may not be highly assumption-dependent. But this is not self-evident and, accordingly, it is important to expand our theoretical outlook on the evolution of aging and life span. Such an effort will lead us to a greater understanding, whether it be one in which the predictions of present theory are revealed to be robust or one in which present theory is seen to be just one part of an expanded scientific explanation of aging and life span evolution.

The construction of additional theories for the evolution of aging and life span can be best motivated by several questions that highlight important aspects of the biology in question.

The first question is "What is the neutral theory of aging and life span?" Such a theory would make predictions about aging and life span without resorting to natural selection as the determinative agent. Instead, patterns of aging and life span evolution would be attributed to nonselective agents, such as genetic drift. The development of this kind of neutral theory has been extremely important in other areas of evolutionary biology, most notably in the study of molecular evolution (e.g., Kimura 1983). The main reason is that competing hypotheses improve the analysis, simply because the need to decide among hypotheses generally requires more detailed data. The benefit of multiple hypotheses is further accentuated when causal and noncausal hypotheses compete in evolutionary biology. Yet, as described above, present theory for the evolution of aging and life span relies almost exclusively on natural selection within populations as the agent of evolutionary change. One reason is the notion that vital rates are the very objects that define evolutionary success since they concern changes in numbers of individuals. So, the reasoning goes, natural selection must always govern life-history evolution. However, the same claim could be made about other life-history traits—such as the ratio of sons and daughters produced by families—that are thought to sometimes be selectively neutral (Kolman 1960). Accordingly, it is important to understand to what extent the evolution of aging and life span can occur independent of the action of natural selection. I outline below some examples of this kind of evolution.

The second question is "What is the stochastic theory of aging and life span?" As noted above, the use of deterministic theories to describe evolutionary dynamics in variable environments can be misleading. What then does an "honest" stochastic theory of aging and life span look like? Does it lead to new ways of understanding aging and life span evolution? Does it change the insights gained from deterministic theory?

The third question is "Do population dynamics play a role in the evolution of aging and life span?" Here, the term "population dynamics" refers to changes in the size of populations over the long term. Most present theory about aging and life span has the common population-genetic assumption that evolutionary change can be described solely in terms of change in allele and genotype *frequencies*. The underlying population *numbers* are not needed. Yet, changes in population size driven by evolutionary change within populations could affect life histories differentially depending on their associated aging patterns and life span. Accordingly, there is a need for a theory of aging and life span evolution in which population dynamics are explicitly incorporated.

New directions

All of the questions outlined above can be addressed either partially or fully by an analytical framework that has been extensively developed in the last 20 years or so (see Tuljapurkar 1989 for a general introduction). In this framework, growth of a population in a temporally variable environment is modeled as a random-matrix product

$$N_{t+1} = X_{t+1} N_t = X_{t+1} X_t \ldots X_1 N_0 \tag{1}$$

where N_t denotes a vector whose ith element represents the number of individuals of the ith life-history stage (age or some other class) at time t, and X_t denotes a projection matrix appearing at time t. The elements of each matrix specify age- or stage-specific vital rates of a life history. The matrix of vital rates appearing at any one time represents the vital rates manifested in a given environmental state. The number of such environmental states can be finite or infinite. Nonzero elements of the projection matrix have an average and a variance that are evaluated with respect to all environmental states.

This analytical framework is based upon standard concepts in classical demographic theory (Keyfitz 1977) and incorporates them into a stochastic framework that draws heavily upon the theory of random-matrix products (see Cohen et al. 1986 for mathematical background). This framework is not complete, as it does not include some plausibly important aspects of most populations. For example, it does not explicitly account for mating dynamics. Nonetheless, this framework is a substantial advance in terms of revealing important aspects of stochastic population dynamics.

One important aspect of the framework just described is that any plausible (or nonplausible) relationship between vital rates can be accounted for. Thus, the negative correlations between early and late vital rates that are central to the antagonistic pleiotropy theory can be represented as a negative relationship between the values of, say, any pair of vital rates expressed at time t or, alternatively, as a negative relationship between the values of any given vital rate or pair of vital rates over time.

A number of important results relate to the population dynamics described by the random-matrix product when the values of the vital rates are assumed to be independent of population size. Such a "density-independent" assumption is valid for many populations, although clearly not for all. Of central importance is the fact that a population has an expected asymptotic growth rate (Tuljapurkar and Orzack 1980). So, instead of there being an ever-changing growth rate because any one of possibly many different projection matrixes affects population dynamics at a particular time, the population is expected to attain a long-term growth rate that is independent of the initial population size; this stochastic growth rate is attained as long as some reasonable assumptions are satisfied about the boundedness of environmental variation and about the timing of reproduction. While this convergence occurs in the strict sense only in the long term, it means something in the short term as the realized stochastic growth rate of a population can closely approach the asymptotic stochastic growth rate after, say, a few dozen time units, depending on the length and structure of the life history (Tuljapurkar and Orzack 1980).

A second important result concerning the population dynamics described by the random-matrix product is that the long-run distribution of population size is lognormal if one ignores population extinction (Tuljapurkar and Orzack 1980). Consider populations that "should" increase because the stochastic growth rate is positive. Even so, an ensemble of identically sized starting populations will tend over time to become skewed in size such that most populations become arbitrarily small in size, while a few populations become extremely large. In formal terms, the asymptotic stochastic growth rate is the mean value of the limiting distribution of the logarithm of population size. But this mean population size is very unrepresentative of the typical population size. These results could only stem from a fully stochastic analysis, as opposed to one in which deterministic results are extrapolated to the stochastic realm. In addition, the fact that many populations decrease in size underscores the importance of studying population numbers, as opposed to simply focusing on frequencies.

What, then, does this framework reveal about the evolution of aging and life span? One central point is that a tremendous variety of life histories can have identical or very similar stochastic growth rates (Orzack and Tuljapurkar 1989; Tuljapurkar 1990; Orzack 1997). This potential for non-selective polymorphism of life histories can also be revealed by a deterministic analysis. Nonetheless, most life-history analyses have paid little attention to conditions under which very distinct life histories have identical growth rates, stochastic or otherwise (see examples below). This means that such life histories are selectively neutral with respect to each other. The implication is that genetic drift can affect the dynamics of aging and life span evolution. Natural selection need not be the first and only explanation for evolutionary differentiation of life histories. Even marked differences in the length of life or age at last reproduction among a group of closely related species could possibly not be due to natural selection. The overall point is that neither selection nor neutrality has necessary priority as an explanation for many aspects of life history evolution, including the evolution of aging and of life span.

The analytical framework described above would be of only formal use if it lacked easy applicability and if it were difficult to understand the separate and joint contributions of the demography of the organism and of the environment to the stochastic growth rate. However, the virtue of this analytical framework is that it is relatively easy to understand its structure and to apply it.

How then does the stochastic growth rate depend upon the life history and the environment? These dependencies can be determined exactly for some life histories and types of environmental fluctuations (e.g., Tuljapurkar 1990). More generally, for any suitable life history and environment, one can construct a "small-noise" approximation to the stochastic growth rate (Tuljapurkar 1982). This approximation can be highly accurate (Orzack and Tuljapurkar 1989; Orzack 1997) even though it is composed only of the contributions of the average life history to growth, the one-period variances and covariances of vital rates, and the two-period temporal correlations between vital rates. What is remarkable and fortunate then is that this approximation works even though it leaves out longer-term aspects of the relationship between the life history and the environment and their effects on the dynamics of population change. Nonetheless, it allows one to understand how a change in the mean or variance of a vital rate or in its covariance with another vital rate affects the stochastic growth rate. Such a correct understanding of the "functional anatomy" of the trait is as essential to evolutionary understanding in demography as it is in any other evolutionary investigation.

What does the stochastic growth rate look like?

When there is no temporal autocorrelation to environmental states, that is, when environmental states occur independently of one another from one time unit to the next, the stochastic growth rate is

$$a \approx \ln \lambda_0 - \frac{1}{2\lambda_0^2} \left(\sum_{\alpha}^{\omega} \left(\frac{\partial \lambda_0}{\partial v_i} \right)^2 \sigma_{v_i}^2 + \sum_{\alpha}^{\omega} \sum_{\alpha}^{\omega} \left(\frac{\partial \lambda_0}{\partial v_i} \right) \left(\frac{\partial \lambda_0}{\partial v_j} \right) \text{cov}(v_i, v_j) \right) \qquad i \neq j \qquad (2)$$

where λ_0 is the dominant eigenvalue of the average projection matrix, $\partial \lambda_0 / \partial v_i$ describes the effect on the eigenvalue of changing any vital rate v_i, $\sigma_{v_i}^2$ is the variance of v_i, and $\text{cov}(v_i, v_j)$ is the covariance of vital rates v_i and v_j. All of these quantities are readily computable given the set of matrixes containing the realized values of the vital rates in all environmental states.

The form of this growth rate approximation reflects the dynamics; it is different from the form derived from an analysis based on statics. The latter can give qualitatively and quantitatively incorrect insights into evolution in stochastic environments (Tuljapurkar 1989; Orzack 1993). In addition, the fact that the average demography as well as environmental variation contributes to the stochastic growth rate indicates that predictions about the evolution of aging and life span must account for both of these contributions; at present, predictions, say, that aging should be more marked when the life span is short (Charlesworth 1994) are based only on an analysis of the average demography.

Consider how the application of this analytical framework might reveal insights into life span evolution. One question to be asked in this context is how the temporal dispersion of reproduction evolves in a variable environment. One way to answer this question is to constrain all life histories being compared so that any evolutionary change stems from differences in the timing and pattern of reproduction and not in its absolute amount. One can impose this constraint by making the total amount of net reproduction, $\Sigma l_i m_i$, a constant ($= \Sigma \phi_i$, where l_i and m_i are average values of the survivorship from age 1 to age i and the fertility at age i, respectively); from here on, comparisons among life histories will be framed in terms of the ϕ_i values.

What environmental conditions allow a given set of life histories to be selectively neutral with respect to one another? Equality of stochastic growth rates defines an indifference curve (Orzack and Tuljapurkar 1989). The indifference curve for a stochastic growth rate of zero is of special significance: it reflects ecological balance in that the population has the potential in an average sense to maintain itself, but not to grow. Assume, for example, that there is an age-invariant coefficient of variation (CV) for the net fertility values and an age-invariant correlation between them. Given these assumptions it is straightforward to show that

$$ a \approx \ln \lambda_0 - \frac{1}{2T_0^2} (CV^2 (\sum_\alpha^\omega G_i^2 + \sum_\alpha^\omega \sum_\alpha^\omega G_i G_j)) \qquad i \neq j \qquad (3) $$

where T_0 is generation length, $G_i = \phi_i \lambda_0^{-i}$, and r denotes the age-invariant correlation coefficient (see Orzack and Tuljapurkar 1989 for further details). Indifference curves for the life histories presented in Table 1 are shown in Figure 1. For any given life history (set of average vital rates) and value of r, the CV value shown in Figure 1 is that which satisfies the approximate equality in equation (3). The general shape of the indifference curves is relatively insensitive to whether changes in the values of net reproduction decline or peak with age (as opposed to being constant), whether the stochastic growth rate is larger, whether the largest age at last reproduction is changed, or whether the total amount of reproduction ($\Sigma \phi_i$) is changed (Orzack and Tuljapurkar 1989). The ecological and evolutionary implications of these curves are simply understood by noting that they divide the parameter space: all CV values higher than the one that is part of the curve result in negative stochastic growth rates; conversely, all lower CV values result in positive stochastic growth rates.

What do indifference curves reveal about life span evolution? Consider life histories 1 through 6. These life histories have identical total amounts of net reproduction (see Table 1), but they differ in life span, with individuals possessing life history 6 ($\omega = 10$) having the potential to live ten times as long as individuals with life history 1 ($\omega = 1$). One can see from Figure 1 that increased life span is always advantageous. This can be inferred by comparing the magnitudes of environmental variation (as measured by the CV)

TABLE 1   A set of life histories with identical amounts of average net reproduction over the life span

Life history	ϕ_i	α	ω	$\ln\lambda_0$	T_0	D
1	1.01	1	1	0.0099	1.000	1.000
2	0.505	1	2	0.0066	1.498	0.500
3	0.2525	1	4	0.0040	2.495	0.250
4	0.16833	1	6	0.0028	3.492	0.167
5	0.12625	1	8	0.0022	4.488	0.125
6	0.101	1	10	0.0018	5.485	0.100
7	0.12625	3	10	0.0015	6.492	0.125
8	0.16833	5	10	0.0013	7.496	0.167
9	0.2525	7	10	0.0011	8.499	0.250
10	0.505	9	10	0.0010	9.500	0.500

$\phi_i = l_i m_i$. $\Sigma\phi_i$ is constant for all life histories ($= 1.01$). α is the age at first reproduction. ω is the age at last reproduction, after which individuals die. λ_0 is the dominant eigenvalue of the average projection matrix. T_0 is the generation length. D ($=\Sigma(\phi_i\lambda_0^{-i})^2$) is an index of the dispersion of reproduction throughout the life span. Values of $\ln\lambda_0$, T_0, and D are approximate.

that populations with life histories 1 through 6 can endure and still maintain a stochastic growth rate of zero. For life history 1, this value of the *CV* is approximately 0.14, regardless of the degree of correlation between stochastic fluctuations in the vital rates. In contrast, life history 6 can maintain such a growth rate even when the *CV* is as high as approximately 0.3 to 1.05, depending upon the degree of correlation. This is an instance in which a longer life span results in an evolutionary advantage. Note that positive covariation increases the realized magnitude of environmental variation (because it means that vital rate fluctuations are entrained with one another, thereby increasing the magnitude of the negative term in equation 3). The intensity of selection increases as the realized magnitude of environmental variation decreases; life histories have the most distinct growth rates when variation affecting one part of the life history is independent of variation affecting another (see the indifference curve for $r = 0.0$ in Figure 1).

The underlying reason for the advantage of longer life span is that generation length quadratically discounts the deleterious effects of environmental variability (as indicated by the inverse quadratic term containing T_0 in equation 3). For example, life history 5 ($T_0 \approx 4.488$) has an approximately 20-fold greater discounting of a given amount of variability as compared to life history 1 ($T_0 = 1.0$).

The results shown in Figure 1 also indicate that increased reproductive span can be advantageous or disadvantageous depending on the correlation between vital rate fluctuations. Consider life histories 6 through 10, which share the same age at last reproduction ($\omega = 10$) but which differ in

FIGURE 1   Indifference curves for populations with the life histories in Table 1 for a range of values of the correlation between the fluctuations of net reproduction at age i and j. The stochastic growth rate is zero. Values of the coefficient of variation are derived from equation (3).

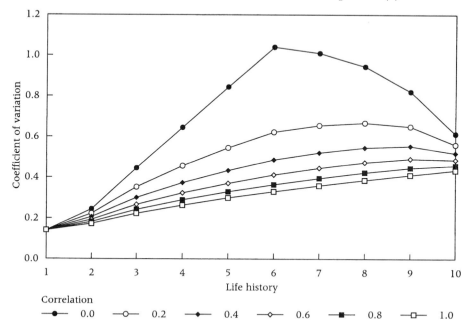

their ages at first reproduction (α). If vital rate fluctuations are independent ($r = 0$), then for a given level of environmental variation it is advantageous to disperse reproduction throughout the life span. This result can be inferred from the fact that life history 6 (maximum dispersion) can maintain a stochastic growth rate of zero in an environment ($CV \approx 1.0$) in which life histories 7 through 10 have negative growth rates. In contrast, when vital rate fluctuations are partially or completely correlated, it can be advantageous to postpone reproduction until the end of life. So, for example, when $r = 0.4$, life history 9 (with reproduction starting at age 7) has a stochastic growth rate of zero in an environment ($CV \approx 0.55$) in which all other life histories have negative growth rates.

These results reveal the potential for interplay between the evolutionary dynamics affecting life span evolution and those affecting patterns of aging. As noted above, present theory describes the evolution of aging for a life history having a given fixed length. Here, one can see how the dynamics affecting the evolution of life span and reproductive span affect the potential for the evolution of aging in a given life history. Consider life history 10, for example, which has the potential to live in environments ($r > 0.8$) when all other life histories go extinct (see Figure 1). This life history has only a limited potential for the evolution of senescence since, of course, only two late age classes reproduce and they are essentially identical in the way changing the vital rate would affect the overall growth rate ($\partial\lambda_0/\partial\phi_9 \approx$ 0.104 and $\partial\lambda_0/\partial\phi_{10} \approx 0.104$). In contrast, life history 6 has more potential for the evolution of senescence because ten age classes reproduce and because they differ more in the way changing the vital rates would affect the overall growth rate (e.g., $\partial\lambda_0/\partial\phi_1 \approx 0.182$ and $\partial\lambda_0/\partial\phi_{10} \approx 0.179$); this difference between these life histories increases as $\ln\lambda_0$ increases.

Consider another question about the evolution of senescence in variable environments that has not been addressed by classical theory. Is senescence allowed to evolve because life histories with longer life spans discount the effects of environmental variability? In lieu of an analytical framework, one might guess that senescence and delayed reproduction are both detrimental since both reduce the growth rate. One way to answer this question is to compare a shorter-lived life history without senescence and a longer-lived life history with senescence. The shorter-lived life history reproduces more quickly; the longer-lived life history discounts variability. When is one of these life histories evolutionarily superior to the other? Contrast a life history that is nonsenescent, with $\alpha = 1$, $\omega = 6$, $\phi_i =$ 0.175, $i = 1,...6$ (net reproduction is constant with age), and a life history that is senescent, with $\alpha = 5$, $\omega = 10$, $\phi_1 = \phi_2 = \phi_3 = \phi_4 = 0.0$, $\phi_5 = 0.2275$, $\phi_6 =$ 0.2065, $\phi_7 = 0.1855$, $\phi_8 = 0.1645$, $\phi_9 = 0.1435$, and $\phi_{10} = 0.1225$ (net reproduction declines after age five). For both life histories, $\Sigma\phi_i = 1.05$ and there are 6 fertile ages. For the first life history $\lambda_0 \approx 1.014$ and $T_0 \approx 3.459$, and for the second life history $\lambda_0 \approx 1.007$ and $T_0 \approx 7.131$; the difference in the value of λ_0 means that the nonsenescent life history has a higher growth rate in a constant environment.

What happens in a stochastic environment? When the coefficient of variation is age-invariant and the vital rates vary independently of each other, equation (3) indicates that the nonsenescent life history has a stochastic growth rate of zero when $CV \approx 1.418$; more environmental variability results in a negative growth rate. But equation (3) indicates that the senescent life history has a stochastic growth rate of zero when $CV \approx 1.996$; this life history can survive in a more variable environment than the nonsenescent life history can. The underlying reason is the difference between the life histories in the values of their generation lengths (T_0). In this analysis, we see that a life history exhibiting senescence and delayed reproduction can be evolutionarily superior to a life history without senescence in a shorter life span because the former is better at discounting the deleterious effects of environmental variability.

The general point of these examples is to illustrate the interaction between the dynamics of the evolution of life span and the dynamics of the evolution of senescence; neither set of dynamics has necessary precedence over the other in regard to governing life-history evolution.

There are further insights to be gained about aging and life span evolution that could come only from stochastic theory. For example, if one views aging as a consequence of a loss of homeostasis (or vice versa), one would expect that older age groups manifest increased temporal variability of vital rates as compared to younger age groups. The evolutionary consequences of this pattern of variation are unknowable with a deterministic analysis, but are readily predicted with the analytical framework described here.

All of the theoretical analyses described above have relied upon the use of the stochastic growth rate as an indicator of the evolutionary potential of a life history. There is good reason for this focus, especially the fact that analytical and numerical analyses of simple genetic dynamics indicate that the stochastic growth rate correctly predicts the fate of rare mutants having a stochastic growth rate distinct from that of the resident population (Tuljapurkar 1982; Orzack 1985). This fact implies that the stochastic growth rate is not an arbitrary construct; it has a natural dynamical meaning.

However, there are additional aspects of the evolutionary dynamics of life histories that are not revealed by these analyses of stochastic growth rates.

As noted above, an indifference curve predicts the set of life histories that are selectively equivalent to one another. This neutrality has important evolutionary implications, especially with regard to the explanation of life-history diversification among closely related populations and species. However, the set of life histories associated with an indifference curve (or any other set) can be assessed with respect to another metric of evolutionary success: extinction probability.

The extinction dynamics of age-structured life histories of the kind discussed here have been analyzed by Lande and Orzack (1988) and Orzack (1993, 1997). This work indicates that the stochastic growth rate is one of two key determinants of the transient and ultimate extinction probabilities of populations with particular life histories (the other is the age distribution of the initial population).

This work makes it possible to refine the insights gained from the analysis of stochastic growth rates described above. So, for example, the lognormality of population size described by Tuljapurkar and Orzack (1980) is seen to apply transiently but to break down eventually as populations decline in numbers, cross an extinction boundary, and cease to exist.

One can use the analytical framework described by Lande and Orzack to show that life histories with a stochastic growth rate of zero are *not* neutral in the transient sense: their extinction probabilities at finite times differ. However, their ultimate extinction probabilities are identical: all will go extinct eventually. The important point is that the differences in transient dynamics can cause differences in the prevalence of life histories with certain kinds of aging patterns and life spans.

The transient dynamics of extinction for populations with life histories 1 through 10 from Table 1 are shown in Figure 2. There are several

FIGURE 2 Transient extinction probabilities for populations with the life histories shown in Table 1. The stochastic growth rate is zero and the extinction boundary is a total population size of one individual. The initial population is assumed to comprise ten newborn individuals. These curves are independent of the covariance between vital rates at any one time (see Orzack 1993 and 1997 for further details). Extinction probabilities for life histories 7, 8, 9, and 10 are so similar that they are plotted together.

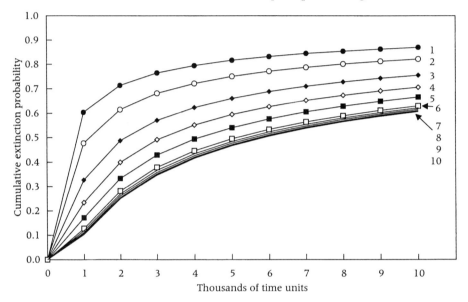

points to be made about these curves. First, life histories with a higher growth rate (λ_0) implied by their average projection matrix have higher finite-time extinction probabilities (Orzack 1997). Second, despite eventual extinction, populations may survive for very long periods of time. For example, as shown in Figure 2, all of the life histories shown in Table 1 have at least a 40 percent probability of being extant after 1,000 time units. One can show that all have a 4 percent chance of being extant at a time equal to 1,000 times the modal lifetime (Orzack 1997). So, for example, life history 10 has a modal lifetime of 880 time units; if the time unit of the analysis is a year, this means that a population has a 4 percent chance of being extant after 880,000 years. Similarly, one can show that a population with life history 1 has this chance of surviving for 90,000 years. Third, a longer life span results in a lower finite-time extinction probability. For example, as shown in Figure 2, life history 6 (with $\alpha = 1$ and $\omega = 10$) has an approximately sixfold smaller probability of extinction after 1,000 time units as compared to life history 1 (with $\alpha = 1$ and $\omega = 1$). Fourth, a longer reproductive span results in a higher finite-time extinction probability (Orzack 1997), although there need be no strong relationship between the two; populations with life histories 6 through 10, with substantially different reproductive spans, differ only slightly in transient extinction probabilities (as shown in Figure 2).

The current focus in explanations of aging and life span on evolution within populations heightens the importance of these results. A number of early claims about aging invoked group selection as an explanation, the idea being that aging might have evolved because it results in the differential survival of populations containing individuals that age more rather than less. The resulting increases in the mortality rate, especially of nonreproductive older age groups, could prevent a population from overexploiting resources and thereby lessen its probability of extinction. Such claims were based on the premise that senescence could not be a product of natural selection within populations because aging is manifestly deleterious to the organism. As described above, subsequent work by a number of evolutionary biologists has elucidated how natural selection within populations *could* explain aging (cf., Williams 1957; Charlesworth 1994). This work, along with a programmatic belief that evolutionary explanations at the level of the individual have precedence over explanations at the level of the group (Williams 1966), has meant that claims about aging that invoke group dynamics have been largely discarded and little attention has been paid to ways in which population dynamics could affect the evolution of aging and life span.

There is good reason to be careful about the application of evolutionary explanations at the level of the group or species. In general, it is not obvious that the turnover among groups is typically substantial enough to account for observed rates of change for many traits. More importantly, it is not always clear how one makes a connection between within-group evolutionary dynamics and between-group evolutionary dynamics.

In the present instance, however, there is a clear connection between the evolutionary dynamics at both of these levels. As described above, the stochastic growth rate, which provides important insights into selection among life histories within populations, also plays an important role in determining extinction dynamics. It is unusual in evolutionary biology to have such a clear analytical understanding of such a connection.

The main point illustrated by the examples described above is that the differential survival of populations and species may be an important determinant of the diversity of aging patterns and life spans that we observe across taxa. Such a claim is meant to expand our conceptual understanding of aging and life span evolution rather than to suggest that older claims about the role of within-population selection are incorrect; instead, it is my belief that they are incomplete.

Empirical work on aging and life span

Just as there are substantial questions to be resolved about our theoretical understanding of aging and life span evolution, so too are there substantial questions to be resolved about our empirical understanding.

As noted above, in the last decade or so there has been a vast increase in our knowledge of the molecular, cellular, and genetic mechanisms that contribute to aging. Less attention has been paid to mechanisms governing life span per se; such mechanisms need not be the same as those governing aging. Nonetheless, this increased knowledge about aging is impressive. What is unclear is whether our understanding of mechanisms will actually translate into an understanding of how to control the processes of aging, especially in humans.

What the last decade has not witnessed is a substantial increase in our understanding of the biology of aging and life span at the broadest biological levels. Although comparative predictions about aging and life span date back at least several decades (e.g., Williams 1957), we still have no comprehensive information on the distribution of these traits across taxa. We know about some of the broader patterns, such as the tendency for birds to have substantially longer life spans as compared to similarly sized mammals (see Carey and Judge 2000 for a compilation of data). Nonetheless, there is little information about the phylogenetic patterns that aging and life span exhibit; what little information we have in this regard provides mixed support for present evolutionary theories of aging (see Promislow 1991; Gaillard et al. 1994). Many important questions about large-scale aspects of aging and life span evolution remain unaddressed. For example, it is not clear whether there is a tendency for older species to have a higher rate of aging. All other things being equal, such an association is predicted by current theory in that an older species has had more time in which natural selection could cause vital rates to "age." This is an ideal question for the application of molecular techniques, which can be used to infer relative ages of related species. More generally, it is not clear whether there is an association between rate of aging and rates of species origination and extinction. I have described theoretically how life span can affect extinction rate; data are needed to refute or substantiate this connection.

Finally and perhaps most importantly, we need a better understanding of what our evolutionary knowledge reveals about aging and life span in humans. There are numerous questions to be answered in this context. One that arises from the analyses presented here of the distinction between the evolutionary dynamics of aging and those of life span is whether this distinction changes our understanding of disease dynamics and thereby influences the design of treatment programs.

More generally, conflicting claims remain to be resolved about the relevance of evolutionary analyses to aging research and geriatric medicine. Some evolutionary biologists have made strong claims about the necessity of incorporating evolutionary insights into aging and life span research, including the design and implementation of treatments for geriatric illness. Rightly or wrongly, this claim has not been generally accepted by clinicians, for whom geriatric medicine is a matter of understanding mechanisms only, without attention to underlying evolutionary causes. The analytical framework discussed above may help achieve at least a partial resolution of this conflict.

Note

This work has been partially supported by NSF award SES 99-06997.

References

Carey, J. R., P. Liedo, D. Orozco, and J. W. Vaupel. 1992. "Slowing of mortality rates at older ages in large medfly cohorts," *Science* 258: 457–461.

Carey, J. R. and D. S. Judge. 2000. *Longevity Records: Life Spans of Mammals, Birds, Reptiles, Amphibians and Fish*. Odense: Odense University Press.

Cohen, J. E., H. Kesten, and C. M. Newman. 1986. *Random Matrices and Their Applications*. Contemporary Mathematics volume 50. Providence, RI: American Mathematical Society.

Charlesworth, B. 1994. *Evolution in Age-structured Populations*. New York: Cambridge University Press.

Charlesworth, B. and K. A. Hughes. 1996. "Age-specific inbreeding depression and components of genetic variance in relation to the evolution of senescence," *Proceedings of the National Academy of Science* 93: 6140–6145.

Finkel, T. and N. J. Holbrook. 2000. "Oxidants, oxidative stress and the biology of ageing," *Nature* 408: 239–247.

Gaillard, J.–M., D. Allaine, D. Pontier, N. G. Yoccuz, and D. E. L. Promislow. 1994. "Senescence in natural populations of mammals: A reanalysis," *Evolution* 48: 509–516.

Hamilton, W. D. 1966. "The moulding of senescence by natural selection," *Journal of Theoretical Biology* 12: 12–45.

Holmes, D. J. and S. N. Austad. 1995. "Birds as animal models for the comparative biology of aging: A prospectus," *Journal of Gerontology: Biological Sciences* 50A: B59–B66.

Hughes, K. A. and B. Charlesworth. 1994. "A genetic analysis of senescence in *Drosophila*," *Nature* 367: 64–66.

Keyfitz, N. 1977. *Applied Mathematical Demography*. New York: John Wiley & Sons.

Kimura, M. 1983. *The Neutral Theory of Molecular Evolution*. Cambridge: Cambridge University Press.

Kirkwood, T. B. L. and R. Holliday 1979. "The evolution of ageing and longevity," *Proceedings of the Royal Society of London B* 205: 531–546.

Kirkwood, T. B. L. and S. N. Austad. 2000. "Why do we age?" *Nature* 408: 233–238.

Kolman, W. A. 1960. "The mechanism of natural selection for the sex ratio," *American Naturalist* 94: 373–377.

Lande, R. and S. H. Orzack. 1988. "Extinction dynamics of age-structured populations in a fluctuating environment," *Proceedings of the National Academy of Sciences* 85: 7418–7421.

Le Bourg, E. 2001. "A mini-review of the evolutionary theories of aging: Is it time to accept them?" *Demographic Research* 4: 1–28.

Leroi, A. M., A. K. Chippindale, and M. R. Rose. 1994. "Long-term laboratory evolution of a genetic life-history trade-off in *Drosophila melanogaster*. 1. The role of genotype-by-environment interaction," *Evolution* 48: 1244–1257.

Luckinbill, L. S., R. Arking, M. J. Clare, W. C. Cirocco, and S. A. Buck. 1984. "Selection for delayed senescence in *Drosophila melanogaster*," *Evolution* 38: 996–1003.

Medawar, P. B. 1952. *An Unsolved Problem of Biology*. London: Lewis.

Orzack, S. H. 1985. "Population dynamics in variable environments V. The genetics of homeostasis revisited," *American Naturalist* 125: 550–572.

———. 1993. "Life history evolution and population dynamics in variable environments: Some insights from stochastic demography," in J. Yoshimura and C. Clark (eds.), *Adaptation in a Stochastic Environment*. New York: Springer-Verlag, pp. 63–104.

———. 1997. "Life history evolution and extinction," in S. Tuljapurkar and H. Caswell (eds.), *Structured Population Models in Marine, Terrestrial, and Freshwater Systems*. New York: Chapman and Hall, pp. 273–302.

Orzack, S. H. and E. Sober. 1994. "Optimality models and the test of adaptationism," *American Naturalist* 143: 361–380.

———. 2001. "Adaptation, phylogenetic inertia, and the method of controlled comparisons," in S. H. Orzack and E. Sober (eds.), *Adaptationism and Optimality*. New York: Cambridge University Press, pp. 45–63.

Orzack, S. H. and S. Tuljapurkar. 1989. "Population dynamics in variable environments VII. The demography and evolution of iteroparity," *American Naturalist* 133: 901–923

Partridge, L. and K. Fowler. 1992. "Direct and correlated responses to selection on age at reproduction in *Drosophila melanogaster*," *Evolution* 46: 76–91.

Pearl, R. 1928. *The Rate of Living, Being an Account of Some Experimental Studies on the Biology of Life Duration*. New York: Knopf.

Promislow, D. E. L. 1991. "Senescence in natural populations of mammals: A comparative study," *Evolution* 45: 1869–1887.

Roff, D. A. 1992. *The Evolution of Life Histories*. New York: Chapman and Hall.

Rose, M. R. and B. Charlesworth. 1981. "Genetics of life history in *Drosophila melanogaster*. II. Exploratory selection experiments," *Genetics* 97: 187–196.

Rubner M. 1908. *Das Problem der Lebensdauer und seine Beziehungen zu Wachstum und Ernährung*. Munich: Oldenbourg.

Stadtman, E. R. 1992. "Protein oxidation and aging," *Science* 257: 1220–1224.

Taylor, A. D. 1992. "Deterministic stability analysis can predict the dynamics of some stochastic population models," *Journal of Animal Ecology* 61: 241–248.

Tuljapurkar, S. 1982. "Population dynamics in variable environments, III. Evolutionary dynamics of r-selection," *Theoretical Population Biology* 21: 141–165.

———. 1989. "An uncertain life: Demography in random environments," *Theoretical Population Biology* 35: 227–294.

———. 1990. "Delayed reproduction and fitness in variable environments," *Proceedings of the National Academy of Science* 87: 1139–1143.

Tuljapurkar, S. D. and S. H. Orzack. 1980. "Population dynamics in variable environments, I. Long-run growth rates and extinction," *Theoretical Population Biology* 18: 314–342.

Vaupel, J. W. et al. 1998. "Biodemographic trajectories of longevity," *Science* 280: 855–860.

Williams, G. C. 1957. "Pleiotropy, natural selection, and the evolution of senescence," *Evolution* 11: 398–411.

Williams, G. C. 1966. *Adaptation and Natural Selection*. Princeton: Princeton University Press.

Wright, S. 1931. "Evolution in Mendelian populations," *Genetics* 16: 97–159.

Zwaan, B., R. Bijlsma, and R. F. Hoekstra. 1995. "Direct selection on life span in *Drosophila melanogaster*," *Evolution* 49: 649–659.

Ecological Correlates of Life Span in Populations of Large Herbivorous Mammals

JEAN-MICHEL GAILLARD
ANNE LOISON
MARCO FESTA-BIANCHET
NIGEL GILLES YOCCOZ
ERLING SOLBERG

Recently, much knowledge has been gained about variation in mammalian life-history traits such as litter size, gestation length, birth mass, weaning mass, age at first breeding, breeding interval, and life span (see Roff 1992; Stearns 1992 for reviews). Life span (length of life) is one of the most influential life-history traits related to individual fitness, from both demographic and evolutionary viewpoints. Differences in life span account for much of the variation in individual fitness in many vertebrate populations (Clutton-Brock 1988; Newton 1989). Among species, the variability in life span is expected to be related to body size (Peters 1983), phylogeny (Harvey and Pagel 1991), and ecological characteristics (Calder 1984). Within species, the main causes of variation in life span should be both tradeoffs between reproduction and survival (Stearns 1992) and individual differences in phenotypic quality (Clutton-Brock 1991).

The objective of this chapter is to present a comprehensive picture of levels and trends of variation in mammalian life span at both intra- and interspecific levels. Improved description and analysis might also be useful to refine current theories of demographic tactics and evolutionary tradeoffs. From a literature survey focused on case studies of large mammalian herbivores, we offer in what follows:

—*A comparative analysis of life span among mammalian species*. Previous analyses have often focused on differences in life span among species. Thus, comparative studies of maximum longevity (Sacher 1978), life expectancy at birth (Read and Harvey 1989), and adult life expectancy (Gaillard et al. 1989) performed over a large range of mammalian species have pointed out that allometry and phylogeny have a marked effect on mammalian life span.

—An analysis of senescence in survival of large mammalian herbivores. Most current theories of life-history evolution predict a decrease in performance with increasing age (see Rose 1991 for a review). However, two processes could prevent survival from decreasing with age. First, a high rate of mortality in natural populations might allow too few individuals to reach the age beyond which survival would decrease with age. Second, if low-quality individuals tend to die younger than high-quality individuals, the average quality of individuals may increase with age. Heterogeneities among individuals of different ages may then mask evidence of senescence.

—An analysis of sexual differences in senescence in relation to the intensity of sexual selection. Current theories of sexual selection predict that males of the most polygynous and sexually dimorphic species should show both lower survival and a greater rate of senescence than males of species that are weakly polygynous and without sexual dimorphism. This prediction is based on the assumption that the intensity of sexual selection, sexual dimorphism, and polygyny are correlated (see Andersson 1994 for a review). Because the intensity of sexual selection should not influence senescence in females, we expect to find increasing male-biased senescence patterns with increasing intensity of sexual selection.

—An analysis of the influence of phenotypic quality on individual life span. Individuals of high phenotypic quality have repeatedly been reported to be better able to survive harsh environmental conditions (see Lomnicki 1978; Van Noordwijck and De Jong 1986; Clutton-Brock 1991 for reviews). We thus expect individuals of high phenotypic quality to outlive individuals of low phenotypic quality.

—An analysis of the tradeoff between reproductive output and life span. Life-history theories predict that mature individuals allocate energy either to maintenance (e.g., body reserves that increase survival chances during adverse conditions) or to reproduction. Because individuals can obtain only a limited amount of energy, they cannot maximize both reproductive output and survival prospects (Law 1979). Therefore, individuals that invest heavily in reproduction should suffer fitness costs by having lower survival chances than individuals that invest less heavily in reproduction. We thus expect a negative relationship between reproductive output and life span in large mammalian herbivores. Alternatively, behavioral adaptations (*sensu* Tuomi et al. 1983) and/or differences between individuals in phenotypic quality may mask such a tradeoff between reproduction and life span.

Comparative analysis of life span among mammalian species

We used data on adult life expectancy collected in Gaillard et al. (1989) on 78 mammalian species belonging to eight orders.[1] As expected, adult life expectancy consistently increased with body mass (mean slope within order of 0.191± 0.030, Figure 1). The allometric exponent, moreover, was not far from 0.25, the exponent expected for biological time (Calder 1984). For a given body mass, however, adult life expectancy differed markedly among mammalian orders. Cetaceans (whales and dolphins), lagomorphs (hares and rabbits), insectivores (shrews and hedgehogs), and rodents (squirrels) all had short adult life expectancy for their body mass (with intercepts of –0.760, –0.582, –0.416, and –0.118, respectively). On the other hand, primates (apes and monkeys) and bats had long adult life expectancy (with intercepts of 0.912 and 0.699), while carnivores and ungulates had intermediate values (with intercepts of 0.014 and 0.030). These findings support the strong influence of body mass and phylogeny in accounting for variation of life span among vertebrate species. Indeed, similar positive allometric relationships seem to hold for all groups of vertebrates (see Gaillard et al. 1989 on birds; Shine and Charnov 1992 on reptiles; and Rochet et al. 2000 on fishes). Among mammals, habitat type might also shape life span. For a given mass, flying mammals (bats) had the longest adult life expectancy, marine mammals (cetaceans) had the shortest life span, and all the terrestrial orders showed intermediate values. However, such an ecological correlate is more likely to reflect differential selective pressures on body mass rather than direct selection on life span. If a large body mass is se-

FIGURE 1 Allometric relationship between adult life expectancy and female body mass in eight orders of mammals

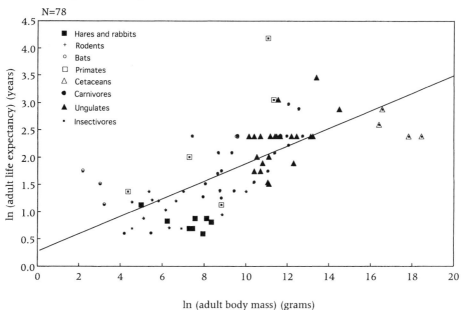

lected against in bats because of flying constraints and is selected for in cetaceans because of diving advantages (lung capacity being isometrically related to body mass; Peters 1983), we should expect to obtain the results reported here in absence of a direct selection on life span. More detailed analyses including other marine (e.g., seals) and flying (e.g., flying squirrels) mammals will provide some tests.

Senescence in fitness components of large mammalian herbivores

Large mammalian herbivores as a model for studying variation in life span

Obtaining reliable information on mammalian life span in the wild is difficult and time consuming. Most previous analyses were based on life tables that were usually based on the ages of animals found dead. Poor assessment of age from tooth wear (Vincent et al. 1994; Hamlin et al. 2000) and obvious sampling biases caused by differential decay rates among age and sex classes (Spinage 1972) make most life tables unreliable. In addition, the estimation of mortality rates from life tables constructed from age at death requires assumptions about age structure and environmental variability that are rarely met (see Caughley 1977 for further details).

Here we used reliable estimates of population parameters by limiting our analyses to populations for which long-term monitoring of individually marked animals was available. Over the last decade, such detailed studies have been published on a few populations of large herbivores, providing the opportunity to quantify patterns of senescence in both survival and reproduction in relation to sex, phenotypic quality, and reproductive output (Gaillard et al. 1993; Loison et al. 1994; Byers 1997; Jorgenson et al. 1997; Toïgo et al. 1997; Cransac et al. 1997; Loison et al. 1999a; Coulson et al. 2001). We avoided the problem of inaccurate age classification by limiting our analyses to animals whose age was known with precision because they were marked within two years of birth. As far as possible, we repeated the analyses by using recent developments of capture-mark-recapture methods (Lebreton et al. 1992), which avoid biased survival estimates by incorporating differences in resighting probabilities (Nichols 1992). Large herbivores included in the following analyses belong to the Order *Artiodactyla* with two main families, deer (*Cervidae*) and wild sheep and mountain goat (*Bovidae*). These species show many similarities in life histories (Gaillard et al. 2000a). Their populations are strongly age-structured with three main stages. (1) A juvenile stage lasts from birth to 1 year of age, during which mortality is very high (up to 90 percent) because of predation, disease, climatic harshness, or food limitation. (2) An adult stage can be subdivided into a transition year from 1 to 2 years of age, when mortality decreases markedly and the first conception may occur, and a prime-age stage (between 2 and 7 years of age, but up to 10 years for the larger species) during which mortality is very low (typically less than 10 percent) and reproductive output peaks, with most females conceiving once a year. Litter size varies from 1 to 3 depending on species. (3) A senescent stage is expected to occur after 7 to 10 years of age, when mortality is presumed to increase and reproductive output to decrease.

Evidence of senescence in survival

Although most biologists agree that senescence is a general property of living organisms, the pervasiveness of senescence in the wild has been challenged. Wild animals suffer high age-independent mortality through predation, disease, or food limitation, so that very few individuals may reach the age from which their ability to survive decreases because of senescence (Comfort 1979). Promislow (1991) and Sibly et al. (1997) reported that senescence in survival was pervasive among mammals, but their conclusions have been criticized (Gaillard et al. 1994) because they ignored biases in the life tables they analyzed. We searched the literature for age-dependent survival probabilities estimated from known-age individuals monitored over many years to test whether senescence occurred in survival of large mammalian herbivores.

We found such data for 12 populations belonging to ten species: two populations of roe deer (*Capreolus capreolus*) monitored by the Office National de la Chasse et de la Faune Sauvage in two French reserves (Chizé and Trois Fontaines; Gaillard et al. 1993), two populations of bighorn sheep (*Ovis canadensis*) in the Canadian Rocky Mountains (Ram Mountain and Sheep River; Jorgenson et al. 1997), one population of mountain goats (*Oreamnos americana*) also in the Canadian Rocky Mountains (Caw Ridge; Festa-Bianchet unpubl. data), one population of pronghorn (*Antilocapra americana*) in Montana (National Bison Range; Byers 1997), one population of moose (*Alces alces*) in Sweden (Västerbotten county; Ericsson et al. 2001), one population of ibex (*Capra ibex*) in the French Alps (Belledone; Toïgo 1998), one population of red deer (*Cervus elaphus*) in Scotland (Rum; Clutton-Brock et al. 1982), one population of Soay sheep (*Ovis aries*) in Scotland (Hirta; Catchpole et al. 1998), one population of mouflon (*Ovis ammon*) in a French reserve managed by the Office National de la Chasse et de la Faune Sauvage in Massif Central (Caroux; Cransac et al. 1997), and one population of isard (*Rupicapra pyrenaica*) in a French reserve also managed by the Office National de la Chasse et de la Faune Sauvage in the Pyrenees (Orlu; Loison et al. 1999a). Species varied markedly in size (from about 25 kg for female Soay sheep, roe deer, and isard to >400 kg for male moose), in habitat type (forest for roe deer, open highlands for pronghorn and red deer, and mountain for ibex, bighorn, isard, and mountain goat), and in level of polygyny (low for roe deer and high for red deer, ibex, and bighorn sheep). To avoid confounding effects of growth or reproduction, we analyzed mortality rates from 2 years of age onward, when most growth has been completed and sexual maturity has been reached.

The most widely known theory of senescence suggests that mortality should increase exponentially with age (Gompertz 1825). To test the validity of the Gompertz model for large mammalian herbivores, we calculated sex- and age-specific mortality rates from the available survival estimates. The Gompertz model predicts a linear relationship between log-transformed mortality rates and age.

In general, our results supported the Gompertz model (see Figure 2 for a representative example): as shown by using simple linear regressions on age-specific mortality estimates, the rate of mortality increase was consistently positive in males (Figure 3A) and females (Figure 3B). The mortality increase with age was statistically significant in 13 cases and marginally significant (0.05<P<0.10) in two cases. Only mountain goat males and Soay sheep of both sexes appear to deviate from this general pattern, because

FIGURE 2   An example of relationships between log-transformed mortality rates and age in a population of large herbivores (female roe deer at Chizé)

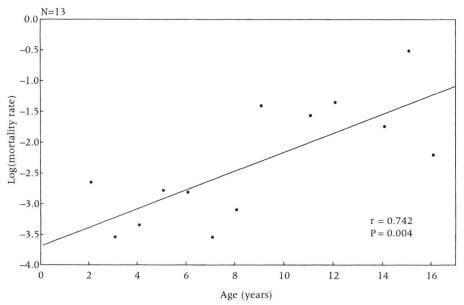

NOTE: According to the the Gompertz model (represented by the solid line), the log-transformed mortality rate should increase with age.

FIGURE 3   Rate of mortality increase (measured as the slope of the regression of log-transformed mortality rates on age) (± 1 standard error) for males (A) and females (B) in 12 populations of large herbivorous mammals

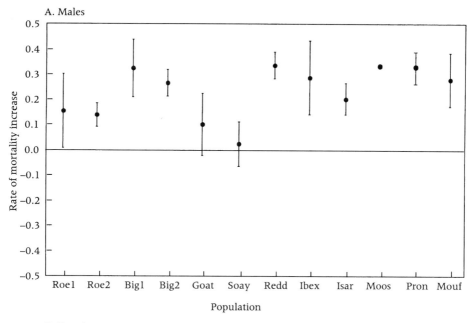

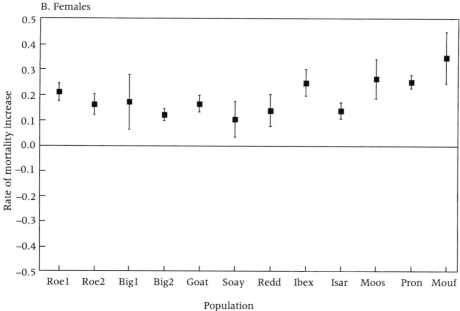

NOTE: Roe1: Roe deer at Trois Fontaines; Roe2: Roe deer at Chizé; Big1: Bighorn sheep at Sheep River; Big2: Bighorn sheep at Ram Mountain; Goat: Mountain goat at Caw Ridge; Soay: Soay sheep on Hirta; Redd: Red deer on Rum; Ibex: Ibex at Belledone; Isar: Isard at Orlu; Moos: Moose at Västerbotten; Pron: Pronghorn at National Bison Range; Mou: Mouflon at Caroux.

they showed much lower mortality rates between 4 and 7 years of age. Our results thus show clear evidence of senescence in survival among free-living large herbivores.

Inspection of the deviations from the Gompertz predictions suggests that prime-age females have lower mortality rates than expected: the observed mortality rate between 3 and 8 years of age was above the regression line in 67 percent of cases (35 times out of 52), which deviates significantly from the 50 percent expected. The low mortality of prime-age females may result from a life-history strategy of risk avoidance. Indeed, prime-age females of large herbivores curtail their reproductive effort to ensure their own survival when facing harsh environmental conditions (Festa-Bianchet and Jorgenson 1998; Gaillard et al. 2000a and references therein).

Are sex differences in senescence related to the intensity of sexual selection?

To answer this question, we used the sex-specific mortality rates calculated previously. As expected for polygynous species, the initial mortality rates (i.e., the intercept of the regression of log-transformed mortality rates on age) of males were higher than those of females in 9 of the 12 populations studied (Figure 4). In both sexes, the initial mortality rate decreased with increasing body mass (ln (initial mortality) = –0.221 – 0.741 ln (adult body mass), r = 0.516, P = 0.086; and ln (initial mortality) = –0.505 – 0.830 ln (adult body mass), r = 0.716, P = 0.009 in males and females respectively). Such decreasing mortality with increasing body mass supports the overall positive influence of size on life span previously reported in most vertebrate groups. Sex differences in initial mortality, however, were not related to the intensity of sexual selection measured either according to the level of polygyny (i.e., the breeding group size *sensu* Clutton-Brock et al. 1982) or according to sexual dimorphism in size (see Andersson 1994 for a review and Loison et al. 1999b for an example in large mammalian herbivores). The male/female ratio of initial mortality did not differ between the four populations of weakly polygynous species (i.e., with a breeding group size between 3 and 5, mean mortality ratio of 2.594 ± 0.478) and the eight populations of highly polygynous species (i.e., with a breeding group size over 5, mean mortality ratio of 1.495 ± 1.256). For instance, the male/female mortality ratio exceeded 2 for both populations of the weakly dimorphic and polygynous roe deer, whereas it was only 0.5 for the highly dimorphic and polygynous Alpine ibex. Likewise, we found no evidence for an increasing male/female ratio of initial mortality with sexual dimorphism in size (r = 0.124, P = 0.700). On the contrary, relative male mortality tended to decrease with increasing sexual dimorphism in size (slope of –0.462 ± 1.166). Clearly, factors other than sexual selection are shaping sexual variation of initial mortality in populations of large herbivores.

FIGURE 4 Initial mortality rate (measured as the antilog-transformed intercept of the regression of log-transformed mortality rates on age) for males (filled squares) and females (open squares) in 12 populations of large herbivorous mammals

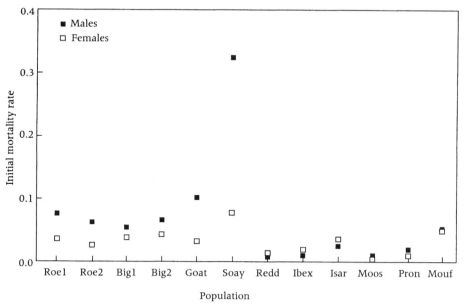

NOTE: See note to Figure 3.

Slight changes occurred when we studied sex differences in the rate of survival senescence (i.e., the slope of the regression of log-transformed mortality rates on age). The senescence rate of male survival was higher than that of females in only 7 out of the 12 populations (Figure 5). While male survival tended to decrease with age faster than female survival in highly polygynous species (mean senescence ratio of 1.456 ± 0.783), and the reverse pattern occurred in weakly polygynous species (mean senescence ratio of 0.857 ± 0.293), the difference was not significant (F = 2.103, df = 1, 10, P = 0.178). The lack of effect of breeding size on the relative senescence rate of male survival was due to the high variability in the rate of survival senescence observed within highly polygynous and strongly dimorphic species. Indeed, the male/female senescence ratio varied from 0.79 in mouflon to 2.52 in red deer. Likewise, we found no evidence of an increasing male/female ratio of senescence rate with sexual dimorphism in size (r = 0.002, P = 0.995). Relative male mortality tended to be constant whatever the sexual dimorphism in size (slope of –0.005 ± 0.713). As with initial mortality, we conclude that factors other than sexual selection are shaping sexual variation of the senescence rate in populations of large herbivores.

Our analyses show that in populations of large mammalian herbivores, females consistently outlived males in the study populations. Contrary to

FIGURE 5　Rate of senescence (measured as the slope of the regression of log-transformed mortality rates on age) for males (filled squares) and females (open squares) in 12 populations of large herbivorous mammals

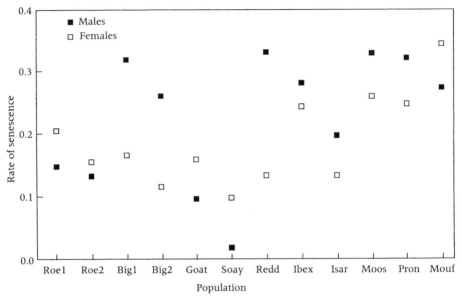

NOTE: See note to Figure 3.

our expectation, however, we did not find support for a higher male-skewed mortality in the most highly polygynous and dimorphic species. In addition to sexual selection, other factors such as ecological conditions may influence the magnitude of male-skewed mortality in populations of large herbivores (Gaillard et al. 2003; Toïgo and Gaillard 2003).

Phenotypic quality and individual life span

To test for a relationship between phenotypic quality and individual life span, we used the long-term (>20 years) data sets on known-aged animals available for roe deer at Trois Fontaines and bighorn sheep at Ram Mountain. Among large herbivores, heavier individuals typically survive better during the juvenile and yearling stages, reproduce earlier, and have larger litter sizes than lighter individuals (see Gaillard et al. 2000a for a review). Body mass has also been reported to increase with habitat quality (Pettorelli et al. 2001) and dominance rank (Clutton-Brock 1991). We therefore used adult body mass as a measure of phenotypic quality. Because we measured adult body mass between 6 and 12 years of age in bighorn sheep and between 4 and 10 years of age in roe deer (Festa-Bianchet et al. unpubl. data; Bérubé et al. 1999), we discarded all females that did not reach the age of 4 years (roe deer) or 6 years (bighorn sheep), leaving us with a sample of 92 bighorn sheep and 86 roe deer.

As previously reported (Bérubé et al. 1999; Gaillard et al. 2000b), we found significant positive relationships between length of life and adult body mass in both species (Linear regressions: r = 0.380, P = 0.0002 for bighorn sheep; r = 0.245, P = 0.023 for roe deer; see Figure 6). However, variation in

FIGURE 6 Relationships between length of life and adult body mass of females in two populations of large herbivores: A. Bighorn sheep at Ram Mountain (Canada) B. Roe deer at Trois Fontaines (France)

A. Bighorn sheep

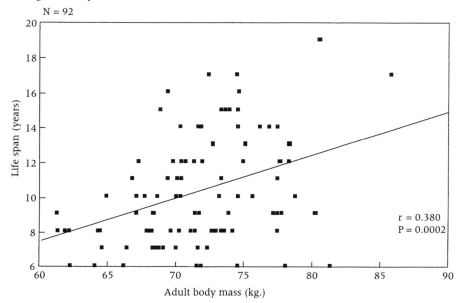

B. Roe deer

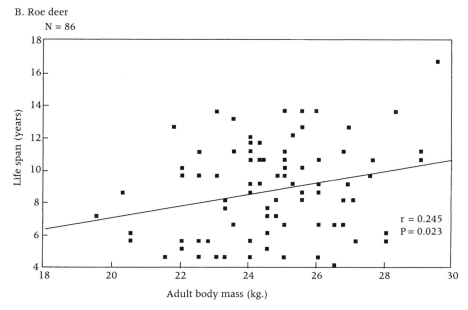

adult body mass accounted for only 14 percent (bighorn sheep) and 6 percent (roe deer) of variation in length of life. While very large body mass appeared necessary for a long life span in both species (no bighorn sheep less than 65 kg lived more than 10 years and no roe deer less than 22 kg lived more than 9 years, while the only bighorn sheep heavier than 82 kg lived 16 years and all four roe deer heavier than 28 kg lived more than 10 years), length of life was highly variable over most of the range of body mass observed in both species (Figure 6). Therefore it appears that the correlation between adult mass and length of life is mostly the result of short life expectancy for the smallest individuals and long life expectancy for the very largest ones. For individuals whose body mass is within about 70 percent of the population average, differences in mass seem to have little effect on life expectancy, and factors other than adult body mass will determine their length of life. Our analyses show, however, that within populations of large herbivores, some large individuals live longer lives while some small individuals live shorter lives than medium-sized individuals. This provides evidence of individual heterogeneities in populations of large herbivores.

The tradeoff between reproductive output and life span

To test the assumption of an evolutionary tradeoff between reproductive output and length of life, we used data from the roe deer population of Trois Fontaines. From the intensive monitoring of marked females between September and December every year since 1976, the mean reproductive success of prime-age females from 2 to 7 years of age was calculated as the average yearly reproductive success. Between one and six yearly values were available for 59 females.

We found no evidence of a negative relationship between prime-age reproductive success and length of life (Figure 7). Contrary to our expectation, mean reproductive success of females during prime age tended to increase weakly with length of life (linear regression: r = 0.198, P = 0.132), similar to results obtained for bighorn sheep (Bérubé et al. 1999). Thus, rather than trading early reproduction for longer life, some female roe deer combined a high reproductive success with great longevity, whereas other females raised few offspring during a short life span. These results support previous claims that differences in quality among individuals are likely to mask evolutionary tradeoffs (Van Noordwijk and De Jong 1986; Clutton-Brock 1991). To test the possibility that individual heterogeneities in body mass may have masked a tradeoff between prime-age reproduction and longer life, we corrected length of life for adult body mass by examining the residuals of the regression of longevity on adult body mass.

FIGURE 7   **Relationship between mean reproductive success per prime-age female (2–7 years of age) (measured as the mean number of live fawns between September and December) and length of life in roe deer females at Trois Fontaines (France)**

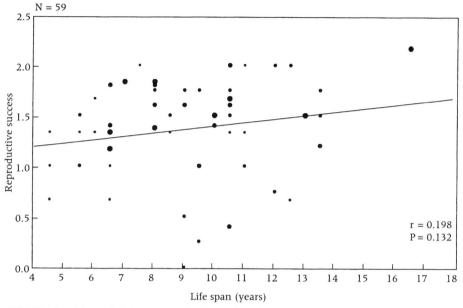

NOTE: The size of the symbols is proportional to the number of years over which mean reproductive success was calculated (from 3 years [smallest filled circles] to 6 years [largest filled circles]).

Accounting for the positive influence of body mass on longevity decreased the strength of the correlation between prime-age reproductive success and longevity (linear regression: r = 0.126, P = 0.361), but the relationship remained positive (Figure 8). We therefore cannot detect any tradeoff between reproductive output and length of life in roe deer, even after accounting for the influence of phenotypic quality. Although (1) body mass may be only a rough proxy of phenotypic quality (it is, however, closely correlated to habitat quality: Pettorelli et al. 2001; and to dominance rank: Clutton-Brock 1991) and (2) only careful experiments can control for phenotypic quality, our results appear to contradict current theories on the evolution of life histories (Williams 1966). Notwithstanding the limitations of the approach we used, at least three explanations can be proposed to account for the absence of a tradeoff. First, several adaptations, such as increasing feeding time or changes in behavior to expend less energy, have evolved in large herbivores to reduce the fitness costs of reproduction (see Tuomi et al. 1983 for elaboration). Second, the occurrence of tradeoffs could depend on the ecological context, being limited to poor environmental conditions. Trois Fontaines provides a highly favorable environment for roe deer, with adequate food supply, yearly removals of

FIGURE 8 **Relationship between mean reproductive success per prime-age female (2–7 years of age) (measured as the mean number of weaned fawns between September and December) and length of life corrected for adult body mass in roe deer females at Trois Fontaines (France)**

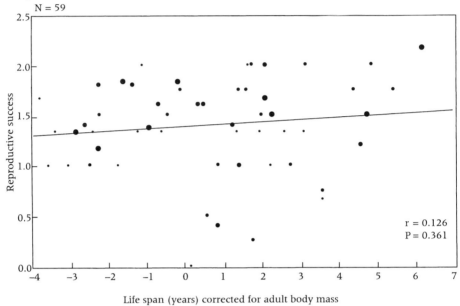

Life span (years) corrected for adult body mass

NOTE: The correction of life span allows us to account for the possible confounding effect of phenotypic quality. The size of the symbols is proportional to the number of years over which mean reproductive success was calculated (from 3 years [smallest filled circles] to 6 years [largest filled circles]).

unmarked deer that limit density, and a mild climate. Under such conditions, high reproductive effort during prime age may not decrease the chance of a long life span. Finally, evolutionary tradeoffs may occur at the level of resource acquisition rather than resource use, leading to "increasing returns" (Houle 1991; Dobson et al. 1999) for individuals that are better adapted to a given environment. Whatever the reasons, the absence of a tradeoff between reproduction and survival that we found for roe deer will increase variance in individual fitness, possibly providing a higher potential for selection among individuals. The ecological and genetic correlates of this high heterogeneity among individuals clearly warrant further investigation in contrasting species and populations, because they likely hold clues to variability in life expectancy.

Conclusions

On the basis of the analyses performed above using high-quality data from long-term monitoring of populations of large herbivorous mammals, we can tentatively propose a new picture of levels and trends of variation in mammalian life span.

First, body mass seems to have an overwhelming influence on life span whatever the object of analysis. We found consistently positive relationships between body mass and life span or related measures among populations of large herbivorous mammals, in both sexes and among individuals of a given population. We may therefore conclude that to survive longer, heavier is better.

Second, it has long been questioned whether senescence in survival really occurs in natural populations of mammals (Comfort 1979; Promislow 1991; Gaillard et al. 1994). From the present work, we can assert that survival consistently decreases with increasing age in both sexes in natural populations of large herbivores. Although predation, disease, human activities, climatic harshness, and population density all affect individuals in natural populations of large herbivores, initial mortality rates during adult stages were remarkably low: less than 10 percent in 10 of 12 populations for males and in all 12 populations for females (Figure 4). Such low mortalities allowed enough individuals to reach ages at which senescence is expected to occur and thereby allowed us to detect decreases in survival with age. We conclude that senescence in survival is pervasive among large mammals. Whether senescence occurs in survival of those small mammals in which initial mortality is expected to be high remains unknown and deserves further study.

Third, contrary to expectations from theories of sexual selection, there is no evidence that strong sexual selection involving either a marked sexual dimorphism in size or a large size of breeding groups decreases the relative male life span. Although the populations studied here included both weakly and highly polygynous species, we found no evidence of higher relative initial mortality of males or greater relative rate of senescence in males in highly polygynous/strongly dimorphic species than in weakly polygynous/ weakly dimorphic ones. Females consistently outlive males in populations of large herbivores, but the intensity of sexual selection leads to no additional sex difference in life span. Because males in most populations of large mammals are more susceptible to environmental harshness than females (Clutton-Brock et al. 1982), we conclude that variation in ecological context rather than variation in intensity of sexual selection accounts for most sex differences we observed in life span and in related measures of large herbivores (Toïgo and Gaillard 2003).

Lastly, although the tradeoff between reproduction and survival is a central tenet of evolutionary theories of life-history variation (Stearns 1992; Roff 1992), we found no evidence for such a tradeoff from our analysis of a roe deer population for which appropriate data were available. Such a negative result cannot be accounted for by the confounding effect of heterogeneities among individuals because we controlled for such heterogeneities by using body mass as a proxy for individual quality. Therefore, rather than dismissing a negative result as is too often done, we challenge the usual assumption according to which a tradeoff between reproduction and survival occurs in all living organisms. Biological mechanisms such as increases in feeding time (i.e., higher energy acquisition), changes in behavior with decreasing movements (i.e., lower energy expenditure), and reproductive pause or cessation when body condition decreases (i.e., a selfish tactic) have all been reported in large mammalian herbivores and could provide ways by which individuals limit fitness costs of reproduction with payoffs for future survival. Further work should assess whether the absence of a tradeoff between reproduction and survival reported here is widespread among mammals in which reproduction does not occur below some body mass threshold.

Note

1 Although species do not represent independent data points, we performed our analyses by fitting usual linear models without accounting for phylogenetic relationships among species. Our choice was motivated by (1) the similar results obtained from analyses on raw data (as performed here) and analyses including corrections for phylogenetic dependence (such as independent contrasts, see Garland et al. 1992) recently reported (see Fisher and Owens 2000 and Nagy and Bradshaw 2000), and (2) recent criticisms of the usefulness of phylogenetic methods such as independent contrasts (Björklund 1997; Price 1997; Ricklefs and Starck 1996) mainly based on the strong assumptions made by such methods on evolutionary changes of traits (Harvey and Rambaut 2001).

References

Andersson, M. B. 1994. *Sexual Selection*. Princeton: Princeton University Press.

Bérubé, C., M. Festa-Bianchet, and J. T. Jorgenson. 1999. "Individual differences, longevity, and reproductive senescence in bighorn ewes," *Ecology* 80: 2555–2565.

Björklund 1997. "Are 'comparative methods' always necessary?" *Oikos* 80: 607–612.

Byers, J. A. 1997. *American Pronghorn: Social Adaptations and the Ghosts of Predators Past*. Chicago: Chicago University Press.

Calder, W. A. 1984. *Size, Function and Life History*. Cambridge, MA: Harvard University Press.

Catchpole, E. A. et al. 1998. *An Integrated Analysis of Soay Sheep Survival Data*. Report UKC/IMS/98/32. Canterbury: University of Kent.

Caughley, G. 1977. *Analysis of Vertebrate Populations*. Chichester: Wiley & Sons.

Clutton-Brock, T. H. 1991. *The Evolution of Parental Care*. Princeton: Princeton University Press.

————. 1988. *Reproductive Success: Studies of Individual Variation in Contrasting Breeding Systems*. Chicago: Chicago University Press.

Clutton-Brock, T. H., F. E. Guinness, and S. D. Albon. 1982. *Red Deer: Behavior and Ecology of Two Sexes*. Chicago: Chicago University Press.

Comfort, A. 1979. *The Biology of Senescence*. Edinburgh: Churchill Livingstone.

Coulson, T. et al. 2001. "Age, sex, density, winter weather and population crashes in Soay sheep," *Science* 292: 1528–1531.

Cransac, N. et al. 1997. "Patterns of mouflon (*Ovis gmelini*) survival under moderate environmental conditions: Effects of sex, age and epizootics," *Canadian Journal of Zoology* 75: 1867–1875.

Dobson, F. S., T. S. Risch, and J. O. Murie. 1999. "Increasing returns in the life history of Colombian ground squirrels," *Journal of Animal Ecology* 68: 73–86.

Ericsson, G., K. Wallin, J. P. Ball, and M. Broberg. 2001. "Age-related reproductive effort and senescence in free-ranging moose, *Alces alces*," *Ecology* 82: 1613–1620.

Festa-Bianchet, M., and J. T. Jorgenson. 1998. "Selfish mothers: Reproductive expenditure and resource availability in bighorn ewes," *Behavioral Ecology* 9: 144–150.

Fisher, D. A. and I. F. P. Owens. 2000. "Female home range size and the evolution of social organization in macropod marsupials," *Journal of Animal Ecology* 69: 1083–1098.

Gaillard, J. M., D. Allainé, D. Pontier, N. G. Yoccoz, and D. E. L. Promislow. 1994. "Senescence in natural populations of mammals: A reanalysis," *Evolution* 48: 509–516.

Gaillard, J. M., D. Delorme, J. M. Boutin, G. Van Laere, B. Boisaubert, and R. Pradel. 1993. "Roe deer survival patterns: A comparative analysis of contrasting populations," *Journal of Animal Ecology* 62: 778–791.

Gaillard, J. M., M. Festa-Bianchet, N. G. Yoccoz, A. Loison, and C. Toïgo. 2000a. "Temporal variation in fitness components and population dynamics of large herbivores," *Annual Review of Ecology and Systematics* 31: 367–393.

Gaillard, J. M., M. Festa-Bianchet, D. Delorme, and J. T. Jorgenson. 2000b. "Body mass and individual fitness in female ungulates: Bigger is not always better," *Proceedings of the Royal Society of London B* 267: 471–477.

Gaillard, J. M., A. Loison, and C. Toïgo. 2003. "Does accounting for variation in life history traits provide a step towards more realistic population models in wildlife management? The case of ungulates," in M. Festa-Bianchet and M. Apollonio (eds.), *Animal Behavior and Wildlife Management*. Washington, DC: Island Press.

Gaillard, J. M., D. Pontier, D. Allainé, J. D. Lebreton, J. Trouvilliez, and J. Clobert. 1989. "An analysis of demographic tactics in birds and mammals," *Oikos* 56: 59–76.

Garland, T. J., P. H. Harvey, and A. R. Ives. 1992. "Procedures for the analysis of comparative data using phylogenetically independent contrasts," *Systematic Biology* 41: 18–32.

Gompertz, B. 1825. "On the nature of the function expressive of the law of human mortality," *Philosophical Transactions of the Royal Society of London* 27: 513–517.

Hamlin, K. L. et al. 2000. "Evaluating the accuracy of ages obtained by two methods for Montana ungulates," *Journal of Wildlife Management* 64: 441–449.

Harvey, P. H. and M. D. Pagel. 1991. *The Comparative Method in Evolutionary Ecology*. Oxford: Oxford University Press.

Harvey, P. H. and A. Rambaut. 2001. "Comparative analyses for adaptive radiations," *Philosophical Transactions of the Royal Society of London Series B* 355: 1599–1605.

Houle, D. 1991. "Genetic covariance of fitness correlates: What genetic correlations are made of and why it matters," *Evolution* 45: 630–648.

Jorgenson, J. T., M. Festa-Bianchet, J. M. Gaillard, and W. D. Wishart. 1997. "Effects of age, sex, disease and density on survival of bighorn sheep," *Ecology* 78: 1019–1032.

Law, R. 1979. "Ecological determinants in the evolution of life histories," in R. M. Anderson, B. D. Turner, and L. R. Taylor (eds.), *Population Dynamics*. Oxford: Blackwell Scientific Publications.

Lebreton, J. D., K. P. Burnham, J. Clobert, and D. R. Anderson. 1992. "Modeling survival and testing biological hypotheses using marked animals: A unified approach with case studies," *Ecological Monographs* 62: 67–118.

Loison, A. et al. 1999a. "Age-specific survival in five populations of ungulates: Evidence of senescence," *Ecology* 80: 2539–2554.

Loison, A., J. M. Gaillard, and H. Houssin. 1994. "New insight on survivorship of female chamois (*Rupicapra rupicapra*) from marked animals," *Canadian Journal of Zoology* 72: 591–597.

Loison, A., J. M. Gaillard, C. Pélabon, and N. G. Yoccoz. 1999b. "What factors shape sexual size dimorphism in ungulates?" *Evolutionary Ecology Research* 1: 611–633.

Lomnicki, A. 1978. "Individual differences between animals and natural regulation of their numbers," *Journal of Animal Ecology* 47: 461–475.

Nagy, K. A. and S. D. Bradshaw. 2000. "Scaling of energy and water fluxes in free-living arid-zone Australian marsupials," *Journal of Mammalogy* 81: 962–970.

Newton, I. 1989. *Lifetime Reproduction in Birds*. London: Academic Press.

Nichols, J. D. 1992. "Capture-recapture models," *Bioscience* 42: 94–102.

Peters, R. H. 1983. *The Ecological Implications of Body Size*. Cambridge: Cambridge University Press.

Pettorelli, N., J. M. Gaillard, P. Duncan, J. P. Ouellet, and G. Van Laere. 2001. "Population density and small-scale variation in habitat quality affect phenotypic quality in roe deer," *Oecologia* 128: 400–405.

Price, T. 1997. "Correlated evolution and independent contrasts," *Philosophical Transactions of the Royal Society of London Series B* 352: 519–529.

Promislow, D. E. L. 1991. "Senescence in natural population of mammals: A comparative study," *Evolution* 45: 1869–1887.

Read, A. F., and P. H. Harvey. 1989. "Life history differences among the eutherian radiations," *Journal of Zoology* 219: 329–353.

Ricklefs, R. E. and J. M. Starck. 1996. "Applications of phylogenetically independent contrasts: A mixed progress report," *Oikos* 77: 167–172.

Rochet, M. J., P. A. Cornillon, R. Sabatier, and D. Pontier. 2000. "Comparative analysis of phylogenetic and fishing effects in life history patterns of teleost fishes," *Oikos* 91: 255–270.

Roff, D. A. 1992. *The Evolution of Life Histories*. London: Chapman & Hall.

Rose, M. R. 1991. *Evolutionary Biology of Aging*. New York: Oxford University Press.

Sacher, G. A. 1978. "Longevity and aging in vertebrate evolution," *Bioscience* 28: 497–501.

Shine, R. and E. L. Charnov. 1992. "Patterns of survival, growth, and maturation in snakes and lizards," *American Naturalist* 139: 1257–1269.

Sibly, R. M. et al. 1997. "Mortality rates of mammals," *Journal of Zoology* 243: 1–12.

Spinage, C. A. 1972. "African ungulate life tables," *Ecology* 53: 645–652.

Stearns, S. C. 1992. *The Evolution of Life Histories*. Oxford: Oxford University Press.

Toïgo, C. 1998. *Stratégies biodémographiques et sélection sexuelle chez le bouquetin des Alpes (*Capra ibex ibex*)*. Unpublished Ph.D. thesis, University of Lyon, France.

Toïgo, C. and J. M. Gaillard. 2003. "Causes of sex-biased survival in prime age ungulates: Sexual size dimorphism, mating tactic or environment harshness?" *Oikos*.

Toïgo, C., J. M. Gaillard, and J. Michallet. 1997. "Adult survival of the sexually dimorphic Alpine ibex (*Capra ibex ibex*)," *Canadian Journal of Zoology* 75: 75–79.

Tuomi, J., T. Hakala, and E. Haukioja. 1983. "Alternative concepts of reproductive effort, cost of reproduction, and selection in life history evolution," *American Zoology* 23: 25–34.

Van Noordwijk, A. J. and G. de Jong. 1986. "Acquisition and allocation of resources: Their influence on variation in life history tactics," *American Naturalist* 128: 137–142.

Vincent, J. P., J. M. Angibault, E. Bideau, and J. M. Gaillard. 1994. "Le problème de la détermination de l'âge: Une source d'erreur négligée dans le calcul des tables de vie transversales," *Mammalia* 58: 293–299.

Williams, G. C. 1966. *Natural Selection and Adaptation*. Princeton, NJ: Princeton University Press.

Environment and Longevity: The Demography of the Growth Rate

Marc Mangel

One of the cornerstones of demography is the mortality rate $m(x,t)$ of individuals of age x at calendar time t. However, from the perspective of evolutionary theory in general (Stearns and Hoekstra 2000) or biodemography more specifically (Gavrilov and Gavrilova 1991; Carey 2001; Carey and Judge 2001) the mortality rate is an evolved quantity, the result of natural selection acting on patterns of growth, behavior, and reproduction.

Our understanding of aging has been enormously advanced by genetic studies in the laboratory, to the point that it appears we will be able to understand all aspects of life span and aging with genetic tools. But aging takes place outside the laboratory and is also very much a phenotypic process. Recent studies, showing that the "public mechanisms" affecting longevity (Martin et al. 1996) are common to a wide variety of organisms (Guarente and Kenyon 2000; Strauss 2001; Zhang and Herman 2002), suggest that we need to understand the role of environment to master our understanding of aging and life span. Furthermore, it is likely that our understanding will be advanced the most when we are able to link proximate (mechanistic) and ultimate (fitness/selection–related) approaches to life-history questions, thereby "dissecting" the life history (Kirkwood 1992; Thorpe et al. 1998).

In this chapter, I develop a model that links environment, growth rate, and life span in the context of a Darwinian framework. In the model, the mortality rate emerges as the result of adaptive processes associated with growth and reproduction. These connections have important demographic implications. For example, it is likely that most organisms evolved in a world with fluctuating food supply, so that deeply embedded in the evolutionary history of organisms lies the ability to survive periods of food shortage. Such survival mechanisms could include adjustments to the growth rate or to reproductive output. This may be one of the causes of the association between caloric restriction and extension of life span, as suggested by Masoro and Austad (1996).

The approach that I use is based on life-history theory (Roff 1992; Stearns 1992). This field of evolutionary biology focuses on natural selection, constraint, adaptation, and Darwinian fitness. R. A. Fisher (1958 [1930]) described life-history theory as asking how environment and age affect the allocation of resources to reproduction. McNamara (1993) described life-history theory through a set of interconnected questions: What is the intrinsic rate of increase of a population that follows a given strategy and how does one find it? What is the current reproductive value of an organism and how does one calculate it? How does one decompose reproductive value into current and future reproductive success? How much effort should an organism put into current reproduction at the expense of future reproduction? How is individual behavior linked to population growth rates? What should an organism maximize over its lifetime?

The work outlined here is motivated in part by the remarkable longevity of the Pacific rockfish, which I describe next.

Longevity and growth in Pacific rockfish

The rockfish (*Sebastes* spp.) are remarkable because of their longevity (Table 1; also see Reznick et al. 2002 and the web site maintained by the Pacific Fisheries Management Council «http://www.psmfc.org/habitat/rockfish_lifespan.html»). Many rockfish species live in the California Current of the Pacific Ocean, which flows eastward toward North America and then southward along the coast. This environment is characterized by seasonal vari-

TABLE 1   Maximum estimated ages of some commercially important rockfish *Sebastes*

Species	Common name	Maximum age (years)
S. aleutianus	Rougheye rockfish	140
S. alutus	Pacific ocean perch	90
S. borealis	Shortraker rockfish	120
S. brevispinis	Silvergray rockfish	80
S. crameri	Darkblotched rockfish	47
S. entomelas	Widow rockfish	58
S. flavidus	Yellowtail rockfish	64
S. paucispinis	Bocaccio	36
S. pinniger	Canary rockfish	75
S. proriger	Redstripe rockfish	41
S. reedi	Yellowmouth rockfish	71
S. variegatus	Harlequin rockfish	43
S. zacentrus	Sharpchin rockfish	45

SOURCE: Leaman and Beamish (1984).

ability (Chelton et al. 1982; Lynn and Simpson 1987) and long-term fluctuations (MacCall 1996; Watanabe and Nitta 1999) in which different environmental regimes persist at times on the order of decades before an abrupt
transition occurs in environmental characteristics (particularly a switch between warm and cold water temperature and differential levels of food).
Seasonally unpredictable environments are associated with high variation
in spawning success, with such success being often associated with long life
span. Murphy (1968) and Mann and Mills (1979) demonstrated a strong
correlation between variation in spawning success (maximum/minimum
observed reproductive success) and reproductive life span; see Figure 1. Long
reproductive life span must perforce be correlated with long life span, and
we are thus led to wonder how long life span is achieved.

   In general, rockfish grow slowly. The growth and maturity of fish are
traditionally measured using the von Bertalanffy equation describing size
at age (Haldorson and Love 1991). This description involves asymptotic size
(L_∞), growth rate (k), and a parameter that describes initial size (t_0).[1] The
growth rates for rockfish are low ($k = 0.08$ to 0.19 per year) compared to
shorter-lived fish of similar size (for example, $k = 0.3$ to 0.4 for cods; Leaman
and Beamish 1984). The description of the life history is completed by inclusion of mortality rate M.

**FIGURE 1   Relationship between variation in spawning success
(fecundity)(often environmentally induced) and reproductive
life span for a variety of species of fish**

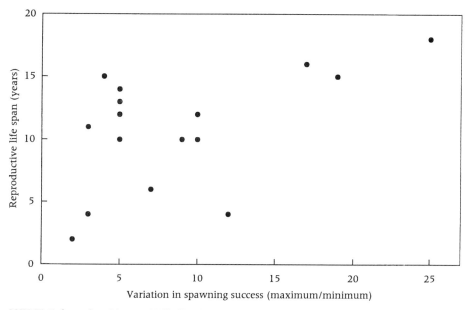

SOURCE: Redrawn from Mann and Mills (1979).

The fishery scientist R. J. H. Beverton explored the relationship between asymptotic size, growth constant, mortality rate, and life span
(Beverton and Holt 1959; Beverton 1963, 1987, 1992) and created a theory
relating growth, maturity, and longevity (or GML). Beverton's theory is reviewed by Mangel and Abrahams (2001); the work reported here is an extension of the line of research started by Beverton.

The importance of facultative growth (metabolic choice) for demography

Work with the nematode worm *C. elegans* (Wolkow et al. 2000; Cowen 2001;
van Voorhies 2001) suggests that metabolism and longevity are separately
regulated, but clearly connected. In common with terrestrial organisms, fish
use oxygen to fuel growth and reproduction. In the California Current, oxygen concentration varies by a factor of nearly 10 (Lynn and Simpson 1987;
Boehlert and Kappenman 1980); faster growth occurs in the northern parts
of the range of the fish, where water temperatures are lower and oxygen
content is higher. Growth rate and reproductive success generally decline as
dissolved oxygen declines (Stewart 1967). Indeed, some authors (Pauly 1981;
Bakun 1996) have proposed that oxygen may be an alternative currency to
food in the ecological budget of fish and that understanding the role of oxygen is crucial to understanding all aspects of fish life history (Pauly 1998).

It has long been understood from both empirical (Pettersson and Bronmark 1993) and theoretical (Clark and Mangel 2000; Mangel and Stamps
2001) perspectives that organisms commonly make tradeoffs between mortality risk and foraging success. Ricklefs et al. (1994) hypothesized that rapid
growth may be inversely correlated with functional maturity and muscle
development, and Ricklefs and Scheuerlein (2001) have shown that in zoo
animals the best explanatory variable for the initial mortality rate and the
rate of aging in a Weibull model is postnatal growth. Here, I explore the
tradeoff between rapid growth, internal damage resulting from oxidative
processes, fitness consequences (von Schantz et al. 1999), and the demography of life span (see Figure 2). I assume that faster growth increases the
prevalence of reactive oxygen species (ROS). There is evidence that ROS
can damage proteins, DNA, and lipids (Tolmassof 1980; Stadtman 1992;
Ames 1993; Barja et al. 1994; Pollack and Leeuwenburgh 1999; Ozawa 1995;
Shigenaga 1994; Lane et al. 1996; Brewer 2000; van Voorhies 2001; Van
Remmen and Richardson 2001). There is also evidence of direct associations between oxidative damage and life span (Sohal et al. 1993; Agarwal
and Sohal 1994; Harshman and Haberer 2000) and between longevity and
low levels of free radical production *in vivo* (Barja et al. 1994).

Furthermore, reproduction is associated with increased use of oxygen.
In some cases, gestating fish used about 35 percent more oxygen per mass

FIGURE 2 Maximum longevity in 46 species of fishes in the family *Scorpaenidae* by maximum depth

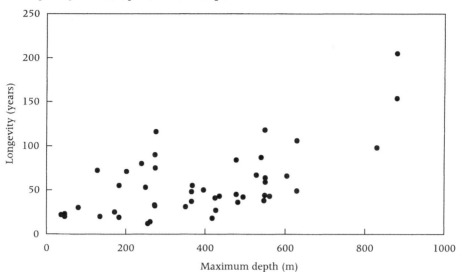

SOURCE: Redrawn from Cailliet et al. (2001).

than nongestating fish (Boehlert et al. 1991). Hopkins et al. (1995) found that metabolic rates of incubating yellowtail rockfish (*Sebastes flavidus*) were 82 percent higher than those of spent females and 101 percent higher than those of nonreproductive males. Whether these differences represent oxygen consumption by embryos or females' providing resources to embryos, they surely represent a potential source of oxidative stress.

Oxidative stress is intimately linked to reproductive fitness (von Schantz et al. 1999); indeed, Sibly and Calow (1986) concluded that evolution proceeded toward maximization of ATP production, but not necessarily minimization of heat per ATP produced. There is evidence for the anticipation of oxidative damage (Novoseltsev et al. 2000), individual regulation of oxygen consumption (Quetin and Ross 1989), and within-individual variation in oxidative stress according to calendar time, age, and physiological or reproductive state (Rikans and Hornbrook 1997; Zielinski and Pörtner 2000; Blount et al. 2001; Filho et al. 2001; Royle et al. 2001; Salmon et al. 2001). I define responses that vary within individuals according to state or time as facultative (Mangel and Clark 1988; Clark and Mangel 2000). Because the facultative response refers to growth rate, which is intimately tied to metabolism, I will also refer to it as metabolic choice.

In order to understand how facultative growth and life span are related, I return to the von Bertalanffy description of growth and dissect it into components. One assumes that:

—Weight ($W(t)$) and length ($L(t)$) are allometrically related by $W(t) = cL(t)^3$.

—The increment in weight is determined by the difference between anabolic and catabolic terms, in which the anabolic term is proportional to surface area (hence $L(t)^2$), food availability, and oxygen; and the catabolic term is proportional to mass (hence $L(t)^3$) and oxygen that the fish uses.

—Asymptotic size is determined by the weight at which anabolic and catabolic terms balance. Larger values of the metabolic choice parameter (a) lead to faster growth, larger size at age, and smaller age at maturity.[2] If predation is size dependent (a common feature in the marine environment; McGurk 1986, 1996), then faster growth will lead to higher survival rates from the viewpoint of predation.

To capture the effects of ROS, I introduce another state variable that measures damage accumulated at age[3]—measured, for example, in terms of lipid peroxidation (Matsuo and Kaneko 1999), oxidized amino acids, or DNA (Leeuwenburgh et al. 1999). Since oxidative damage accumulates as a function of metabolism, the dynamics of damage are related to metabolism. The reason for introducing oxidative damage caused by ROS is that survival from one age to the next depends upon both size (bigger size and thus faster growth are better) and accumulated damage (slower growth is better). Thus, there is an interplay between the short-term and long-term survival processes.[4]

This formulation is sufficient to compute the survival to a specific age for an individual using a fixed value of the metabolic choice parameter. In general, predation-related survival is an increasing function of the metabolic choice parameter, and damage-related survival is a decreasing function of the parameter. Thus, a fixed high metabolic rate allows the fish to grow rapidly, escaping size-dependent predation and reaching age at maturity sooner. But a fixed high metabolic rate also leads to increased levels of oxidative damage, thus decreasing long-term survival. Survival, and thus life span, depends upon the balance of these factors. In Figure 3, I show life span (defined as the age at which survival from birth falls to less than one in 10 million) when the metabolic parameter is fixed throughout an individual's life. We see, for example, that in the base case maximum life span is about 48 years, achieved by using a fixed metabolic parameter of 0.6; when the repair rate is doubled, the corresponding values are 79 years and $a = 1.2$. Thus, although doubling the repair rate allows individuals to grow at a faster rate, with concomitant benefits, it does not double life span.

On the other hand, whereas fast growth may be advantageous at small size (to help the organism escape size-dependent predation), there is little advantage to fast growth at larger sizes and a definite cost. We are thus led to ask what pattern of growth (and thus ROS production) maximizes the

FIGURE 3   Life span (age at which survival probability drops to less than 1 in 10 million) among fish as a function of a fixed metabolic parameter *a*. When there is a balance between predation-related survival and damage-related survival, life span peaks for an intermediate value of the metabolic parameter.

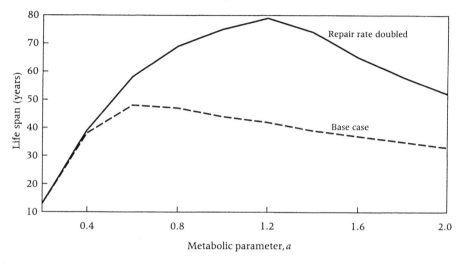

long-term number of descendants (equivalent to *r* of the Euler–Lotka equation) and what is the effect on life span?

To characterize reproduction, I assume that a fish is capable of reproduction once its size exceeds 60 percent of asymptotic size (Beverton 1992) and that the amount of reproduction is determined allometrically by length. Reznick et al. (2002) argued that this increasing fecundity with age separates fish from birds or mammals and has allowed the evolution of such remarkably long lives. The optimal pattern of growth can be determined by the method of stochastic dynamic programming (Mangel and Clark 1988; Clark and Mangel 2000; McNamara 2000). The output of such an analysis is a "decision matrix" that gives the optimal value of the parameter of metabolic choice for each size and damage level. Age dependence is an implicit rather than explicit relationship. Thus, aging, senescence, and longevity are emergent properties in this approach (Waldorp 1992; Jazwinski et al. 1998; Goodwin 2001; Mangel 2001), rather than constitutive properties of the model.

The theory leads to the prediction of age-dependent growth and damage, through the parameter of metabolic choice (Figure 4a), and thus the dynamics of length (Figure 4b), survival (Figure 4c), and mortality (Figure 4d). The theory thus provides a method for computing life span and for deriving the mortality rate as a result of the adaptive processes of growth and reproduction. The life span, again defined as the age at which survival

FIGURE 4   A theory that allows facultative growth (metabolic choice) allows prediction of the optimal age-dependent value of the metabolic parameter (panel a) and the length (panel b), survival (panel c), and mortality (panel d) dynamics. This theory thus allows the prediction of the mortality rate as a result of adaptive processes of growth and reproduction.

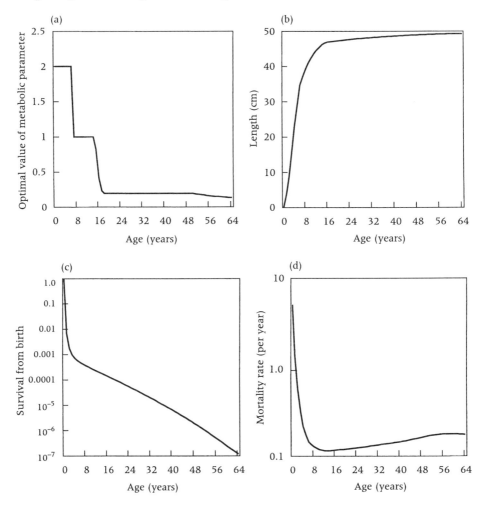

probability from birth drops below one in 10 million, is 65 years; with a doubled repair rate the life span is 88 years. The precise factors affecting life span include size-dependent mortality, damage, and food availability. Because mortality is size dependent, small fish experience high predation rates, but these decline as the fish grow older. In all of these cases, facultative growth leads to longer life span than occurs with a fixed growth parameter. It is important to recognize, however, that life span, like the mortality rate itself, is the outcome of natural selection maximizing expected reproductive success, rather than a direct target of natural selection per se.

Discussion

The models introduced here are conceptual ones, intended to facilitate exploration of the relationship between environment and longevity and to illustrate how the perspective of evolutionary ecology allows us to predict the mortality rate as a function of growth and reproduction. To convert these models to applied, calculational tools (*sensu* Mangel et al. 2001) requires extensions that include temperature dependence and density dependence of growth (Ylikarjula, Heino, and Dieckmann 1999; Lorenzen 2000; Mangel and Abrahams 2001).

Even so, a strong qualitative prediction of this theory is an ontogenetic shift with age toward slower growth (and thus to generally deeper and oxygen-poorer water). That is, a behavioral trait is predicted on the basis of physiological processes in a demographic context. Various species of fish fit this pattern. Chilipepper rockfish (*Sebastes goodei*) and splitnose rockfish (*Sebastes diplorproa*) show an ontogenetic shift toward deeper water and lower oxygen concentrations as they get bigger (Vetter and Lynn 1997). Shortspine thornyhead (*Sebastolobus alascanus*), longspine thornyhead (*Sebastolobus altivelis*), and dover sole (*Microstomus pacificus*) have the same pattern but move even deeper and into the oxygen-minimum zone (Vetter et al. 1994; Vetter and Lynn 1997). Similarly, juvenile sablefish (*Anoplopoma fimbria*) (Pallas) are found at the surface in oxygen- (and predator-) filled regions of the ocean, but adult sablefish are found in the subsurface zone of low oxygen concentration (Bakun 1996; Sogard and Olla 1998). The study of these patterns, known as bathymetric demography (Jacobson and Hunter 1993; Jacobson and Vetter 1996), has important implications for fishery management. In this regard fish, because they live in a much-enlarged three-dimensional world, allow exploration of demographic ideas that terrestrial organisms do not.

It is also clear that not all long-lived marine species live in deep water (sturgeon and yelloweye rockfish, for example, do not) and that several short-lived species are found in deep water (the lantern fish, for example). According to Donald Gunderson (personal communication), data on thornyheads suggest that oxygen, temperature, and depth do little to explain differences in mortality across species or regions. These differences may be a result of the balance of production of ROS and repair. However, the arguments presented here could be adapted to other mechanisms that connect rate of growth and damage.

The models show the importance of considering the organism in its environment, the role of natural selection in shaping life span in the context of environment, and the need for broad integration of different levels of biological organization for a full understanding of longevity.

Notes

For conversations or correspondence on various aspects of this chapter, I thank James Childress, Donald Gunderson, Alec MacCall, Jon Miller, Timothy Moreland, George Somero, and Russell Vetter. For comments on the manuscript, I thank Donald Gunderson, Kai Lorenzen, Nicholas Royle, and Nicholas Wolf. This work was started when I held the position as William R. and Lenore Mote Eminent Scholar Chair in Fisheries Ecology and Enhancement at Florida State University and the Mote Marine Laboratory in winter 2000.

1 If $L(t)$ denotes the length of a fish at age t, the von Bertalanffy description of growth is $L(t) = L_\infty(1-\exp(-k(t-t_0)))$. This equation is the solution of the differential equation

$$\frac{dL}{dt} = k(L_\infty - L(t))$$

and satisfies the difference equation

$$L(t + 1) = L(t)\exp(-k) + L_\infty(1 - \exp(-k))$$

as can be verified directly from substitution into the differential equation.

2 In a discrete time formulation the weight dynamics are

$$W(t + 1) - W(t) = aY[O_2]L(t)^2 - ma[O_2]L(t)^3$$

where a is the parameter of metabolic choice and can be interpreted as the fraction of ambient oxygen flowing across the gills (Waller et al. 2000) or a correlate of depth; Y is food availability; $[O_2]$ is the concentration of oxygen in the water; and m determines catabolic costs. Length dynamics in discrete time follow directly from this since it is assumed that weight is proportional to length cubed.

3 The discrete time dynamics for damage in the nonreproductive period are

$$D(t + 1) = D(t) + \kappa ma[O_2]L(t)^3 - \rho$$

where κ is a parameter converting metabolism to damage and ρ is the damage repair rate. When the fish is reproductively mature, additional damage accumulates, associated with the numbers and quality of offspring (more or higher-quality offspring imply higher rates of accumulation of damage).

4 I assume here that the two processes are independent, so that given $L(t) = 1$ and $D(t) = d$

$$\text{Prob}\{\text{survive to age } t+1\} =$$
$$\exp\{-\mu / 1\}\exp\{-(d / d_c)^2\}$$

where μ and d_c are parameters characterizing size- and damage-dependent mortality respectively. Given the dynamics of size and damage, this probability can be iterated from $t = 0$, when survival probability is unity, into the future. The mortality rate is then computed from the survival probability in the standard way.

References

Agarwal, S. and R. S. Sohal. 1994. "DNA oxidative damage and life expectancy in house-flies," *Proceedings of the National Academy of Sciences USA* 91: 12332–12335.

Ames, B. N., M. K. Shigenaga, and T. M. Hagen. 1993. "Oxidants, antioxidants and the degenerative disease of aging," *Proceedings of the National Academy of Sciences USA* 90(17): 7915–7922.

Bakun, A. 1996. *Pattern in the Ocean: Oceanic Process and Marine Population Dynamics.* La Jolla, CA: California Sea Grant.

Barja, G., S. Cadenas, C. Rojas, M. Lopez-Torres, and R. Perez-Campo. 1994. "A decrease of free radical production near critical targets as a cause of maximum longevity in mammals," *Comparative Biochemistry and Physiology* 108B: 501–512.

Beverton, R. J. H. 1963. "Maturation, growth, and mortality of clupeid and engraulid stocks in relation to fishing," *Rapport Conseil International Exploration de la Mer* 154: 44–67.

———. 1987. "Longevity in fish: Some ecological and evolutionary considerations," in A. D. Woodhead and K. H. Thompson (eds.), *Evolution of Longevity in Animals.* New York: Plenum, pp. 161–185.

———. 1992. "Patterns of reproductive strategy parameters in some marine teleost fishes," *Journal of Fish Biology* 41 (Supplement B): 137–160.

Beverton, R. J. H. and S. J. Holt. 1959. "A review of the life-spans and mortality rates of fish in nature, and their relationship to growth and other physiological characteristics," *CIBA Foundation Colloquium on Ageing* 54: 142–180.

Boehlert, G. W. and R. F. Kappenman. 1980. "Variation of growth with latitude in two species of rockfish (*Sebastes pinniger* and *S. diploproa*) from the Northeast Pacific Ocean," *Marine Ecology-Progress Series* 3: 1–10.

Blount, J. D. et al. 2001. "Antioxidants, showy males and sperm quality." *Ecology Letters* 4: 393–396.

Boehlert, G. W., M. Kusakari, and J. Yamada. 1991. "Oxygen consumption of gestating female *Sebastes schlegeli*: Estimating the reproductive costs of livebearing," *Environmental Biology of Fishes* 30: 81–89.

Brewer, G. J. 2000. "Neuronal plasticity and stressor toxicity during aging," *Experimental Gerontology* 35: 1165–1183.

Cailliet, G. M. et al. 2001. "Age determination and validation studies of marine fishes: Do deep-dwellers live longer?" *Experimental Gerontology* 36: 739–764.

Carey, J. R. 2001. "Insect biodemography," *Annual Review of Entomology* 46: 79–110.

Carey, J. R. and D. S. Judge. 2001. "Principles of biodemography with special reference to human longevity," *Population: An English Selection* 13: 9–40.

Chelton, D. B., P. A. Bernal, and J. A. McGowan. 1982. "Large-scale inter annual physical and biological interaction in the California Current," *Journal of Marine Research* 40: 1095–1125.

Clark, C. W. and M. Mangel. 2000. *Dynamic State Variable Models in Ecology. Methods and Applications.* New York: Oxford University Press.

Cowen, T. 2001. "A heady message for life span regulation," *Trends in Genetics* 17: 109–113.

Filho, D. W. et al. 2001. "Seasonal changes in antioxidant defenses of the digestive gland of the brown mussel (*Perna perna*)," *Aquaculture* 203: 149–158.

Fisher, R. A. 1958 [1930]. *The Genetical Theory of Natural Selection* 2nd edition [1st edition]. New York: Dover Press.

Gavrilov, L. A. and N. S. Gavrilova. 1991. *The Biology of Life Span: A Quantitative Approach.* Chur, Switzerland: Harwood Academic Publishers.

Goodwin, B. 2001. *How the Leopard Changed Its Spots. The Evolution of Complexity.* Princeton, NJ: Princeton University Press.

Guarente, L. and C. Kenyon. 2000. "Genetic pathways that regulate ageing in model organisms," *Nature* 408: 255–262.

Haldorson, L. and M. Love. 1991. "Maturity and fecundity in the rockfishes, *Sebastes* spp., a review," *Marine Fisheries Review* 53(2): 25–31.

Harshman, L. G. and B. A. Haberer. 2000. "Oxidative stress resistance: A robust correlated response to selection in extended longevity lines of *Drosophila melanogaster*?" *Journal of Gerontology: Biological Sciences* 55A: B415–B417.

Hopkins, T. E., M. B. Eldridge, and J. J. Cech. 1995. "Metabolic costs of viviparity in yellowtail rockfish, *Sebastes flavidus*," *Environmental Biology of Fishes* 43: 77–84.

Jacobson, L. D. and J. R. Hunter. 1993. "Bathymetric demography and management of Dover sole," *North American Journal of Fisheries Management* 13: 405–420.

Jacobson, L. D. and R. D. Vetter. 1996. "Bathymetric demography and niche separation of thornyhead rockfish: *Sebastolobus alascanu* and *S. altivelis*," *Canadian Journal of Fisheries and Aquatic Sciences* 53: 600–609.

Jazwinski, S. M., S. Kim, C-Y. Lai, and A. Bengura. 1998. "Epigentic stratification: The role of individual change in the biological aging process," *Experimental Gerontology* 33: 571–580.

Jensen, A. 1996. "Beverton and Holt life history invariants result from optimal trade-off of reproduction and survival," *Canadian Journal of Fisheries and Aquatic Sciences* 53: 820–822.

Kirkwood, T. B. L. 1992. "Comparative life spans of species: Why do species have the life spans they do?" *American Journal of Clinical Nutrition* 55: 1191S–1195S.

———. 1977. "Evolution of ageing," *Nature* 270: 301–304.

Kirkwood, T. B. L. and M. R. Rose. 1991. "Evolution of senescence: Late survival sacrificed for reproduction," *Philosophical Transactions of the Royal Society of London B*: 15–24.

Lane, M. A. et al. 1996. "Calorie restriction lowers body temperature in Rhesus monkeys, consistent with a postulated anti-aging mechanism in rodents," *Proceedings of the National Academy of Sciences USA* 93(9): 4159–4164.

Leaman, B. M. 1991. "Reproductive styles and life history variables relative to exploitation and management of *Sebastes* stocks," *Environmental Biology of Fishes* 30: 253–271.

Leaman, B. M. and R. J. Beamish. 1984. "Ecological and management implications of longevity in some northeast Pacific groundfishes," *International North Pacific Fisheries Commission Bulletin* 42: 85–97.

Leeuwenburgh, C., P. A. Hansen, J. O. Holloszy, and J. W. Heinecke. 1999. "Oxidized amino acids in the urine of aging rats: Potential markers for assessing oxidative stress in vivo," *American Journal of Physiology (Regulatory Integrative Comparative Physiology)* 276: R128–R135.

Lorenzen, K. 2000. "Population dynamics and management," in M. C. M. Beveridge and B. J. McAndrew (eds.), *Tilapias: Biology and Exploitation*. Dordrecht: Kluwer Academic Publishers, pp. 163–226.

Lynn, R. J. and J. J. Simpson. 1987. "The California Current system: The seasonal variability of its physical characteristics," *Journal of Geophysical Research* 92: 12947–12966.

MacCall, A. 1996. "Patterns of low-frequency variability in fish populations of the California current," *California Cooperative Fishery Investigations (CalCOFI) Reports* 37: 100–110.

Mangel, M. 2001. "Complex adaptive systems, aging, and longevity," *Journal of Theoretical Biology* 213: 559–571.

Mangel, M. and M. V. Abrahams. 2001. "Age and longevity in fish, with consideration of the ferox trout," *Experimental Gerontology* 36: 765–790.

Mangel, M. and C. W. Clark. 1988. *Dynamic Modeling in Behavioral Ecology*. Princeton, NJ: Princeton University Press.

Mangel, M., O. Fiksen, and J. Giske. 2001. "Theoretical and statistical models in natural resource management and research," in T. M. Shenk and A. B. Franklin (eds.), *Modeling in Natural Resource Management. Development, Interpretation, and Application*. Washington, DC: Island Press, pp. 57–72.

Mangel, M. and J. Stamps. 2001. "Trade-offs between growth and mortality and the maintenance of individual variation in growth," *Evolutionary Ecology Research* 3: 583–593.

Mann, R. H. K. and C. A. Mills. 1979. "Demographic aspects of fish fecundity," *Symposia of the Zoological Society of London* 44: 161–177.

Martin, G. M., S. N. Austad, and T. E. Johnson. 1996. "Genetic analysis of ageing: Role of oxidative damage and environmental stresses," *Nature Genetics* 13: 25–34.

Masoro, E. J. and S. N. Austad. 1996. "The evolution of the antiaging action of dietary restriction: A hypothesis," *Journal of Gerontology: Biological Sciences* 51A: B387–B391.

Matsuo, M. and T. Kaneko. 1999. "Lipid peroxidation," in B. P. Yu (ed.), *Methods in Aging Research*. Boca Raton, FL: CRC Press, pp. 571–606.

McGurk, M. D. 1986. "Some remarks on 'Model of monthly marine growth and natural mortality for Babine Lake sockeye salmon (*Oncorhynchus nerka*)' by Furnell and Brett," *Canadian Journal of Fisheries and Aquatic Sciences* 43: 2535–2536.

———. 1996. "Allometry of marine mortality of Pacific salmon," *Fishery Bulletin* 94: 77–88.

McNamara, J. M. 1993. "State-dependent life history equations," *Acta Biotheoretica* 41: 165–174.

————. 2000. "A classification of dynamic optimization problems in fluctuating environments," *Evolutionary Ecology Research* 2: 457–471.

Murphy, G. I. 1968. "Pattern in life history and environment," *American Naturalist* 102: 391–403.

Novoseltsev, V. N., J. R. Carey, P. Liedo, J. Novoseltseva, and A. I. Yashin. 2000. "Anticipation of oxidative damages decelerates aging in virgin female medflies: Hypothesis tested by statistical modeling," *Experimental Gerontology* 35: 971–987.

Ozawa, T. 1995. "Mitochondrial DNA mutations associated with aging and degenerative diseases," *Experimental Gerontology* 30: 269–290.

Pauly, D. 1980. "On the interrelationships between natural mortality, growth parameters, and mean environmental temperature in 175 fish stocks," *Journal of the International Council for the Exploration of the Seas* 39(2): 175–192.

————. 1981. "The relationships between gill surface area and growth performance in fish: A generalization of von Bertalanffy's theory of growth," *Meeresforschung* 28: 251–282.

————. 1998. "Tropical fishes: Patterns and propensities, " *Journal of Fish Biology Supplement A* 53: 1–17.

Pettersson, L. B. and C. Bronmark. 1993. "Trading off safety against food: State dependent habitat choice and foraging in crucian carp," *Oecologia* 95: 353–357.

Pollack, M. and C. Leeuwenburgh. 1999. "Molecular mechanisms of oxidative stress in aging: Free radicals, aging, antioxidants and disease," in C. K. Sen, L. Packer, and O. Hanninen (eds.), *Handbook of Oxidants and Antioxidants in Exercise.* Amsterdam: Elsevier Science B.V., pp. 881–923.

Preston, S. H., P. Heuveline, and M. Guillot. 2001. *Demography: Measuring and Modeling Population Processes.* Oxford: Blackwell.

Quetin, L. B. and R. M. Ross. 1989. "Effects of oxygen, temperature and age on the metabolic rate of the embryos and early larval stages of the Antarctic krill *Euphausia superba* Dana," *Journal of Experimental Marine Biology and Ecology* 125: 43–62.

Reznick, D., C. Ghalambor, et al. (2002). "The evolution of senescence in fish," *Mechanisms of Ageing and Development* 123: 773–789.

Ricklefs, R. E., R. E. Shea, and I-H. Choi. 1994. "Inverse relationship between functional maturity and exponential growth rate of avian skeletal muscle: A constraint on evolutionary response," *Evolution* 48(4): 1080–1088.

Ricklefs, R. E. and A. Scheuerlein. 2001. "Comparison of aging-related mortality among birds and mammals," *Experimental Gerontology* 36: 845–857.

Rikans, L. E. and K. R. Hornbrook. 1997. "Lipid peroxidation, antioxidant protection and aging," *Biochimicia Biophysica Acta* 1362: 116–127.

Roff, D. A. 1992. *The Evolution of Life Histories.* New York: Chapman and Hall.

Royle, N. J., P. F. Surai, R. J. McCartney, and B. K. Speake. 1999. "Parental investment and yolk lipid composition in gulls," *Functional Ecology* 13: 298–306.

Royle, N. J., P. F. Surai, and I. R. Hartley. 2001. "Maternally derived androgens and antioxidants in bird eggs: Complementary but opposing effects?" *Behavioral Ecology* 12: 381–385.

Salmon, A. B., D. B. Marx, and L. G. Harshman. 2001. "A cost of reproduction in *Drosophila melanogaster*: Stress susceptibility," *Evolution* 55(8): 1600–1608.

Shigenaga, M. K., T. M. Hagen, and B. N. Ames. 1994. "Oxidative damage and mitochondrial decay in aging," *Proceedings of the National Academy of Sciences USA* 91(23): 10771–10778.

Sibly, R. M. and P. Calow. 1986. *Physiological Ecology of Animals.* Oxford: Blackwell Scientific Publications.

Sogard, S. M. and B. L. Olla. 1998. "Behavior of juvenile sablefish, *Anoplopoma fimbria* (Pallas), in a thermal gradient: Balancing food and temperature requirements," *Journal of Experimental Marine Biology and Ecology* 222: 43–58.

Sohal, R. S., S. Agarwal, A. Dubey, and W. C. Orr. 1993. "Protein oxidative damage is associated with life expectancy of houseflies," *Proceedings of the National Academy of Sciences USA* 90: 7255–7529.

Stadtman, E. R. 1992. "Protein oxidation and aging," *Science* 257: 1220–1224.

Stearns, S. C. 1992. *The Evolution of Life Histories*. New York: Oxford University Press.

Stearns, S. C. and R. F. Hoekstra. 2000. *Evolution*. New York: Oxford University Press.

Stewart, N. E., D. L. Shumway, and P. Doudoroff. 1967. "Influence of oxygen concentration on the growth of juvenile largemouth bass," *Journal of the Fisheries Research Board of Canada* 24: 475–494.

Strauss, E. 2001. "Growing old together," *Science* 292: 41–43.

Thorpe, J. E., M. Mangel, N. B. Metcalfe, and F. A. Huntingford. 1998. "Modelling the proximate basis of salmonid life-history variation, with application to Atlantic salmon, *Salmo salar* L.," *Evolutionary Ecology* 12: 581–600.

Tolmassof, J. M., T. Ono, and R. G. Cutler. 1980. "Superoxide dismutase: Correlation with life-span and specific metabolic rate in primate species," *Proceedings of the National Academy USA* 77(5): 2777–2781.

Van Remmen, H. and A. Richardson. 2001. "Oxidative damage to mitochondria and aging," *Experimental Gerontology* 36: 957–968.

van Voorhies, W. A. 2001. "Metabolism and life span," *Experimental Gerontology* 36: 55–64.

Vetter, R. D. and E. A. Lynn. 1997. "Bathymetric demography, enzyme activity patterns, and bioenergetics of deep-living scorpaenid fishes (genera *Sebastes* and *Sebastolobus*): Paradigms revisited," *Marine Ecology Progress Series* 155: 173–188.

Vetter, R. D., E. A. Lynn, and A. S. Costa. 1994. "Depth zonation and metabolic adaptation in Dover sole, *Microstomus pacificus*, and other deep-living flatfishes: Factors that affect the sole," *Marine Biology* 120: 145–159.

von Schantz, T., S. Bensch, M. Grahn, D. Hasselquist, and H. Wittzel. 1999. "Good genes, oxidative stress and condition-dependent signals," *Proceedings of the Royal Society of London Series B* 266: 1–2.

Waldorp, M. M. 1992. *Complexity*. New York: Touchstone Books.

Waller, U., E. Black, D. Burt, C. Groot, and H. Rosenthal. 2000. "The reaction of young coho *Oncorhynchus kisutch* to declining oxygen levels during long-term exposure," *Journal of Applied Ichthyology* 16: 14–19.

Watanabe, M. and T. Nitta. 1999. "Decadal changes in the atmospheric circulation and associated surface climate variations in the northern hemisphere winter," *Journal of Climate* 12: 494–510.

Wolkow, C. A., K. D. Kimura, M-S. Lee, and G. Kuvkun. 2000. "Regulation of *C. elegans* life-span by insulinlike signaling in the nervous system," *Science* 290: 147–150.

Ylikarjula, J., M. Heino, and U. Dieckmann. 1999. "Ecology and adaptation of stunted growth in fish," *Evolutionary Ecology* 13: 433–453.

Zhang, Y. and B. Herman. 2002. "Ageing and apoptosis," *Mechanisms of Ageing and Development* 123: 245–260.

Zielinski, S. and H.-O. Pörtner. 2000. "Oxidative stress and antioxidative defense in cephalopods: A function of metabolic rate or age?" *Comparative Biochemistry and Physiology Part B* 125: 147–160.

Life Span in the Light
of Avian Life Histories

ROBERT E. RICKLEFS
ALEX SCHEUERLEIN

Our purpose is to consider the comparative study of aging in natural and captive populations as a potential source of insight into the future of human aging. In conducting this assessment in the context of life-history comparisons, we consider only birds and mammals to restrict our analyses to organisms whose physiology is similar to that of humans. We show that the basic evolutionary theory of aging is well supported, and we explore variation in aging among species to characterize its relationship to factors in the environment and to other aspects of the life histories of organisms. Finally, we draw some inferences from comparative studies about possibilities for changes in human aging in the future.

The life history of an individual consists of a set of optimized tradeoffs related to its allocation of time, energy, materials, and body components to different functions (Roff 1992, 2002; Stearns 1992). Individuals have limited resources at their disposal, and resources devoted to one function cannot be used for another. The allocation of resources among functions can be mapped onto evolutionary fitness (Figure 1). This mapping occurs through the interaction of the genotype with the environment to determine survival and reproductive success at each age, which are summarized in a life table. Fitness can be defined retrospectively for an individual as the number of descendants it leaves in future generations. Fitness can be estimated prospectively from the life table entries, as we see below. Evolutionary fitness ties particular genotypes to variation in reproductive success. Variation in evolutionary fitness within a population is the basis for evolutionary response to the environment. We presume that the pattern of aging of an individual is in part genetically controlled and influences individual fitness, and that allocation of resources to extend potential life span reduces other aspects of the individual's performance and thus also affects its fitness. These relationships represent a tradeoff that defines the range of possible form and function of individuals—the phe-

FIGURE 1   The connection between the genotype and the life table involves several steps with progressively increasing environmental influence; variation in the reproductive success of different genotypes shapes the gene pool of the population through natural selection and evolutionary response, thereby closing the circle of adaptation; the life table expression of the genotype may also be influenced by density-dependent feedbacks on the conditions and resources of the environment

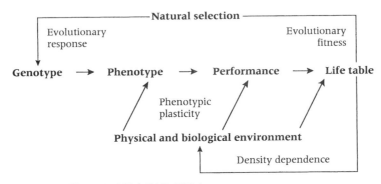

SOURCE: After Arnold (1983); Ricklefs (1991, 2000a).

notype. Every point along this tradeoff continuum may be represented by a fitness value (Figure 2).

We presume that evolution maximizes fitness and that any particular value of a tradeoff is selected according to its contribution to evolutionary fitness, as shown in Figure 2. A population's demography, as defined by the life table, may influence the optimum value of a tradeoff between fecundity and survival at different ages. With respect to aging, the fitness contribution of a life-history trait that extends life depends on the proportion of the population that lives long enough to experience the trait, as well as its reproductive success after that point (Abrams 1991, 1993). Thus, where individuals survive at a high rate through most of their lives, many achieve old age and the fitness value of extending life further is relatively high. Of course, this applies only when older individuals produce offspring or contribute post-reproductively to the survival and reproductive success of their own offspring (individual fitness: Hawkes et al. 1998; Alvarez 2000) or those of close relatives (inclusive fitness: Alexander 1974). Where individuals rarely survive through young adulthood, few reach old age and the fitness value of extended life span is small (Figure 3). Thus, selection to extend life span is strong in populations with high survival. Accordingly, the central prediction of evolutionary theory is that the rate of aging should be directly related to the mortality rate of adults from causes unrelated to aging (Edney and Gill 1968; Hamilton 1966; Medawar 1952; Williams 1957). As we see below, this prediction is supported by comparisons of the rate of increase in

FIGURE 2   The constrained relationship, for example between fecundity and potential life span (left), is optimized by maximizing fitness; each point along the constraint curve is a potential phenotype; the fitness value of phenotypes may vary between environments, leading to different optimized points

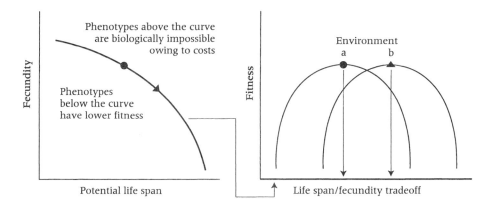

aging-related mortality in natural populations. If senescence could be postponed without cost, potential life span would evolve to be infinite because fitness increases with increasing life span, all else being equal. That life span is finite shows that the evolutionary modification of life span is constrained.

FIGURE 3   Lower extrinsic mortality results in more individuals surviving to old age and therefore potentially greater strength of selection on traits that influence fitness through contributions to fecundity or further survival at old age

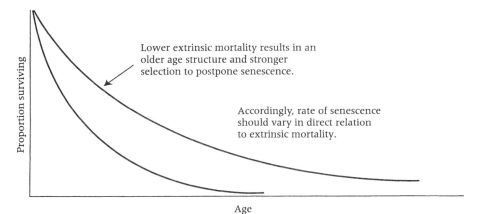

Mechanisms and tradeoffs in aging

The mechanisms of senescence are thought to include (1) wear and tear induced by the environment and by life activity itself, (2) accumulation of deleterious mutations with delayed onset of expression, or at least delayed onset of their influence on survival and reproductive success, and (3) pleiotropic genes that have beneficial effects early in life and deleterious effects at older ages. The means that organisms employ to postpone senescence are not well understood, but certainly include mechanisms that prevent and repair damage (Finkel and Holbrook 2000; Perez-Campo et al. 1998; Promislow 1994). These mechanisms may impose energetic or other costs, which would lead to tradeoffs between modification of senescence and other life-history traits that affect fitness in opposite ways.

Constrained relationships, such as the one shown in Figure 2, may embody mechanisms controlling the rate of senescence. As yet, we have little insight into the nature of such constrained relationships. George Williams (1957) introduced a general term, "antagonistic pleiotropy," to describe constrained relationships, and this has been elaborated by Michael Rose (1991) and others. Antagonistic pleiotropy implies a genetic link between two functions or aspects of performance having contrasting effects on evolutionary fitness (e.g., Weinstein and Ciszek 2002). Originally, antagonistic pleiotropy was thought to result from genes whose expression extended life span but also reduced early reproductive success. Relatively few such genes have been identified (Caruso et al. 2000; Ricklefs and Finch 1995; Rose 1991; Silbermann and Tatar 2000; Stearns and Partridge 2000), but antagonistic pleiotropy also can arise by way of functional rather than genetic connections (Ricklefs 2000a). For example, if aging resulted from wear and tear, mechanisms that prevented or repaired damage might prolong life span at a cost of resources allocated to reproduction, even though the deterioration associated with aging itself did not have a direct genetic cause (Kirkwood 1990; Kirkwood and Holliday 1979). This idea, known as the disposable soma theory, implies that the body (soma) is maintained only for the purpose of propagating the germ line.

One of the ways in which biologists have searched for insights into tradeoffs between life-history traits is to look for correlations between variables among species (e.g., Sacher 1959). This approach presupposes that the relationships between life-history variables observed among species in some way reflect the constrained relationships that pertain to tradeoffs within a species. Because the purpose of comparative analyses is to gain insight into the ways that aging mechanisms might be modified to prolong life span, it is important to tie comparisons to particular explanations by making predictions that are unique to each hypothesis. For example, any explanation relating prolonged life span to mechanisms that also extend development could be rejected by failing to find a relationship between life span and development periods. Unfortunately, this approach has several difficulties.

First, correlations may arise fortuitously among interrelated variables. The life history encompasses many fundamental tradeoffs that are interconnected in some way through functional relationships within the organism or through various environmental feedbacks in the population, such as density dependence (Figure 1; Ricklefs 2000a). Thus, correlations between variables do not necessarily imply causal or evolved relationships.

Second, variation among species need not parallel variation within populations (i.e., tradeoffs expressed among alternative phenotypes). The reason for this is that the tradeoff relationship, such as shown in Figure 2, may vary among populations owing to differences in body plan or life style or to differences in environmental influences on life-history traits. Thus, for example, one may observe a positive correlation between life span and fecundity among species over a gradient of habitat productivity from generally stressful to generally favorable conditions, whereas the tradeoff between these traits within a population (thus holding environment constant) might be negative (Figure 4).

Finally, a particular theory might make multiple predictions and therefore be consistent with a wide range of observations or experimental results. For example, in a population with senescence, selection favors increased reproductive investment toward the end of life. The argument is that if an individual has a low probability of surviving to reproduce in the future, it may as

FIGURE 4 The relationship between optimized traits among populations may not parallel variation among phenotypes within populations; at left, under better conditions, both constrained traits have higher values; at right, the optimized traits are consistent with constraint but in fact represent shifts in the constraint functions rather than different optimization points on the same constraint function

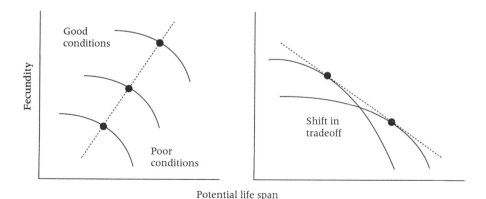

well increase investment in the present in spite of increasing its risk of death. As a result, reproductive success toward the end of life might actually increase owing to a response to selection rather than decrease as a result of physiological deterioration. Thus, a simple correlation between reproductive success and longevity might be uninformative or even misleading.

As we search for insights among patterns of correlation between life-history traits, we must be aware of several potential limitations of the approach and, in particular, we must not view correlations as confirming a functional relationship at the organismic level. In general, comparative studies are most useful when a hypothesis based on functional relationships predicts a pattern of correlation that we can either confirm or reject.

Comparative studies

The most general prediction of life-history theory concerning aging is that the "rate" of senescence should vary in direct relation to the age-independent risk of mortality (Charlesworth 1994; Edney and Gill 1968). This prediction arises because the strength of selection on factors that prolong life span varies in direct relation to the proportion of individuals who survive to old age. Theories of aging involving the optimization of tradeoffs also predict an inverse relationship between life span and reproductive success or survival rate. This points up a difficulty with comparative approaches. We have already seen that life span should be positively correlated with survival of young adults, but a tradeoff between early and late survival would predict the opposite relationship. The inconsistency arises because extrinsic factors, for example predation and accidental death, contribute to variation in survival rate independently of mechanisms that influence the rate of aging.

Empirical patterns among species suggest pervasive relationships between maximum life span, reproductive rate, body size or a particular aspect of body size, and development rate (Austad 1997; Calder 1983, 1984). These relationships may reflect allometric physiological connections, perhaps mediated through growth and metabolism, or they may reflect shifts in optimization related to the different demographic environments experienced by large and small species (Ricklefs 2000a). When body size is removed from such analyses by considering residuals from regressions against body size, life histories of organisms still appear to be organized on a single fast–slow continuum interrelating variation in life span, development period, metabolism, and reproductive rate (Bennett and Harvey 1985, 1987; Calder 1984; Promislow and Harvey 1990; Ricklefs 2000a; Sæther 1988).

Tests of the fundamental evolutionary prediction of aging models, that the rate of aging should be positively related to the rate of mortality of young adults, have met with varying success. Austad (1993) demonstrated a correlation between both physiological and demographic indexes of rate of aging and extrinsic mortality in insular and continental populations of the opossum *Didelphis virginianus*. Other studies have found that the rate of aging is lower where individuals are protected from external mortality factors (e.g., Dudycha 2001; Keller and Genoud 1997). These studies suggest the possibility that life span may evolve in response to variation in the environment, and that life span and the rate of aging might be phenotypically flexible as well. In general, short life span is associated with high mortality rates of young adults, as predicted. For example, under protected conditions in captivity, mice and ungulates, which suffer severe mortality in nature, never achieve the longevity of primates or, for that matter, most birds. Promislow (1991) attempted to place this observation on a quantitative footing by estimating the rate of aging by the rate parameter γ of the Gompertz aging model. However, he failed to find a relationship between γ and the minimum rate of mortality among adults. Ricklefs (1998) characterized the rate of aging by an index ω derived from the age-dependent mortality term of the Weibull aging model and found a positive relationship between ω and the estimated mortality rate of young adults among many species of birds and mammals. In this case, different ways of quantifying the rate of aging produced contradictory results. Therefore, it is clear that we need to consider how aging is measured demographically and how to interpret these measures in light of the questions that we have posed.

Methodological considerations

Quantifying aging

Our first task is to define what we mean by the rate of aging. We restrict ourselves to demographic measures of aging, sometimes referred to as actuarial senescence (Holmes and Austad 1995b). We assume that there is a general correlation between actuarial senescence and physiological decline in function. Most models of aging-related death relate increased vulnerability to disease, accidental death, and other causes to deteriorating physiological condition (e.g., Strehler and Mildvan 1960).

The most commonly used measure of the rate of aging is the maximum reported life span. This measure has the potential disadvantage of being determined to some degree by the age-independent rate of mortality and the number of individuals sampled. When the mortality rate does not increase with age and individuals potentially are immortal, a sample of individual ages at death will nonetheless have a maximum value, which would be the maximum reported life span. Moreover, errors in reporting introduce considerable variation in estimated life span (Carey and Judge 2000). However, when aging does occur and the mortality rate rises rapidly with increasing age, maximum reported life span may reasonably parallel the rate of aging in an underlying model of the increase in mortality with age (Finch 1990; Scheuerlein and Ricklefs unpubl.).

Botkin and Miller (1974) used maximum reported life span to infer that birds actually expressed the effects of aging in natural populations. Previously, it was thought that natural (extrinsic) mortality of birds was so high that few individuals reached "old age" (Lack 1954; Medawar 1952). This was supported by the observation from band return studies of natural populations that the adult mortality rate did not decline with increasing age. Botkin and Miller showed that a non-aging model of reported life span, assuming a constant exponential decline in survivorship with age, substantially overestimated maximum longevity. From this discrepancy, Botkin and Miller inferred that an increasing mortality rate with age must be responsible for the difference (see also Curio 1989).[1] Nevertheless, the misconception remains that natural populations lack aging individuals (Hayflick 2000; Kirkwood and Austad 2000).

A more informative index to the rate of aging than maximum reported life span can be obtained by fitting a model that incorporates the rate of increase in mortality with age to data on ages at death within a sample of individuals. Two types of models have been used for the most part (Gavrilov and Gavrilova 1991; Wilson 1994). The first, typified by the Gompertz model, represents the rate of aging by the exponential rate of increase γ of an initial mortality m_0. Thus, the mortality rate at age x is

$$m_x = m_0 e^{\gamma x}.$$

The rate of aging may also be expressed as the mortality rate doubling time, which is calculated as $\log_e 2/\gamma$ (Finch, Pike, and Witten 1990; Sacher 1977). In the second type of model, the aging component of mortality adds to an initial, or baseline, rate. Thus, according to the Weibull aging function,

$$m_x = m_0 + \alpha x^\beta.$$

In this equation, β describes the shape of the aging-related mortality curve and α is a scaling factor that describes its magnitude.[2]

Ricklefs (1998) devised an index from the parameters α and β of the Weibull function that has units of $1/\text{time}$:

$$\omega_W = \alpha^{1/(\beta+1)}.$$

Thus, ω is a rate.[3] Although the Gompertz γ also has units of $1/\text{time}$, the rate of mortality at any given age is also a function of the initial mortality rate, m_0. For comparison with the Weibull rate of aging (ω_W), Ricklefs and Scheuerlein (2001) devised an ad hoc index including both m_0 and γ, which has units of $1/\text{time}$:

$$\omega_G = \sqrt{m_0 \gamma}.$$

In general, simulated data for actuarial aging show that maximum recorded life span and the parameters of Gompertz and Weibull functions are highly correlated with each other and therefore provide adequate descriptions of the rate of aging in most cases (Ricklefs and Scheuerlein 2002).

Aging models and the biology of life span

Regardless of their similar abilities to describe empirical data, the Gompertz and Weibull equations imply different mechanisms for the increase in mortality with age. In the case of the Gompertz model, the mortality rate of young adults (m_0) is multiplied by a factor that increases exponentially with age to obtain the mortality rate at age x. Thus, the model suggests that older individuals become progressively more vulnerable to causes of death that also afflict young adults. Presumably, these causes are for the most part extrinsic and comprise primarily death by predation and accident, including effects of inclement weather. Increasing vulnerability to extrinsic causes of death results from deterioration of physiological function with age. Strehler and Mildvan (1960) explained the exponential rise in the mortality rate of the Gompertz model in terms of a linear increase in vulnerability to aging factors combined with a sensitivity threshold that was exponentially distributed among individuals in a population.

In the Weibull model, aging-related mortality is added to the baseline (extrinsic) level, implying that different mechanisms cause the increasing rate of death in old age. Thus, the Weibull model would be consistent with catastrophic causes of death from acute illness or system failure (cardiovascular disease, carcinoma) in older individuals. These are intrinsic causes of death that would likely kill regardless of extrinsic mortality factors, even though the immediate cause of death may nonetheless be extrinsic. That is, a terminally ill individual almost certainly would have an elevated risk of death from predation, inclement weather, social strife, and other external causes. Several mathematical derivations have developed the idea, which may serve as a model of intrinsic mortality, that a sequence of randomly occurring events (e.g., mutations, radiation damage) is required to initiate a fatal cancerous growth or other failure. This theory shows that a series of $\beta + 1$ such events produces a Weibull mortality function with shape parameter β (Armitage and Doll 1954, 1961).

With increasing age, humans suffer higher mortality rates from both extrinsic and intrinsic causes of death, although the aging-related rate of increase is lower for accidental death than it is for spontaneous intrinsic causes, such as cancer and cardiovascular disease (e.g., Horiuchi and Wilmoth 1997). The few data available for nonhuman populations, such as rhesus macaques, show a similar rapid increase in geriatric disease with advancing age (Uno 1997). Although few studies address changes in the rate of death from extrinsic causes, comparative analyses of natural and captive populations shed some light on the relative balance of extrinsic and intrinsic mortality factors in aging-related mortality.

The proportion of aging-related deaths

Another insight that can be gained from fitting aging functions to data concerns the proportion of deaths that can be attributed to aging-related causes (P_S). This can be calculated from the parameters of the Weibull model by the expression

$$P_S = \int_{x=0}^{\infty} \alpha x^\beta l_x dx,$$

where l_x is the proportion of individuals who survive to age x. This expression integrates over age the proportion of deaths that occur in excess of the fraction expected from constant mortality at rate m_0. This index is proportional to the potential strength of selection on factors that delay senescence. Ricklefs (1998) used Weibull functions fitted to empirical data to show that the proportion of senescent deaths in wild populations is high—up to 50 percent—when m_0 is low and a large proportion of the population attains old age. Because a large proportion dies at old age, reducing α should result in a large increase in fitness. Failure to reduce aging-related mortality, as observed in populations with long life spans, implies that there is limited genetic variation for mechanisms to delay aging at old age or that the associated cost of these mechanisms is too high.

Aging and evolutionary fitness

In order to place senescence in an evolutionary context, it is necessary to determine how changes in the rate of actuarial senescence influence fitness (Abrams 1991, 1993). We have already seen that the strength of selection on life span depends on the proportion of the population reaching a certain age, more specifically on the future reproductive potential of a cohort upon reaching that age. Any gain in fitness caused by an increase in life span must be sufficient to balance the cost of the mechanisms that extend life. The effect on fitness (λ) of changes in survival and reproduction at a particular age can be evaluated by the Euler–Lotka equation, also referred to as the "characteristic equation" of a population, which relates fitness to the life table variables of age-specific survival (l_x) and age-specific fertility (b_x)—

$$1 = \sum_{x=a}^{z} \lambda^{-x} l_x b_x$$

—where a and z are the ages at the onset and termination of reproduction, and age-specific survival is the product of annual survival probabilities s_x up to age $x - 1$. Each term of the sum is the realized production of offspring at a given age (x) discounted by the rate of population growth (λ).[4] To illustrate the use of the equation to evaluate changes in fitness resulting from changes in life span in as simple a manner as possible, we use a life table in which fertility (B) and adult survival rate (S) are constant, and we express senescence as a truncation of the life table at life span z (Figure 5). The age at maturity is a, and survival to maturity is S_a. Accordingly, the Euler–Lotka equation becomes

$$1 = \sum_{x=a}^{z} \lambda^{-x} S_a S^{x-a} B,$$

which may be solved for

$$\lambda = S + S_a B \lambda^{1-a} - S_a B S^{z-a+1} \lambda^{-z}.$$

This equation can then be differentiated to determine the effect of changes in life span (z) or adult mortality rate ($1 - S$) on evolutionary fitness (λ).[5]

The differential forms of this equation show how changes in life table entries influence fitness. These are illustrated in Figure 6 for changes in life span (z) and adult mortality rate ($1 - S$) as a function of the maximum life span. First, we see that a change in life span influences fitness most when life span is short, that is, when a large proportion of individuals are still alive toward the end of the truncated life span and thus will benefit from

FIGURE 5 Diagram of a simplified life table in which fertility (B) and mortality rate (M) are constant after the age at maturity (a), to which proportion S_a survive, and the life span is truncated at age z.

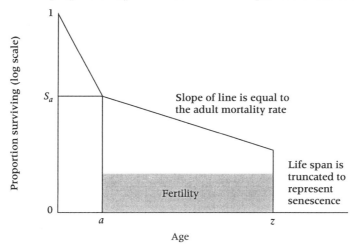

FIGURE 6 Relative (percentage) change in fitness (λ) resulting from changes in life span and mortality rate, plotted for different values of adult mortality rate (M) as a function of the maximum life span (z)

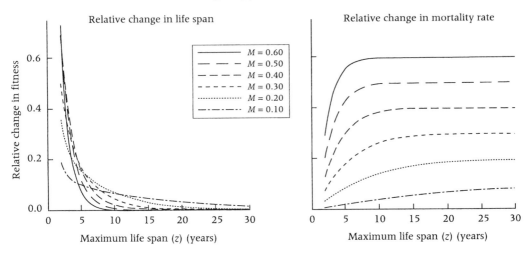

the extension. For the same reason, the benefits of prolonging life at older ages are greater for populations with lower adult mortality rates. A decreasing mortality rate benefits populations with high adult mortality rates more than it does those with low adult mortality rates, as one would expect. For this discussion, the telling point is that variation in the mortality rate has a much greater influence on fitness than does variation in life span, especially beyond a modest maximum life span (for birds) of about ten years. This means that selection on patterns of aging is more likely to modify aging-related mortality at all ages, for example by changing the α parameter of the Weibull function, than it is to extend life span alone. Increasing z in this simple model would be equivalent to simultaneously increasing β and decreasing a. In comparative studies, we find that β is independent of initial mortality m_0 and ω, and that variation in the pattern of aging among species involves solely changes in α (Ricklefs and Scheuerlein 2001).[6]

How does one analyze and interpret life-history variation?

Correlations between the rate of aging and other life-history traits may provide insights into the mechanisms of extending life. Comparative analyses are fraught with problems, however. The most important of these are spurious correlations arising from third variables and lack of evolutionary independence of the data. The first problem may be addressed using analytical techniques that account for the correlations among independent variables: multiple regression, path analysis, and multivariate analysis. These analytical tools seek unique correlations between two variables that are independent of correlations with other variables, or they combine independent variables into derived axes that incorporate their intercorrelations. The problem of evolutionary independence has been addressed by a number of techniques based on phylogenetic relationships among taxa that isolate independent aspects of their evolutionary history. In most cases, however, it may be reasonable to proceed with correlations based on terminal taxa in a phylogeny, that is, with raw species values.[7]

Another issue in evolutionary analysis that is not addressed by analyzing either contrasts or terminal taxa is the evolutionary lability versus conservatism of life-history traits. Some traits vary substantially among closely related species and presumably are responsive to variation in selection. Other traits are much more conservative, with a large part of their variation concentrated in comparisons among higher taxonomic categories. This situation may reflect rapid evolution of life-history patterns in conjunction with the early expansion of a group followed by a subsequent period of relative stasis to the present (Nealen and Ricklefs 2001; Ricklefs and Nealen 1998). The distribution of variation in traits and covariation between traits with respect to evolutionary history can be approached by hierarchical nested analysis of variance. This procedure partitions variation into components associated with a hierarchy of taxonomic categories (Bell 1989; Derrickson and Ricklefs 1988; Stearns 1983), such as families within orders and species within genera.[8] We apply this analytical technique below.

Can one apply experimental approaches to life-history variation?

Most experimental work on the evolution of aging has been carried out with laboratory organisms, most notably *Drosophila melanogaster* (Partridge and Barton 1993; Rose and Graves 1990). The general approach is to select on variation in life table traits and to observe correlated evolutionary responses. Laboratory populations tend to respond rapidly to selection on life span or on fertility during early or late portions of the life span. In such experiments, increased life span is typically associated with increased ability to tolerate stress (Harshman et al. 1999) but also with reduced fertility early in life (Rose 1984). In many laboratory populations, individual genes with striking effects on life span have been identified. One must be careful, however, in placing such cases in the context of life span in natural populations. One should ask, for example, whether the extension of life span in laboratory populations exceeds that of individuals in nature or remains within the biological potential of natural populations (Linnen, Tatar, and Promislow 2001; Sgrò and Partridge 2000). Most domesticated and laboratory populations have been selected for rapid maturation and high early fertility, which would under most circumstances reduce potential life span. In this case, variation in laboratory populations and, particularly, large effects of individual genes, as well as the response of laboratory populations to selection on life span, may have limited application to understanding limits to life span in natural populations (Clark 1987; Promislow and Tatar 1998).

Natural experiments

In some cases, it may be possible to observe closely related natural populations under different environmental conditions to detect the evolutionary response of life span to variation in environmental conditions. Such natural experiments may offer reasonable control over the intrinsic biological characteristics of the organism (such control is more difficult to achieve in broader comparative studies); however, environmental variables may be more difficult to control. Such natural experiments nonetheless provide insights into the evolutionary flexibility of the life span and, presumably, the aging process. In one case already mentioned, Austad and Fischer (1991) compared populations of opossums separated over several thousand years in mainland and island environments and found that the rate of actuarial senescence was lower on the island, where predators were absent and individuals suffered lower mortality from extrinsic factors. Austad also found that a physiological index of aging—the cross-linking of collagen fibers—occurred more slowly among island opossums.

Comparison of survival curves in populations of the water flea *Daphnia pulex* in lakes and ponds that differ in the intensity of predation showed parallel variation in longevity consistent with a response of aging to selective factors in the environment (Dudycha 2001). Several authors have emphasized the disparity in life span between workers and reproductive castes of social insects as indicative of environmental controls over the rate of aging; but because the genotypes of queens and workers do not differ, these differences clearly do not reflect evolutionary responses (Keller and Genoud 1997). Nonetheless, they indicate the strong direct environmental influence on aging, presumably mediated through rate of activity or differences in hormonal and physiological state.

Another type of experiment that can elucidate mechanisms responsible for aging is the establishment of captive populations, whereby extrinsic mortality factors are reduced considerably compared to natural populations. We discuss the results of such comparisons below.

Patterns of life-history variation involving life span

Until recently, most empirical studies of variation in longevity were based on maximum reported life span. When organisms exhibit an increase in mortality rate with age and a natural population is sampled for a sufficient period to cover the maximum natural life span, simulated data show that the maximum reported life span is highly correlated with aging parameters of Gompertz and Weibull models of aging (Ricklefs and Scheuerlein 2002; Scheuerlein and Ricklefs unpubl.). Problems arise primarily when studies are too brief. For example, among the maximum reported life spans in Carey and Judge's (2000) thorough compilation, those for red-legged kittiwakes (4 years) and Townsend's shearwater (5.1 years) are barely long enough for these seabirds to attain maturity.[9] Although the data must be viewed cautiously, life span analyses nevertheless reveal general patterns in the rate of aging among animals.

Body mass

Biologists have recognized for many years that maximum potential life span in birds and mammals is closely related to body size (Comfort 1979; Finch 1990). Large animals live longer (Austad and Fischer 1991; Calder 1983, 1984, 1985; Sacher 1959). This has been attributed to several causes. First, many physiological parameters, such as metabolism, development rate, and fecundity, vary with body size. Large animals have a slow rate of living, and long life may express this general property. Reduced metabolism may also result in reduced concentrations of reactive forms of oxygen, thereby slowing the aging process (e.g., Barja and Herrero 2000). Another consequence of large body size is a lower extrinsic mortality rate. Larger animals are better able to avoid encounters with predators, and their bodies are also better buffered against vagaries in the environment.

Second, it is well understood that greater survival in the face of extrinsic factors increases the strength of selection on life span, leading to a demographic or life-history connection between body size and rate of aging. Several authors have examined this demographic connection directly by relating aging parameters of Gompertz or Weibull functions to the estimated extrinsic mortality rate. As mentioned previously, Promislow (1991) compared the aging parameter γ of the Gompertz equation to an estimate of the minimum mortality rate of young adults but found no relationship between them. In contrast, Ricklefs (1998) found a strong relationship between the derived aging parameter ω and the fitted initial mortality rate (m_0) of the Weibull equation. This discrepancy raises a fundamental issue concerning the proper definition of the rate of aging. Using the Gompertz γ, one defines aging as the exponential rate of increase in mortality with age. Using ω derived from either the Gompertz or Weibull equation, one defines aging according to the magnitude of the mortality rate at a particular age. In the case of the Gompertz function, the magnitude of the mortality rate depends on both m_0 and γ, hence it is difficult to separate their effects.

Multiple regression shows that the Weibull ω varies with the initial mortality rate independently of the relationship of both to body size, thereby making the demographic connection between extrinsic mortality and rate of aging more plausible and the direct allometric physiological connection between size and m_x less plausible. This is further substantiated by comparisons between birds and mammals. For individuals of a given body size, birds have longer life spans than mammals, perhaps by an average factor of 2–3 times (Calder 1983; Holmes and Austad 1995a). When birds and mammals are analyzed together, the rate of aging ω is related to the initial mortality rate m_0 independently of body size or of differences between birds and mammals (Ricklefs 1998). Accordingly, the difference in longevity between birds and mammals of the same body size is statistically associated with the greater ability of birds to escape mortality factors, as shown by their lower values of m_0. This observation strengthens the idea that the rate of aging is evolutionarily flexible rather than being tied closely to other aspects of physiology, such as the rate of metabolism (Holmes and Austad 1995a, b).

Brain mass

Life span has been related to other attributes besides body mass. Brain mass, in particular, has captured the attention of comparative biologists, perhaps because the brain is considered the pacemaker of life in general and is the organ that most resists aging-related change. Sacher (1959, 1978) found a good correlation between life span and brain mass for mammals, and Allman et al. (1993) have explored this relationship in more detail for primates. However, the brain–life span relationship was questioned by Economos (1980a, b), who pointed out that brain mass is not uniquely highly correlated with longevity, given the strong intercorrelations, for example, between brain mass and body mass. Why mass itself should exercise a dominant influence over life span is also not clear. Ricklefs and Scheuerlein (2001), comparing the Weibull parameter ω to both body mass and brain mass in birds in a multiple regression, found that only brain mass was uniquely related to rate of aging. Clearly, however, in comparisons involving both birds and mammals, brain mass does not explain as much of the variation in life span or rate of aging as extrinsic mortality. Thus, it is likely that the correlation observed within either class between brain mass and life span reflects some aspect of the ecology of species that is expressed independently in both brain mass and extrinsic mortality.

Genome size

Another attribute that has been linked to longevity is the size of the genome. Although the two do not appear to be related in mammals (see Finch 1990), Monaghan and Metcalfe (2000) have recently made such a claim for birds based on independent contrasts calculated from phylogenetic relationships among avian families (Sibley and Ahlquist 1990). Ricklefs and Scheuerlein (2001) and Morand and Ricklefs (2001) failed to find a relationship between the Weibull aging parameter ω and genome size in a correlation among terminal taxa at the species level. However, a hierarchical analysis of covariance based on Monaghan and Metcalfe's data reveals a strong correlation between life span and genome size at the level of families within orders but no correlation at the level of species within genera (Table 1). Even though the family-level correlation is significant, the amount of the total variance that resides at this taxonomic level is relatively small, hence it is hard to attach a biological meaning to the correlation. Indeed, when the data were partitioned into equal halves, only one of the two data sets retained a significant correlation between life span and genome size at the family level. This suggests that the significant correlation may have resulted from a chance association of data rather than an underlying mechanistic connection between the two.

A broader life-history perspective

Ricklefs and Scheuerlein (2001) investigated life-history correlations of the rate of aging (ω) in birds and mammals separately. The rate of aging was generally inversely related to body size, brain size (tested only for birds), and rate of postnatal development (significant only for mammals), and was directly related to the lengths of incubation (birds) or gestation periods (mammals). When the independent variables were treated together in a multiple regression, only brain mass (birds) and postnatal growth rate (mam-

TABLE 1 Distribution of variance components in life span, genome size, and their correlation among birds over nested taxonomic levels

Taxonomic level	Life span	Genome size	Correlation
Order	0	0	0.00
Family	26	6	1.59
Genus	74	94	0.02
Total	100	100	—

NOTE: Life span is the maximum recorded value and is subject to considerable error; correlations exceeding 1.0 are possible because variance components are estimates.
SOURCE: Data are from Monaghan and Metcalfe (2000) as analyzed by Morand and Ricklefs (2001); based on 67 species.

TABLE 2　Distribution of variance components in life-history traits and aging parameters of birds over nested taxonomic levels

Taxonomic level	Mass	Brain	Incubation period	Growth rate	Initial mortality (m_0)	Rate of aging (ω)
Order	42	20	47	80	10	25
Family	1	0	41	0	21	0
Genus	52	76	9	20	49	61
Species	5	4	4	0	21	14
Total	100	100	100	100	100	100

NOTE: Variance components are estimates based on the hierarchical structure of mean squares in a nested analysis of variance. When the data are poorly balanced, as in analyses of this type, mean squares may decrease from one level to the next higher level, resulting in a negative variance component, which is arbitrarily set to zero.
SOURCE: Data from Ricklefs and Scheuerlein (2001), based on 53 species.

mals) were uniquely related to rate of aging. Hierarchical analysis of variance for life-history and aging traits of birds reveals differences in the distribution of variation related to the evolutionary responsiveness of each of the traits (Table 2). In particular, about 80 percent of the variance in embryonic and postnatal growth rates resides at the level of orders and families, reflecting body size differences between large taxonomic groups and differences in mode of development (i.e., altricial versus precocial) that influence growth rate (Starck and Ricklefs 1998). Variance in mass itself is about equally distributed between higher and lower taxonomic levels, whereas the variance in initial mortality and rate of aging is more heavily weighted toward the taxonomic level of genus. For the smaller number of species for which brain masses are available, variance in brain mass is concentrated at the level of genus, as is the variance in rate of aging.

Variance component correlations relating the rate of aging ω to other life-history traits show consistent significant correlations between rate of aging and extrinsic mortality at the level of orders and genera within families, but correlations with body mass and incubation period are weaker (Table 3).

TABLE 3　Variance component correlations for the rate of aging (ω) and initial mortality rate (m_0), mass, and incubation period in birds

Taxonomic level	ω versus mass	ω versus m_0	Mass versus m_0	ω versus incubation period
Order	−0.41	1.10	0.59	−0.67
Family	0.00	0.00	−0.80	0.00
Genus	−0.84	0.76	−0.92	−0.56
Species	−0.19	0.33	−0.51	−0.31

SOURCE: Data from Ricklefs and Scheuerlein (2001).

Comparisons between wild and captive populations

When individuals are brought into captivity, the rate of extrinsic mortality is reduced considerably, except perhaps for the incidence of stress-related and contagious disease. If the increase in mortality with age resulted from increasing vulnerability to extrinsic mortality factors, then one would expect animals in captivity to exhibit lower mortality rates at older ages than they do in the wild. If aging-related mortality were caused by intrinsic factors that kill independently of external causes (even though the latter may hasten intrinsically caused death), then aging-related mortality would not differ between captive and wild populations. Comparing wild and captive populations is complicated by the fact that ages at death are often estimated by different methods, and mortality rates of young adults in captivity may have additional causes related to stress and contagion that are not prominent in the wild. Thus, such comparisons must be made with caution.

Using phylogenetically controlled comparisons, Ricklefs (2000b) showed that captive populations of birds had lower initial mortality (m_0) but comparable rates of aging (ω) vis-à-vis wild populations. This suggested that the Weibull model of aging, which separates extrinsic and intrinsic causes of death in additive terms, provides a reasonable demographic description of the aging process. A similar comparison of taxonomically matched wild and captive populations of mammals suggested in contrast that the Weibull rate of aging decreases in captivity by an amount expected of the multiplicative Gompertz model (Ricklefs and Scheuerlein 2001). This result was particularly striking for mammals of open savanna habitats, for which survival rate must be closely tied to speed and endurance. For such species, physiological decline would seemingly lead to higher vulnerability to extrinsic mortality factors, particularly predation. Predators themselves should be similarly afflicted by physiological decline, and it is perhaps relevant that the lion *Panthera leo* shows a pattern of reduced aging-related mortality in captivity, as do several ungulates that are potentially its prey.[10]

Discussion

Comparative analyses of the rate of aging in the context of life histories have provided insights into the aging process while leaving many issues unresolved. In the case of birds, which are long lived compared to mammals of the same size, comparisons between wild and zoo populations suggest that the Weibull model provides a better description of the aging process than the Gompertz model. Causes of aging-related deaths are apparently unrelated to extrinsic mortality factors; accordingly, an additive model of extrinsic and intrinsic mortality, such as the Weibull function, is appropriate. Comparison of rates of aging with body size and the extrinsic (initial) mortality rate indicates that the rate of aging responds to selection based on the demography of a population rather than on physiological characteristics of individuals. Thus, for the most part, evolutionary theories of aging are supported by comparative studies. Moreover, comparisons between natural and captive populations indicate that birds in the wild do not die because of increasing vulnerability to extrinsic causes. This suggests that individual birds maintain high fitness late into life and that the increase in mortality with age is due to the increasing probability of catastrophic causes of death, such as cancer and stroke. The increasing proportion of aging-related deaths in populations with a large proportion of old individuals further suggests that there are inherent biological limitations on the control of intrinsic aging-related mortality.

Correlations of the rate of aging with life-history attributes other than extrinsic mortality are not particularly informative because of the strong correlations among independent variables. By considering mammals and birds together, however, one can break this pattern of intercorrelation because the different life styles of birds and mammals result in contrasting patterns of extrinsic mortality. Thus, although the rate of aging is related to body size and other correlated life-history attributes within both birds and mammals, analyses involving both groups clearly tie the rate of aging to extrinsic mortality. Accordingly, the rate of aging appears to be responsive to selection on traits that extend life span or modify the rate of increase in mortality. Because the rate of aging varies in response to the demographic environment of a population, we may infer that postponing aging imposes a fitness cost, although the mechanisms involved are not understood. Because the proportion of aging-related mortality increases with maximum potential life span, this cost evidently increases as senescence is delayed further. Moreover, the difference in the rate of aging between birds and mammals of the same body size is not due to some fundamental biological difference between the two groups of organisms. Rather, it is related to the fitness benefits of delayed aging, which are lower for mammals owing to their higher extrinsic mortality rates.

The nature of the fitness cost of postponing aging (Abrams 1991) is not evident in comparative analyses. In general, increasing body size is associated with a reduced rate of aging, reduced extrinsic mortality, lower fecundity resulting from smaller brood size and longer development, and a higher pre-reproductive survival rate. If postponing senescence imposes a cost in terms of early fecundity or survival, this might be obscured either by contrasting or supplementing contributions to fertility and survival from other factors. For example, there are many reasons, related to lower metabolic intensity and slower life processes, why larger organisms might have lower fecundity than smaller organisms besides the costs of delayed senescence.

Comparative analyses of life histories might provide more information concerning mechanisms of delayed senescence if they were applied in conjunction with mechanistic models of control of the aging process (Abrams and Ludwig 1995). For example, Ricklefs (1992, 1993) suggested that variation in the development period among birds of the same body size might involve mechanisms to prolong life through production of a better immune system or a larger, more complex nervous system. In this case, one could search for a direct correlation between the rate of aging and the development rate, where the latter was sufficiently uncoupled from body size to permit separation of the two effects statistically. To the extent that such analyses have been performed for birds and mammals, they are consistent in that a longer development period is associated with lower extrinsic mortality and longer life span independently of body size (Ricklefs 1993). Clearly, slow development reduces fitness because it extends the period of exposure of highly vulnerable offspring to extrinsic mortality factors. The size of this fitness impact can be estimated from mortality rates of parents and offspring during the development period. If prolongation of life span were achieved by reducing fertility, one might be able to use comparative analyses to identify associated adaptations of organisms that secondarily decrease the reproductive rate.

What the comparative analysis of aging requires most at this point is an unambiguous quantitative definition of the rate of aging; mechanistic models of aging to inform comparative analyses; and comparative data specifically related to aging and its causes. General insights have been gained by comparative life-history analyses of aging; however, at present these lack the specificity to help characterize mechanisms of aging by their effects on life-history traits. More-focused gathering of data and application of comparative methods should help considerably in this regard.

We conclude with four general observations about life span—its nature, evolution, theoretical study, and future prospects.

(1) What are the nature and fundamental properties of life span? Comparative studies of life tables do not address physiological processes directly. What they can say is that senescence leads to a continuous increase in the mortality rate with age and that differences between populations reflect primarily the magnitude rather than the pattern of this increase. Although a species may have a maximum potential life span, it is clear that few individuals reach this age. From the standpoint of the evolution of the aging pattern, factors that influence adult mortality throughout life determine the overall pattern of aging-related mortality. An important issue is whether the demographic expression of aging results from increasing vulnerability to extrinsic causes of death due to reduced physiological capacity, or from increasing frequency of catastrophic intrinsic mortality factors, such as carcinomas and cardiovascular disease. In birds, the relative constancy of the

aging-related component of mortality when extrinsic mortality varies between captive and natural populations weighs in favor of intrinsic causes. The data for mammals are less conclusive and may reflect a basic difference in the way mammals age.

(2) How does life span evolve in the context of the life history? Comparative studies make quite clear that the primary selective factor is the level of extrinsic mortality. The role of other factors, such as parental care and sociality (Allman et al. 1998; Carey and Judge 2001), over and above extrinsic mortality, has not been evaluated quantitatively in a comparative context. Because most of the variance in the rate of aging resides at the level of genera within families, this observation suggests that the rate of aging is relatively flexible from an evolutionary standpoint, but also that most of the variance in extrinsic mortality rates is expressed at this level.

(3) Why is there no theory of life span per se, notwithstanding the dominant focus on aging? From an evolutionary standpoint, one answer is that the fitness effect of variation in life span is much smaller than the effect of changes in mortality throughout adult life (Figure 6). This is because so few individuals in natural populations reach the maximum potential life span. This does not mean that life span per se is not responsive to selection; the evolutionary response also depends on genetic variation for a trait within a population. Comparative studies provide no evidence, however, that maximum potential life span varies independently of changes in aging-related mortality throughout adult life.

(4) What is the future of the human life span? The increasing proportion of aging-related mortality in populations with older age structures suggests that natural populations have approached biological constraints limiting the potential length of life. Average human life span clearly has increased greatly in recent years, with many of the gains coming from reduced mortality rates in the oldest age classes (Robine et al. 1997). It is clear that these changes have resulted from manipulation of the environment rather than changes in the genetic makeup of the human population. Thousands of generations of selection on natural populations of long-lived animals, such as elephants and albatrosses, have failed to reduce aging-related mortality to low levels. Thus, it is likely that future progress in extending average human life span will require either further improvement in the conditions of life conducive to survival at old age or manipulations of the human genome that exceed the genetic repertoire of natural populations. A more important lesson from the study of aging in natural populations of birds is that individuals can maintain high personal fitness to an advanced age as part of the normal progress of aging. This lesson suggests that it may be more fruitful to focus our attention on realizing and enhancing for humans the potential natural quality of life of elderly individuals rather than trying to extend life per se.

Notes

This study was supported by NIH R03 AG16895-01 and R01 AG20263-01.

1 The failure of banding studies to detect aging-related mortality can be attributed to several factors. First, most banding studies with large sample sizes were restricted to small species with short life spans and high adult mortality rates, such as tits (*Parus*) and flycatchers (*Ficedula*). As shown by Ricklefs (1998), only a small proportion of the mortality in populations of this type is associated with the expression of aging. Second, techniques for analyzing demographic data in early studies of band returns were not well developed and may have had little power to detect age dependency of the survival rate. Recent development of maximum likelihood models has made demographic analysis much more sensitive (Lebreton et al. 1992). Third, recovery of ages at death from natural populations, unlike cohort analysis, is not well suited to analysis of age-specific changes in the survival rate. This is because, in populations that are sampled through recoveries of dead individuals, the declining sizes of successively older age classes are balanced by an increasing mortality rate with age. Thus, the changes in age structure associated with aging are partly or completely compensated by change in the mortality rate, often resulting in a distribution of ages at death sampled from a natural population that resembles a constant exponential (that is, non-senescing) decline.

2 Additional models incorporate additional additive and multiplicative terms (e.g., the Gompertz–Makeham function) or a leveling off at high age (e.g., logistic function) to match observations on humans and on laboratory populations of flies (Carey et al. 1992; Fukui et al. 1993; Horiuchi and Wilmoth 1998; Vaupel et al. 1998). We have not used functions with an upper asymptote because too few data are available for natural populations of birds and mammals to reveal aging plateaus, and we wished to keep the number of aging parameters small to facilitate comparisons among taxa. Note also that the Gompertz–Makeham function,

$$m_x = m_0 + A_0 e^{\gamma t},$$

incorporates an additive term for initial mortality rate, but, as in the Gompertz equation, the aging-related mortality term remains an exponential increase in a portion (A_0) of the initial mortality. Thus, it does not provide the independence between initial and aging-related mortality that the Weibull function does.

3 Specifically, ω is the value of the aging-dependent mortality term αx^β obtained when $x = 1/\omega$. From the logarithmic form of the expression, $\log\omega = \log\alpha /(\beta + 1)$, one can see that for a given value of β, the value of $\log\omega$ is directly proportional to $\log\alpha$. Empirically, β averages about 3 in bird and mammal populations.

4 Application of this equation involves many assumptions, most notably that the life table of the population is constant over time and that males and females are demographically indistinguishable. When survival and fertility vary stochastically over time, selection generally favors longer life (lower adult mortality) over reproduction (Hastings and Caswell 1979; Schaffer 1974). For a constant life table, the change in fitness resulting from a change in any life table entry, i.e., fertility or survival rate at a particular age, can be obtained by differentiating the equation to find $d\lambda/ds_x$ or $d\lambda/db_x$ (Hamilton 1966). However, as Abrams (1991, 1993) has pointed out, genetic changes influencing the aging process typically have an age at onset and affect life table values at all subsequent ages. These considerations complicate the application of the Euler–Lotka equation to the evolution of aging. Nevertheless, it has heuristic value for exploring the effects of changes in life table variables on evolutionary fitness, which can be illustrated by the simple approach taken here.

5 Because populations are maintained in an approximate equilibrium by density-dependent factors, this equation can be simplified by assuming $\lambda = 1$, at which point recruitment of young individuals into the breeding population ($S_a B$) can be replaced by

$$S_a B = \frac{1 - S}{1 - S^{z-a+1}}.$$

6 The estimated fitness value of extending average life span also shows how much

mechanisms for extending life cost in terms of life table values. In the simple model presented above, costs may be expressed as an increase in age at maturity (a), decrease in prereproductive mortality (increase in S_a), decrease in fertility (B), or any combination of these. For example, suppose life span were extended by some mechanism that lengthened the development period. This would incur a cost in terms of the reduced survival (expressed as lower B or S_a) resulting from a longer period of exposure of the more vulnerable offspring to extrinsic mortality factors. In tropical birds, for example, daily rates of nest predation commonly exceed 5 percent, hence extending development even for a few days, as so many tropical species do, can be very costly.

7 The technique of independent contrasts used to correct for phylogenetic relationship (Garland et al. 1992; Harvey and Pagel 1991) requires estimation of the character values or states of ancestral nodes and thus shifts the problem of independence to a problem of estimation. Restricting comparisons to terminal pairs of species (sister-taxon comparisons) eliminates the problem of independence and does not rely on estimation of ancestral states, but results in a considerable reduction of sample size. For example, in a sample of 20 taxa, the number of independent contrasts is 19, while the maximum number of sister-taxon comparisons is 10 or fewer. However, results based on independent contrasts and results based on terminal taxa rarely differ qualitatively and generally are similar quantitatively (Price 1997; Ricklefs and Starck 1996).

8 One assumption of this approach is that taxonomic ranks are homogeneous throughout a phylogeny. In the case of birds, Sibley and Ahlquist (1990) have defined taxonomic categories on the basis of genetic divergence estimated by DNA hybridization. Genetic differences are expressed as difference in the melting point temperatures of homoduplexed and heteroduplexed DNA ($\Delta T_H 50$, °C). Thus, for example, families include taxa descending from a single ancestral node at a depth of 9–11 °C.

9 Most data on maximum life span for natural populations of animals come from birds, which have been the subjects of banding programs for many years. When birds are fitted with an individually numbered leg band as chicks or in the first year of life, when age can be assessed by plumage, age at death can be determined. Ages at death in mammal populations are based primarily on patterns of growth or tooth wear and are therefore less accurate than those for birds. In some cases, long-term studies of individually recognizable mammals have provided more accurate demographic information in natural populations (Gaillard et al. 1994).

10 One nagging feature of captive populations is the relatively high "extrinsic" mortality rate in the absence of most extrinsic mortality factors. We suggest that relatively high values of m_0 are related to the stress of the zoo environment, which may impair immune system function and render zoo animals more vulnerable to infection and other stress-related causes of mortality in young adults. The higher rate of initial mortality may also result in a lower apparent rate of aging as an artifact. This result might be particularly pronounced if the high initial mortality rate of captive populations actually decreased with age. We have tested the idea that stress might be related to high initial mortality in captive populations by comparing captive populations of wild species with those of domesticated species, which presumably have been selected for reduced stress response to the conditions of captivity (Scheuerlein and Ricklefs, unpubl.). Preliminary results show that initial mortality of domesticated populations is very low, but that the rate of aging (ω) differs little from that of comparable wild or captive populations. Thus, it seems plausible that captivity and domestication have reduced extrinsic mortality considerably, but that the pattern of actuarial senescence has changed little. This result is contrary to Austad's (1993) observation that an insular population of opossums isolated from the mainland for about 4,000 years (i.e., similar to the period of domestication of dogs and livestock) exhibited both reduced extrinsic mortality and slower aging-related increase in mortality. The result with domesticated mammals also reinforces the idea that the additive terms of the Weibull function portray the underlying causes of actuarial senescence more realistically than the multiplicative terms of the Gompertz equation.

References

Abrams, P. A. 1991. "The fitness costs of senescence: The evolutionary importance of events in early adult life," *Evolutionary Ecology* 5: 343–360.

———. 1993. "Does increased mortality favor the evolution of more rapid senescence?" *Evolution* 47: 877–887.

Abrams, P. A., and D. Ludwig. 1995. "Optimality theory, Gompertz' law, and the disposable soma theory of senescence," *Evolution* 49: 1055–1066.

Alexander, R. D. 1974. "The evolution of social behavior," *Annual Review of Ecology and Systematics* 5: 325–383.

Allman, J., T. McLaughlin, and A. Hakeem. 1993. "Brain weight and life-span in primate species," *Proceedings of the National Academy of Sciences USA* 90: 118–122.

Allman, J., A. Rosin, R. Kumar, and A. Hasenstaub. 1998. "Parenting and survival in anthropoid primates: Caretakers live longer," *Proceedings of the National Academy of Sciences USA* 95: 6866–6869.

Alvarez, H. P. 2000. "Grandmother hypothesis and primate life histories," *American Journal of Physical Anthropology* 113: 435–450.

Armitage, P. and R. Doll. 1954. "The age distribution of cancer and a multi-stage theory of carcinogenesis," *British Journal of Cancer* 8: 1–12.

———. 1961. "Stochastic models for carcinogenesis," *Proceedings of the Fourth Berkeley Symposium on Mathematical Statistics and Probability* 4: 19–38.

Arnold, S. J. 1983. "Morphology, performance and fitness," *American Zoologist* 23: 347–361.

Austad, S. N. 1993. "The comparative perspective and choice of animal models in aging research," *Aging-Clinical and Experimental Research* 5: 259–267.

———. 1997. "Comparative aging and life histories in mammals," *Experimental Gerontology* 32: 23–38.

Austad, S. N. and K. E. Fischer. 1991. "Mammalian aging, metabolism, and ecology: Evidence from the bats and marsupials," *Journal of Gerontology: Biological Sciences* 46: B47–B53.

Barja, G. and A. Herrero. 2000. "Oxidative damage to mitochondrial DNA is inversely related to maximum life span in the heart and brain of mammals," *FASEB Journal* 14: 312–318.

Bell, G. 1989. "A comparative method," *American Naturalist* 133: 553–571.

Bennett, P. M. and P. H. Harvey. 1985. "Brain size, development and metabolism in birds and mammals," *Journal of Zoology* 207: 491–509.

———. 1987. "Active and resting metabolism in birds: Allometry, phylogeny and ecology," *Journal of Zoology* 213: 327–363.

Botkin, D. B. and R. S. Miller. 1974. "Mortality rates and survival of birds," *American Naturalist* 108: 181–192.

Calder, W. A., III. 1983. "Body size, mortality, and longevity," *Journal of Theoretical Biology* 102: 135–144.

———. 1984. *Size, Function, and Life History.* Cambridge, MA.: Harvard University Press.

———. 1985. "The comparative biology of longevity and lifetime energetics," *Experimental Gerontology* 20: 161–170.

Carey, J. R. and D. S. Judge. 2000. *Longevity Records: Life Spans of Mammals, Birds, Amphibians, Reptiles, and Fish.* Odense, Denmark: Odense University Press.

———. 2001. "Principles of biodemography with special reference to human longevity," *Population: An English Selection* 13: 9–40.

Carey, J. R., P. Liedo, D. Orozco, and J. W. Vaupel. 1992. "Slowing of mortality rates at older ages in large medfly cohorts," *Science* 258: 457–461.

Caruso, C. et al. 2000. "HLA, aging, and longevity: A critical reappraisal," *Human Immunology* 61: 942–949.

Charlesworth, B. 1994. *Evolution in Age-structured Populations*, 2nd edition. Cambridge: Cambridge University Press.

Clark, A. G. 1987. "Senescence and the genetic correlation hang-up," *American Naturalist* 129: 932–940.

Comfort, A. 1979. *The Biology of Senescence*, 3rd ed. Edinburgh and London: Churchill Livingstone.

Curio, E. 1989. "Is avian mortality preprogrammed?" *Trends in Ecology and Evolution* 4: 81–82.

Derrickson, E. M. and R. E. Ricklefs. 1988. "Taxon-dependent diversification of life histories and the perception of phylogenetic constraints," *Functional Ecology* 2: 417–423.

Dudycha, J. L. 2001. "The senescence of *Daphnia* from risky and safe habitats," *Ecology Letters* 4: 102–105.

Economos, A. C. 1980a. "Brain-life span conjecture: A re-evaluation of the evidence," *Gerontology* 26: 82–89.

———. 1980b. "Taxonomic differences in the mammalian life span-body weight relationship and the problem of brain weight," *Gerontology* 26: 90–98.

Edney, E. B. and R. W. Gill. 1968. "Evolution of senescence and specific longevity," *Nature* 220: 281–282.

Finch, C. E. 1990. *Longevity, Senescence, and the Genome*. Chicago: University of Chicago Press.

Finch, C.E., M.C. Pike, and M. Witten. 1990. "Slow mortality rate accelerations during aging in some animals approximate that of humans," *Science* 249: 902–905.

Finkel, T. and N. J. Holbrook. 2000. "Oxidants, oxidative stress and the biology of ageing," *Nature* 408: 239–247.

Fukui, H. H., L. Xiu, and J. W. Curtsinger. 1993. "Slowing of age-specific mortality rates in *Drosophila melanogaster*," *Experimental Gerontology* 28: 585–599.

Gaillard, J. M., D. Allaine, D. Pontier, N. G. Yoccoz, and D. E. L. Promislow. 1994. "Senescence in natural populations of mammals: A reanalysis," *Evolution* 48: 509–516.

Garland, T., Jr., P. H. Harvey, and A. R. Ives. 1992. "Procedures for the analysis of comparative data using phylogenetically independent contrasts," *Systematic Biology* 41: 18–32.

Gavrilov, L. A. and N. S. Gavrilova. 1991. *The Biology of Life Span: A Quantitative Approach*. New York: Harwood Academic Publishers.

Hamilton, W. D. 1966. "The moulding of senescence by natural selection," *Journal of Theoretical Biology* 12: 12–45.

Harshman, L. G., K. M. Moore, M. A. Sty, and M. M. Magwire. 1999. "Stress resistance and longevity in selected lines of *Drosophila melanogaster*," *Neurobiology of Aging* 20: 521–529.

Harvey, P. H. and M. S. Pagel. 1991. *The Comparative Method in Evolutionary Biology*. Cambridge: Cambridge University Press.

Hastings, A. and H. Caswell. 1979. "Role of environmental variability in the evolution of life history strategies," *Proceedings of the National Academy of Sciences USA* 76: 4700–4703.

Hawkes, K., J. F. O'Connell, N. G. B. Jones, H. Alvarez, and E. L. Charnov. 1998. "Grandmothering, menopause, and the evolution of human life histories," *Proceedings of the National Academy of Sciences USA* 95: 1336–1339.

Hayflick, L. 2000. "New approaches to old age," *Nature* 403: 365.

Holmes, D. J. and S. N. Austad. 1995a. "Birds as animal models for the comparative biology of aging: A prospectus," *Journal of Gerontology: Biological Sciences* 50A: B59–B66.

———. 1995b. "The evolution of avian senescence patterns: Implications for understanding primary aging processes," *American Zoologist* 35: 307–317.

Horiuchi, S. and J. R. Wilmoth. 1997. "Age patterns of the life table aging rate for major causes of death in Japan, 1951–1990," *Journals of Gerontology Series A—Biological Sciences and Medical Sciences* 52: B 67–B 77.

———. 1998. "Deceleration in the age pattern of mortality at older ages," *Demography* 35: 391–412.

Keller, L. and M. Genoud. 1997. "Extraordinary lifespans in ants: A test of evolutionary theories of ageing," *Nature* 389: 958–960.

Kirkwood, T. B. L. 1990. "The disposable soma theory of aging," in D. E. Harrison (ed.), *Genetic Effects on Aging*. Caldwell, NJ: Telford Press, pp. 9–19.

Kirkwood, T. B. L. and S. N. Austad. 2000. "Why do we age?" *Nature* 408: 233–238.

Kirkwood, T. B. L. and R. Holliday. 1979. "The evolution of ageing and longevity," *Proceedings of the Royal Society of London—Series B: Biological Sciences* 205: 531–546.

Lack, D. 1954. *The Natural Regulation of Animal Numbers*. Oxford: Clarendon Press.

Lebreton, J. D., K. P. Burnham, J. Clobert, and D. R. Anderson. 1992. "Modeling survival and testing biological hypotheses using marked animals: A unified approach with case studies," *Ecological Monographs* 62: 67–118.

Linnen, C., M. Tatar, and D. Promislow. 2001. "Cultural artifacts: A comparison of senescence in natural, laboratory-adapted and artificially selected lines of *Drosophila melanogaster*," *Evolutionary Ecology Research* 3: 877–888.

Medawar, P. B. 1952. *An Unsolved Problem in Biology*. London: H. K Lewis.

Monaghan, P. and N. B. Metcalfe. 2000. "Genome size and longevity," *Trends in Genetics* 16: 331–332.

Morand, S. and R. E. Ricklefs. 2001. "Genome size, longevity, and development time in birds," *Trends in Genetics* 17: 567–568.

Nealen, P. M. and R. E. Ricklefs. 2001. "Early diversification of the avian brain:body relationship," *Journal of Zoology, London* 253: 391–404.

Partridge, L. and N. H. Barton. 1993. "Evolution of aging: Testing the theory using *Drosophila*," *Genetica* 91: 89–98.

Perez-Campo, R., M. Lopez-Torres, S. Cadenas, C. Rojas, and G. Barja. 1998. "The rate of free radical production as a determinant of the rate of aging: Evidence from the comparative approach," *Journal of Comparative Physiology—B, Biochemical, Systemic, & Environmental Physiology* 168: 149–158.

Price, T. 1997. "Correlated evolution and independent contrasts," *Philosophical Transactions of the Royal Society of London—Series B: Biological Sciences* 352: 519–529.

Promislow, D. E. L. 1991. "Senescence in natural populations of mammals: A comparative study," *Evolution* 45: 1869–1887.

———. 1994. "DNA repair and the evolution of longevity: A critical analysis," *Journal of Theoretical Biology* 170: 291–300.

Promislow, D. E. L. and P. H. Harvey. 1990. "Living fast and dying young: A comparative analysis of life-history variation among mammals," *Journal of Zoology* 220: 417–437.

Promislow, D. E. L. and M. Tatar. 1998. "Mutation and senescence: Where genetics and demography meet," *Genetica* 103: 299–314.

Ricklefs, R. E. 1991. "Structures and transformations of life histories," *Functional Ecology* 5: 174–183.

———. 1992. "Embryonic development period and the prevalence of avian blood parasites," *Proceedings of the National Academy of Sciences USA* 89: 4722–4725.

———. 1993. "Sibling competition, hatching asynchrony, incubation period, and lifespan in altricial birds," *Current Ornithology* 11: 199–276.

———. 1998. "Evolutionary theories of aging: Confirmation of a fundamental prediction, with implications for the genetic basis and evolution of life span," *American Naturalist* 152: 24–44.

———. 2000a. "Density dependence, evolutionary optimization, and the diversification of avian life histories," *Condor* 102: 9–22.

———. 2000b. "Intrinsic aging-related mortality in birds," *Journal of Avian Biology* 31: 103–111.

Ricklefs, R. E. and C. E. Finch. 1995. *Aging: A Natural History*. New York: Scientific American Library.

Ricklefs, R. E. and P. M. Nealen. 1998. "Lineage-dependent rates of evolutionary diversification: Analysis of bivariate ellipses," *Functional Ecology* 12: 871–885.

Ricklefs, R. E. and A. Scheuerlein. 2001. "Comparison of age-related mortality among birds and mammals," *Experimental Gerontology* 36: 845–857.

———. 2002. "Biological implications of the Weibull and Gompertz models of aging," *Journals of Gerontology Series A—Biological Sciences and Medical Sciences* 57: B69–B76.

Ricklefs, R. E. and J. M. Starck. 1996. "Applications of phylogenetically independent contrasts: A mixed progress report," *Oikos* 77: 167–172.

Robine, J.-M., J. W. Vaupel, B. Jeune, and M. Allard. 1997. *Longevity: To the Limits and Beyond*. Berlin: Springer.

Roff, D. A. 1992. *The Evolution of Life Histories*. New York: Chapman and Hall.

———. 2002. *Life History Evolution*. Sunderland, MA: Sinauer Associates.

Rose, M. R. 1984. "Laboratory evolution of postponed senescence in *Drosophila melanogaster*," *Evolution* 38: 1004–1010.

———. 1991. *Evolutionary Biology of Aging*. New York: Oxford University Press.

Rose, M. R. and J. L. Graves, Jr. 1990. "Evolution of aging," *Review of Biological Research in Aging* 4: 3–14.

Sacher, G. A. 1959. "Relation of lifespan to brain weight and body weight in mammals," in G. E. W. Wolstenholme and M. O. Connor (eds.), *CIBA Foundation Colloquia on Aging*. Boston: Little, Brown, pp. 115–141.

———. 1977. "Life table modification and life prolongation," in C. E. Finch and L. Hayflick (eds.). *Handbook of the Biology of Aging*. New York: Van Nostrand, pp. 582–638.

———. 1978. "Evolution of longevity and survival characteristics in mammals," in E. L. Scheider (ed.), *The Genetics of Aging*. New York: Plenum, pp. 151–168.

Sæther, B.-E. 1988. "Pattern of covariation between life-history traits of European birds," *Nature* 331: 616–617.

Schaffer, W. M. 1974. "Optimal reproductive effort in fluctuating environments," *American Naturalist* 108: 783–790.

Sgrò, C. M. and L. Partridge. 2000. "Evolutionary responses of the life history of wild-caught *Drosophila melanogaster* to two standard methods of laboratory culture," *American Naturalist* 156: 341–353.

Sibley, C. G. and J. E. Ahlquist. 1990. *Phylogeny and the Classification of Birds*. New Haven, CT: Yale University Press.

Silbermann, R., and M. Tatar. 2000. "Reproductive costs of heat shock protein in transgenic *Drosophila melanogaster*," *Evolution* 54: 2038–2045.

Starck, J. M. and R. E. Ricklefs. 1998. "Variation, constraint, and phylogeny: Comparative analysis of variation in growth," in J. M. Starck and R. E. Ricklefs (eds.), *Avian Growth and Development: Evolution within the Altricial-Precocial Spectrum*. New York: Oxford University Press, pp. 247–265.

Stearns, S. C. 1983. "The impact of size and phylogeny on patterns of covariation in the life history traits of mammals," *Oikos* 41: 173–187.

———. 1992. *The Evolution of Life Histories*. New York: Oxford University Press.

Stearns, S. C. and L. Partridge. 2000. "The genetics of aging in *Drosophila*," in E. Masoro and S. Austad (eds.), *Handbook of Aging*, 5th ed. San Diego: Academic Press, pp. 345–360.

Strehler, B. L. and A. S. Mildvan. 1960. "General theory of mortality and aging," *Science* 132: 14–21.

Uno, H. 1997. "Age-related pathology and biosenescent markers in captive rhesus macaques," *Age* 20: 1–13.

Vaupel, J. W. et al. 1998. "Biodemographic trajectories of longevity," *Science* 280: 855–860.

Weinstein, B. S. and D. Ciszek. 2002. "The reserve-capacity hypothesis: Evolutionary origins and modern implications of the tradeoff between tumor-suppression and tissue-repair," *Experimental Gerontology* 37: 615–627.

Williams, G. C. 1957. "Pleiotropy, natural selection and the evolution of senescence," *Evolution* 11: 398–411.

Wilson, D. L. 1994. "The analysis of survival (mortality) data: Fitting Gompertz, Weibull, and logistic functions," *Mechanisms of Ageing and Development* 74: 15–33.

Life Span Extension of *Drosophila melanogaster*: Genetic and Population Studies

Lawrence G. Harshman

During the past two decades, genetic studies of model organisms have been the most important tool underlying advances in understanding the biological basis of aging and longevity. *Drosophila melanogaster*, the geneticist's "fruit fly," is a model organism because it has been the focus of genetic studies for more than 90 years. This review argues that studies on *D. melanogaster* will make an especially important contribution to the field of aging and longevity at the intersection of research on genetics, complex traits, and fly populations.

Five approaches have been used to study the genetics of longevity of *D. melanogaster*: (1) laboratory selection, (2) quantitative genetics, (3) transgenic overexpression, (4) mutation analysis, and (5) measurement of gene expression. The first two approaches attempt to decompose longevity as a complex character. The third and fourth approaches start by looking for major gene effects on life span. The fifth approach is emerging as part of a major advance in technology in which the expression of almost all genes in the genome can be measured at one time.

Genetic research on aging and longevity using *D. melanogaster* has been reviewed previously (Arking 1987, 1988; Arking and Dudas 1989; Rose 1991; Curtsinger et al. 1995; Tower 1996; Stearns and Partridge 2001). The present chapter reviews the range of genetic approaches used to study aging and life span (length of life).

Selection experiments in the laboratory

Natural selection has shaped organic evolution whereas artificial selection is a human endeavor, usually with some utilitarian goal. Artificial selection is typically conducted on complex traits that are controlled by multiple genes.

Generation by generation, artificial selection can progressively change the mean value of a complex trait, such as longevity, in a population. The genes that underlie the response to selection, and traits that change as correlated responses to selection, are of interest in the context of understanding aging and longevity.

Figure 1 shows the distribution of a trait in a population of individuals undergoing selection. When individuals from the distribution are non-representatively selected to contribute to the next generation and genetic variation contributes to the trait variation in the population, there can be a genetic-based change in the mean value of the trait in the next generation.

Artificial or natural selection in the laboratory has been used to study a range of traits (Rose 1984; Hoffmann and Parsons 1989; Rose et al. 1990; Huey and Kingsolver 1993; Rose et al. 1996; Gibbs 1999). Selection experiments magnify the difference between selected populations and ancestral or unselected control populations. Large differences are easier to study; selection experiments facilitate investigation by increasing the signal-to-noise ratio in the comparison of selected and control (unselected) populations.

FIGURE 1   The distribution of a trait and population mean (μ)

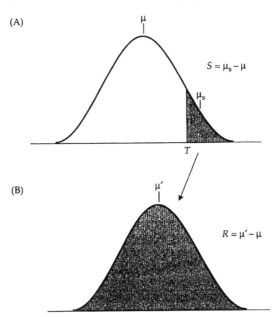

NOTES: (A) The mean of the subset of individuals used to propagate the next generation is μ_s. Selection intensity (S) is a function of the difference between μ and μ_s. (B) The response (R) to selection is defined in terms of the mean trait value in the next generation (μ') and the previous generation (μ).
SOURCE: Hartl and Clark 1997.

   The direct response to selection is the change in the trait targeted for selection. The indirect responses to selection consist of changes in other traits. Indirect responses to selection can be informative because they indicate genetic correlations between traits. Such correlations can reveal traits whose association with the directly selected trait was not anticipated at the beginning of the experiment. Moreover, genetic associations between traits can provide circumstantial evidence about mechanisms underlying the direct response to selection. For example, negative correlations between traits in selection experiments suggest tradeoffs (Rose et al. 1990). Tradeoffs are based on constraints involving energy, space, hormones, or structural biology such that an increase in the expression of one trait results in a decrease in another trait. Tradeoffs are a pivotal consideration in the evolution of longevity and other life-history traits (Williams 1957; Stearns 1989, 1992; Reznick 1985; Zera and Harshman 2001).

   A great deal of *D. melanogaster* genetic research on longevity has been based on artificial selection in the laboratory. The following summary of a subset of results is designed to focus on general outcomes and provide perspective on the pluses and minuses associated with the use of selection experiments to study aging and longevity.

Selection experiments on life span

Although not the first studies of this genre, two artificial selection experiments have had a long-term impact on the field (Rose 1984; Luckinbill et al. 1984). In each case, the mode of selection was to select flies to propagate the next generation using individuals that remained fertile at old ages. For example, an artificial selection experiment by Rose (1984) increased the age of breeding from 4-day-old adults to 28 days (generation 1), then to 35 days (generations 2 and 3), then to 42 days, and ultimately used 70-day-old adults to maintain the selected lines. The important features of the experimental design included a high degree of replication (5 selected and 5 control lines), a relatively large population size in each line to mitigate inbreeding, and a laboratory-adapted ancestral population (Rose 1984; Rose et al. 1996). The control lines were maintained using the same generation time as the ancestral population, meaning that the breeders for the next generation were young adults. Rose (1984) and Luckinbill et al. (1984) substantially increased longevity (by 50 to 100 percent) in their selected lines. Research on their lines highlighted stress resistance as a genetic correlate of longevity.

   Selection experiments provided initial evidence for a genetic relationship between stress resistance and longevity. Relative to control lines, the Rose lines, selected for longevity, were resistant to starvation, desiccation, and ethanol fumes, as well as elevated temperature under some conditions

(Service et al. 1985). The Luckinbill set of lines provided the first evidence for a genetic correlation between longevity and oxidative stress resistance (Arking et al. 1991). This observation was particularly significant because it provided indirect support for the free radical theory of aging (Harman 1956), a predominant biochemical hypothesis that explains aging in terms of oxidative damage to macromolecules in cells (Wallace 1992; Martin et al. 1996; Johnson et al. 1999; Guarente and Kenyon 2000).

Selection experiments also provided early evidence for a genetic association between early-age reproduction and longevity. Females from control populations exhibited a relatively short life span and high early-age egg production, but the converse was observed for the lines selected for extended longevity (Rose 1984). Selected-line females were less fecund early in life regardless of whether they had mated or were virgin (Service 1989). Selected-line males also exhibited reduced reproductive function early in life compared to control-line males (Service and Fales 1993; Service and Vossbrink 1996). During the course of many generations of selection, however, early-age egg production evolved to become higher in the selected than in the control lines in some environments (Leroi et al. 1994). This outcome was interesting from an evolutionary standpoint, but it raises questions about the utility of the selection experiment approach because some results can be inconsistent over time.

Ancillary selection experiments have been used to test the direct and correlated responses to selection. For example, direct selection for starvation resistance resulted in flies with extended longevity (Rose et al. 1992). However, Harshman et al. (1999) selected only for female starvation resistance and did not observe an increase in female or male longevity. Both experiments were highly replicated, but they differed in the intensity of selection and in other design features. In general, the relevant literature does not indicate consistent support for an association between starvation resistance and extended longevity in selection experiments (Harshman et al. 1999). The heterogeneity in outcomes suggests the limitations of selection experiments for understanding mechanisms underlying extended longevity when such experiments are conducted in isolation from other genetic approaches.

Synopsis of results from laboratory selection experiments that extend longevity

Six selection experiments that extended the life span of *D. melanogaster* are summarized here. The first two have already been described (Rose 1984; Luckinbill et al. 1984). The third and fourth artificial selection experiments were also conducted by selection for advanced age of reproduction (Partridge and Fowler 1992; Partridge et al. 1999). The fifth artificial selection

TABLE 1   Life span extension: Correlated responses to selection in six experiments

	SR	Lip	DR	Gly	OR	DT	JV	EF	LF	Met
R	>	>	>	>	>	>	>	<	>	0
L	0	0	0>	0>	>	>	<	<	>	0
P1	na	na	na	na	na	>	<	0	>	na
P2	na	na	na	na	na	0	0	<	0	na
Z	0	0	na	na	na	0	na	<	0	na
S	0	0	>	na	na	>	na	<	0	0

NOTES: The six selection experiments are identified in the table using the following abbreviations: R = Rose (1984), L = Luckinbill et al. (1984), P1 = Partridge and Fowler (1992), P2 = Partridge et al. (1999), Z = Zwann et al. (1995), S = Stearns et al. (2000). Life-history traits typically have to do with development, growth, size, reproduction, and survival. The life-history traits in Table 1 are: DT = development time, JV = juvenile viability (survival rate), EF = early-life female fecundity, and LF = late-life female fecundity. Table 1 also summarizes stress resistance traits: SR = starvation resistance, DR = desiccation resistance, OR = oxidative stress resistance. The metabolic traits in Table 1 are: Lip = lipid abundance, Gly = glycogen abundance, Met = metabolic rate (respiration rate). The symbols in the text of the table indicate whether the selected lines exhibit a greater mean trait value (>) compared to the control lines, a relatively reduced mean trait value (<), marginally greater mean trait value than the control line value (0>), no statistically significant difference between selected and control lines (0), or data are not available for a particular experiment (na).

experiment selected from families that exhibited relatively high mean longevity (Zwann et al. 1995). The sixth experiment was based on differential extrinsic mortality controlled by the investigators, not artificial selection for extended longevity (Stearns et al. 2000). Some lines in this experiment experienced high levels of extrinsic survival and others experienced relatively low levels. Flies from lines that experienced low extrinsic mortality became relatively long-lived.

As described previously, genetic correlations can provide insight into the mechanistic underpinnings of a response to selection. The results of the six selection experiments are summarized in Table 1 in terms of factors that are genetically correlated (indirect responses to selection) with extended longevity. In some cases the correlated responses to selection were largely consistent. Five of six selection experiments documented that relatively low early-age egg production (EF) was associated with a longer life span. This is an example of a consistent response to selection; consistent indirect responses to similar selection experiments are arguably most suitable for continued study (Harshman and Hoffmann 2000a). A useful approach might be to use the robust genetic correlations to guide the search for underlying longevity factors (Gibbs 1999; Zera and Harshman 2001).

Analysis of selection experiments for extended longevity: Studies of consistent correlated responses

Stress resistance, primarily resistance to starvation and desiccation, was identified as a correlate of extended longevity in selected lines (Rose 1984;

Service et al. 1985; Service 1987). However, selection for extended longevity did not result in significantly increased starvation resistance in three of the four longevity selection experiments in which a starvation assay was conducted (Table 1). In one of these experiments, selection on the basis of differential adult mortality did not alter starvation resistance as an indirect response, but the selection response to selection did indicate tradeoffs between early fecundity (egg production), late fecundity, and starvation resistance that was mediated by lipid allocation (Gasser et al. 2000). Desiccation resistance appears to be more consistent as a correlated response to selection for longevity (Table 1). Investigation of desiccation resistance in selection lines has revealed that the selected lines have more glycogen, store more water, and have a reduced rate of water loss, but there is no difference in water content at death in the selected and control lines (Bradley et al. 1999; Graves et al. 1992; Gibbs et al. 1997; Nghiem et al. 2000). Oxidative stress resistance in the selected lines might be particularly important for extended longevity. Flies from Luckinbill extended longevity lines (L in Table 1) have higher levels of mRNA abundance corresponding to a range of antioxidant defense genes that produce enzymes including superoxide dismutase (SOD), glutathione *S*-transferase, catalase (CAT), and xanthine dehydrogenase as well as higher levels of CAT and SOD enzyme activity (Dudas and Arking 1995; Force et al. 1995). These results were corroborated by a reverse selection experiment (Arking et al. 2000a), which produced data indicating a time delay in oxidative damage of proteins and lipids in the extended longevity lines. Antioxidant enzymes (CAT and Cu-Zn SOD) are found on the left arm of the 3rd chromosome, and chromosome substitution studies showed that the proximal part of this chromosome arm is associated with extended longevity in the Luckinbill selected lines (Buck et al. 1993b). Arking et al. (2000a) used the Luckinbill et al. (1984) lines to reverse selection for longevity by taking young-age breeders for a series of generations. As a correlated response, relatively elevated measurements of antioxidant enzymes decreased in control line levels, but there was no significant decrease in 11 other enzymes that play a general role in metabolism not directly linked to antioxidant defense. Oxidative stress resistance is also present in the Rose lines, suggesting oxidative stress resistance might play a general role underlying increased life span in response to selection for extended longevity (Harshman and Haberer 2000).

Reduced early-age reproduction appears to be a consistent correlated response to selection for extended longevity (Table 1). A delay in early-stage egg maturation was found to be responsible for reduced early-age reproduction in the Rose extended longevity lines (Carlson et al. 1998; Carlson and Harshman 1999). As has been described, selection experiments also indicate that increased stress resistance is a consistent correlated response to selection (Table 1). Does this imply a negative relationship between reproduction and stress resistance? As an example, selection for cold resistance in *D. melanogaster* and *Drosophila simulans* was correlated with decreased early-age fecundity (Watson and Hoffmann 1996). In general, experiments that have selected for stress resistance have found that decreased early-age reproduction is a correlated response (reviewed in Zera and Harshman 2001). Support for this relationship also comes from phenotypic manipulation experiments. Stimulation of female egg production in the Rose lines resulted in decreased starvation resistance (Chippindale et al. 1996), and stimulation of egg production resulted in decreased oxidative stress resistance (Wang et al. 2001; Salmon et al. 2001). The loss of oxidative stress resistance as a function of reproduction might be particularly relevant because of the role that oxidative damage putatively plays in aging and longevity.

An extension of metabolic life is another consistent outcome of selection experiments for longevity. The rate of gas exchange by animals, the uptake of O_2 and production of CO_2 (respiration rate), is a measure of metabolic activity. Metabolism generates oxygen radicals that are thought to cause aging. Therefore, aging may be an inevitable byproduct of the metabolic functions of life. As a relevant observation, one or more of the mutations that extend the life span of the worm *Caenorhabditis elegans* reduce the metabolic rate (Van Voorhies and Ward 1999).

Is reduced respiration rate associated with extended longevity in *D. melanogaster* selection experiments? A series of respiration rate studies on one set of extended longevity lines is summarized in Rose and Bradley (1998). When fly respiration was measured in chambers smaller than the cages used for selection, Service (1987) found relatively higher rates at young ages in the control (unselected) lines than in the long-lived selected lines, but not at later ages. Djawdan et al. (1996) found no difference between selected and control lines when respiration was measured in cages used for selection. When adults from both lines were provided with supplementary yeast, the control-line females exhibited a slightly higher metabolic rate than selected-line females (Simmons and Bradley 1997). Djawdan et al. (1997) found no difference in respiration rate when the mass of selected and control flies was adjusted by removing the weight of water, lipid, and carbohydrate. Using the Luckinbill lines (L in Table 1), Arking et al. (1988) demonstrated that selected longevity resulted in appreciably greater lifetime metabolic potential than measured in control lines. The study by Gasser et al. (2000), S in Table 1, included a behavioral activity assay and a measurement of respiration rate; neither was reduced in the longer-lived lines. There is little evidence that extended longevity is genetically correlated with reduced metabolic rate. In selected lines of *D. melanogaster*, longevity does not appear to be a simple byproduct of reduced metabolism.

Analysis of selection experiments for extended longevity: Demography

What is the age-specific pattern of mortality in lines selected for extended longevity? At issue is whether selection has reduced the rate of aging or reduced the initial mortality parameter. Demographic analyses have been conducted on the Rose and Luckinbill lines selected for extended longevity (Service et al. 1998; Pletcher et al. 2000). Service et al. (1998) used the Rose selected lines and found that the age-dependent mortality rate parameter was significantly smaller and the frailty parameter had a significantly smaller variance whereas the age-independent mortality parameter did not differ between selected and control lines. Thus, there was evidence that the Rose selected lines evolved a reduced rate of senescence (Nusbaum et al. 1996; Service et al. 1998). Pletcher et al. (2000) investigated the lines selected for extended longevity and found that differences in baseline mortality, compared to the age-dependent mortality parameter, accounted for most of the difference between selected and control lines. In terms of understanding the potential for life span extension, it is important to know whether selected life span extension is based on a delay in the aging process or on reduced age-specific mortality across a range of ages. One of the difficulties in answering this question arises from the demands of adequate sample size because mortality is uncommon among young flies and few flies remain alive at the oldest ages.

Analysis of selection experiments for extended longevity: The cost of reproduction

The cost of reproduction is a demographic relationship in which current reproduction is associated with a reduction in future reproduction and decreased longevity (Williams 1966; Bell and Koufopanou 1986; Reznick 1985; Stearns 1989, 1992; Partridge and Sibly 1991; Carey et al. 1998). This relationship has been documented in a wide variety of plants and animals (Roff 1992). The cost of reproduction is so widespread that it would be appropriate to consider it to be a general feature of life.

To investigate the cost of reproduction, experiments were conducted on lines selected for extended longevity (Sgro and Partridge 1999). The selected and control lines used were genetically differentiated in terms of early-age egg reproduction (lower in the selected lines) and longevity (greater in the selected lines). Sgro and Partridge abolished egg maturation in females from selected and control lines to determine whether this manipulation removed the survival-rate differences between selected and control lines. They ablated egg production by using irradiation or introducing a sterile mutation. Thus, the hypothesis was tested in parallel experiments, which helps circumvent potentially idiosyncratic results from any one experimental manipulation. When female reproduction was ablated, age-specific survival differences between selected and control lines disappeared (Figure 2). With implications that extend beyond *D. melanogaster*, a delayed survival cost of reproduction could play an important role in defining lifetime survival curves (Carey et al. 1998; Sgro and Partridge 1999).

The study by Sgro and Partridge (1999) produced results indicating that one phenomenon could underlie both the acceleration of mortality (aging) and mortality deceleration at late ages. The leveling off of mortality at late ages was originally documented in medflies by Carey et al. (1992) and in *D. melanogaster* by Curtsinger et al. (1992). The causes of aging and of the deceleration of mortality at advanced ages are two of the most important problems in biodemography. The results of Sgro and Partridge suggest that reproduction could underlie aging and the deceleration of mortality at late ages. The generality of this study could be tested by repeating it with different lines and with sample sizes that allow finer resolution of age-specific mortality patterns, especially at ages when there were few surviving flies.

Perspectives on selection experiments

Genetic correlations are the sine qua non of laboratory selection experiments, but they are problematic. First, what appears to be a genetic correlation could be due to inadvertent independent selection on two traits (coselection). Second, genetic correlations may be based on strong biological connections between traits or very indirect association. Thus, they may or may not provide useful evidence about mechanisms that underlie the response to selection. Third, indirect responses (genetic correlations) to selec-

FIGURE 2 Age-specific female mortality of selected populations (closed circles) and control populations (open circles) of *D. melanogaster*

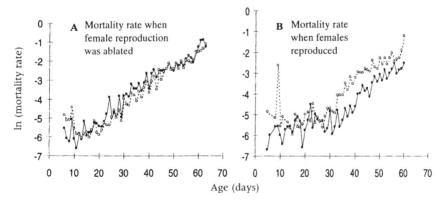

NOTE: When female reproduction was ablated by irradiation (or introduced mutation), the age-specific mortality differences between selected and control lines disappear.
SOURCE: Sgro and Partridge 1999

tion for extended longevity are sometimes inconsistent between experiments (Table 1; Tower 1996; Harshman and Hoffmann 2000a). This inconsistency could stem from a number of causes including inbreeding, inappropriate assay conditions, genetic differences between base populations, variable strength of selection in different experiments, container effects, and multiple mechanisms underlying the selection response. Unfortunately, the causes of heterogeneity between experiments are extremely difficult to sort out. In the face of this vagary, it is best to focus on consistent indirect responses (robust genetic correlations among experiments) as the most important targets for further research to elucidate the mechanisms underlying responses to selection for longevity (Harshman and Hoffmann 2000a; Harshman and Haberer 2000).

Flies from the Rose longevity-selected lines have been found to exhibit no greater life span than flies collected from the field and then tested for longevity before numerous generations have passed (Promislow and Tatar 1998). This fact represents a major problem for selection experiments on longevity because selection for extended longevity in the laboratory may merely involve removal of deleterious mutant forms of genes that have accumulated during laboratory culture (Promislow and Tatar 1998). Alternatively, if populations adapt to the laboratory by evolving a reduction in life span, then laboratory selection for extended longevity is an informative reversal of this initial adaptation. Transfer from a natural population to the laboratory environment and subsequent culture of flies in bottles resulted in reduced stress resistance, higher early-age egg production, and a shortened life span (Sgro and Partridge 2000; Hoffmann et al. 2001), indicating that these traits are interrelated (Hoffmann et al. 2001). Overall, the evidence indicates that these changes are due to natural selection in the laboratory.

Selection experiments have been useful for identifying properties of organisms whose life span has been genetically extended. And they have proven useful for study of such phenomena as the cost of reproduction (Sgro and Partridge 1999) and oxidative stress resistance (Arking et al. 2000b). They have had limited utility, however, in elucidating specific mechanisms (genes, signaling pathways, cell biology, physiology) that directly cause life span extension. Quantitative genetics provides tools that could identify specific genes that cause extended longevity in selection experiments and other populations.

Quantitative genetics

Quantitative traits are continuously distributed in a population, and multiple genes determine the distribution of these traits. For such genetically complex traits, statistical techniques have been developed to characterize genetic correlations and genetic variances, to estimate the number of major genes controlling the distribution of a trait, and to localize regions of chromosomes that control variation in the trait. The regions are known as quantitative trait loci (QTL). An important goal of QTL analysis is to identify the specific genes responsible for extended longevity in populations.

Using very small sequences and the polymerase chain reaction (PCR) of DNA it is possible to amplify regions of DNA from many locations in the genome of *D. melanogaster*. Curtsinger et al. 1998 used this technique (RAPD) to assess DNA variation at approximately 1,000 positions in the genome in relation to life span. Five genomic regions had a significant effect on the life span of males, females, or both sexes. The largest effect was found for a variant DNA region that does not make a protein product but was associated with reduced mortality at all ages.

A more common approach used to QTL map (localize in the genome) employs recombinant inbred lines (RILs). Briefly, flies from two parental lines are crossed and the hybrid offspring are crossed back to one of the parental lines through a series of generations of matings to siblings. As a consequence, regions of chromosomes are introduced from one parental line into an otherwise homogenous background of the second parental line. These introduced segments can cause variation in the trait of interest among the RILs. The location of the introduced segments can be identified by variable molecular markers distributed across the genome. Association of the trait of interest with specific genetic markers provides a way of mapping regions of chromosomes that contribute to population variation in a complex trait (Figure 3).

A series of QTL analyses of longevity have been conducted. One study, based on a set of RILs derived from laboratory lines, assayed the longevity of virgin males and females among the RIL lines (Nuzhdin et al. 1997; Leips and Mackay 2000; Pasyukova et al. 2000; Viera et al. 2000). Nuzhdin et al. identified 5 QTLs with major effects on longevity, but these QTLs affected life span in only one sex or the other. Viera et al. extended the longevity assays conducted on these recombinant inbred lines by determining mean life span at different temperatures and after heat shock. Seventeen QTLs were identified that increased life span, and they were largely specific to each environment and sex. There was evidence for opposing effects such that a QTL associated with increased life span in one environment tended to decrease life span in other environments.

A few studies have been conducted with outbred RILs. Using recombinant inbred lines crossed to an unrelated stock (outbred), Reiwitch and Nuzhdin (2002) found correspondence between female life span QTLs identified in the recombinant inbred lines and outbred lines. This study measured longevity with males and females held in the same container, and in this case there were no sex differences in QTLs. In Leips and Mackay (2000), the recombinant inbred lines were crossed to the ancestral lines used to produce them. The investigators studied the effect of another environment, larval (juvenile) density, on adult longevity QTLs. The six sig-

FIGURE 3   Quantitative trait loci (QTL) mapping by marker–trait association in recombinant inbred lines (RILs)

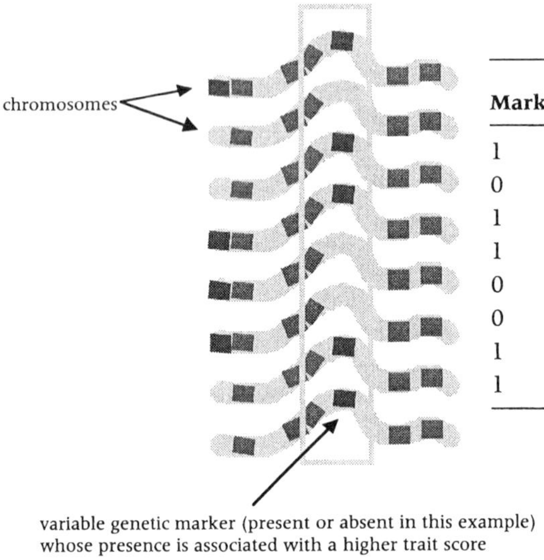

Marker	Trait score	Chromosome (from top)
1	16	1
0	12	2
1	15	3
1	18	4
0	10	5
0	8	6
1	15	7
1	16	8

variable genetic marker (present or absent in this example)
whose presence is associated with a higher trait score

nificant QTLs identified by Leips and Mackay (2000) typically extended life span only at one density or in one sex. Interactions were common among the life span QTLs, but a combination of interacting QTLs did not necessarily result in exceptional life span. Importantly, the QTLs identified by Nuzhdin et al. (1997) and Veira et al. (2000) tended to correspond to the QTLs identified in the study using relatively outbred lines (Leips and Mackay 2000).

QTL analyses have been conducted on recombinant inbred lines derived from the Luckinbill et al. (1984) selection experiment for extended longevity. In these studies, three life span QTLs have been identified by assaying large numbers of males and females held together from each recombinant inbred line (Curtsinger, personal communication). One of the QTLs, on chromosome 3L, confers resistance to paraquat (oxidative stress), and it maps to a region of the genome that includes a gene that metabolizes oxygen radicals (Curtsinger et al. 1998). Unlike the results of Nuzhdin et al. (1997), these life span QTLs are not sex-specific in their effects. Thus, for the two cases in which males and females were held in the same container during the longevity assays, sex-specific QTLs were not found (Reiwitch and Nuzhdin 2002; Curtsinger, personal communication).

Perspectives on quantitative genetic studies of life span

A criticism of recombinant inbred lines for the characterization of QTLs is that inbreeding exposes the effect of deleterious mutations that have accumulated in populations. However, the use of RILs is a legitimate first step in a process that can include outbred lines. Finding the genes that do extend life span in populations is an ultimate goal, and *D. melanogaster* is the best option for such studies among the organisms that are used as models for genetic research. The genetic toolbox available for *D. melanogaster* and genomics/bioinformatics methods will expedite identification of the genes that extend the life span of flies.

QTL analyses of extended life span in *D. melanogaster* selection experiments could be an especially informative avenue of research because metabolic rate apparently has not been reduced as a result of artificial selection for longevity. This implies that there are forms of genes that have the potential to extend life span without sacrificing normal metabolism and activity. In general, quantitative genetics has the potential to vastly improve the power to detect genes that cause extended longevity in *D. melanogaster* populations. Transgenic overexpression and mutation analysis have identified genes that can cause extended longevity in the laboratory.

Transgenic overexpression of candidate genes

Candidate genes are those that might have an anticipated effect on the trait of interest. Candidate genes in the present context are those that might extend longevity. Foreign genes can be introduced into the genome of *D. melanogaster* as candidate transgenes. The goal is to alter the level of specific gene expression for the purpose of investigating gene function. Generally, transgenes are introduced into other genomes using modified transposable elements ("jumping genes") that can insert themselves into a recipient genome (Figure 4). For this purpose, a gene of interest can be combined with a modified DNA element (P element) that can integrate into a genome. A population of such recombinant molecules is injected into specific cells in developing embryos. These cells are destined to become the germ line that produces eggs and sperm and eventually adult flies. After recombinant P element vectors are injected into embryonic germ line cells (pole cells), insertion into random chromosome positions can occur because of co-injection with another vector that can make the enzyme (transposase) needed for recombinant vector integration into the genome. In short, it is possible to introduce foreign genes into the recipient genome, and the foreign genes can be transmitted from generation to generation with the rest of the chromosomal genetic material. Candidate genes that have extended life span by overexpression or by mutation are presented in Table 2. In general, transgenes that extend life span are associated with stress resistance.

Two enzymes, found in a wide variety of organisms, are known to have a mode of action that nullifies oxygen radicals. Superoxide dismutase (SOD) scavenges superoxide anions and produces hydrogen peroxide as a

FIGURE 4   P element–mediated introduction of foreign genes

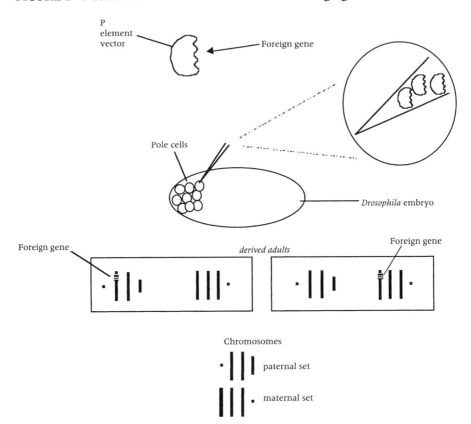

NOTE: A P element is a short segment of DNA that can allow an exogenous carrier molecule with a transgene (foreign gene) to insert at random locations in the genome.

byproduct. Hydrogen peroxide, also an oxidizing agent, is broken down by catalase (CAT) into molecular water and oxygen. A transgenic extra copy of a *D. melanogaster CAT* gene resulted in increased enzyme activity, but no increase in longevity (Orr and Sohal 1992). Transgenic over-expression of a *SOD1* (CuZn SOD) gene resulted in increased resistance to oxidative stress agents and modestly increased life span in one study, but not in another (Reveillaud et al. 1991; Orr and Sohal 1993). Simultaneous introduction of *CAT* and *SOD1* genes was observed to increase life span by approximately 20 percent (Orr and Sohal 1994). However, this result has been called into question because of inadequate controls for the experiment (Tower 1996).

One of the advantages of *D. melanogaster* transgenic methodology is that it is possible to overexpress introduced genes in specific tissues or at specific times during the life cycle. When human *SOD1* was overexpressed

TABLE 2   Genes that have extended the life span of *D. melanogaster* in the laboratory

Genes	Gene product	Possible life span extension effect
Overexpression (high level of gene activity)		
SOD1 (CuZn) and *SOD2* (Mn)	superoxide dismutase (an enzyme)	oxidative stress resistance
hsp70	heat shock protein	high temperature tolerance
DPOSH	signaling protein	stress resistance pathway activation and/or resistance to programmed cell death
Mutation (partial or total loss of gene function)		
mth	G protein–coupled receptor (putative)	multiple stress resistance
Indy	dicarboxylate co-transporter (putative)	reduced intermediary metabolism
InR and *chico* [1]	insulin receptor and insulin receptor docking protein	altered control of metabolism and some stress resistance

exclusively in adult motor neurons of *D. melanogaster*, life span was extended by 40 percent (Parkes et al. 1998). As an aside, mutations in the human *SOD1* gene are the basis of familial amyotrophic lateral sclerosis (commonly known as Lou Gehrig's disease), which is an inherited life-shortening human disease characterized by deterioration of motor neurons. Sun and Tower (1999) overexpressed *D. melanogaster SOD1* in adult *D. melanogaster*, extending life span by 48 percent. In this study, overexpression was not constrained to motor neurons. To the extent that the studies can be compared, the life span–extending effect of *SOD1* overexpression in the whole body (Sun and Tower 1999) was not much greater than that observed by Parkes et al. (1998). This comparison suggests that oxidative damage to motor neurons is a major cause of aging. The *DPOSH* gene, also associated with stress resistance, extends longevity by 14 percent when expressed in the nervous system throughout the life cycle (Seong et al. 2001: Table 2). In general, cells that are not replaced by new cells in adults, such as neurons, are especially vulnerable to oxidative damage and may play an important role in aging. Mitochondria are the major source of oxygen radical production that may cause aging and limit life span (Wallace 1992). The gene for the manganese-dependent superoxide dismutase (*SOD2*) is associated with mitochondria in higher organisms. *SOD2* has been stably introduced into the genome of *D. melanogaster* and overexpressed, thereby extending the life span of *D. melanogaster* adults by approximately 50 percent (Sun et al. 2002).

Oxidative damage, high temperature, and other factors can denature proteins, resulting in deleterious physiological and fitness effects (Feder and Hoffmann 1999). Heat shock proteins (HSP) and molecular chaperones can refold damaged proteins. Thus, investigators tested the hypothesis that a relatively high level of HSP70 production would increase life span (Khazaeli et al. 1997; Tatar et al. 1997). For one test, two P element transgenic strains of *D. melanogaster* were used (Tatar et al. 1997). One of the strains had extra copies of *hsp70* genes. The second strain was very similar in having a remnant P element present at the same location as the extra copy strain, but the extra copies of the *hsp70* genes had been excised. When adults were briefly exposed to high temperature, which stimulates activity of the transgenes, flies from the high copy *hsp70* strain exhibited relatively decreased mortality rates for two weeks and a 4–8 percent increase in life span. If a high level of this heat shock protein is beneficial, why wouldn't heat shock proteins be continuously expressed at high levels? The answer may be that high levels of HSP70 negatively affect other parts of the life cycle. For example, overexpression of *hsp70* can negatively affect larval (juvenile) flies by retarding growth, reducing viability, and reducing egg hatch (Krebs and Feder 1997, 1998; Silbermann and Tatar 2000).

Perspectives on transgenic overexpression

When transgenes are inserted into genomes, their effects can depend on the position of insertion or on differences between genomes into which they are inserted. Overexpression studies have varied in the extent to which they have controlled for such effects to rule out the possibility of artifacts (Kaiser et al. 1997; Stearns and Partridge 2001). The studies by Sun and Tower (1999) and Sun et al. (2002) are exemplary in their controls for position and genetic background effects and their use of a sufficient number of replicate lines.

Mutation analysis

Mutation analysis is a preeminent analytical tool in contemporary biology. In general, new mutations are generated and screened to identify genes that affect a biological trait of interest. The goal is to identify the underlying genes controlling trait manifestation and the role of these genes in the process. Mutations in *D. melanogaster* have identified genes that can increase longevity. In all cases, transposable DNA (P elements) was used to induce the mutations (Figure 5). When a P element moves and reinserts itself elsewhere in the genome it can cause a mutation. As opposed to transgenic overexpression, all of the mutations described in this section reduce the expression of, or totally inactivate, specific genes (such as those included in Table 2).

FIGURE 5   **Transposon insertional mutagenesis**

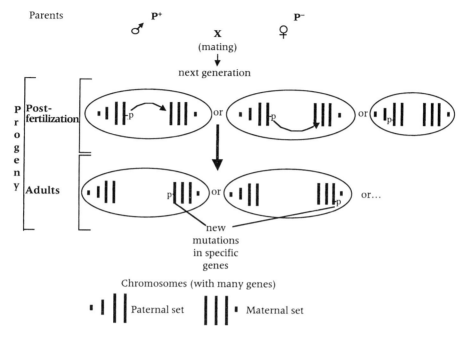

NOTE: A P element transposon is a short segment of DNA that can move within the genome and cause mutations. If a male with one or more P elements is mated to a female with none, then P elements can move to new locations in the chromosomes of the progeny, sometimes causing mutation.

The first *D. melanogaster* longevity-extension mutation was in the *Methuselah* (*mth*) gene (Lin et al. 1998). The *mth* mutation results in partial loss of function of the gene and extends longevity by 35 percent. The *mth* gene produces a protein, presumably a G protein–coupled receptor, that belongs to a family of proteins associated with a range of functions in higher organisms including endocrinology, neurology, and response to external stimuli. The putative G protein product of the *mth* gene is not similar to any of the G proteins of known function in other organisms, and thus the specific function of the *mth* gene is not suggested by comparison with other organisms. There is evidence that the *mth* gene plays a role in regulation of neuromuscular neurotransmitter function (Song et al. 2001). Again, neuromuscular function might be especially important for aging and longevity. The *mth* mutation resulted in increased stress resistance (Lin et al. 1998), and the mutant flies were approximately one-third larger than controls. Mutant flies were substantially more resistant to starvation, high temperature, and oxidative stress.

Mutations that markedly extend life span were also found in the *Indy* gene (Rogina et al. 2000). *Indy* mutations can result in a 50 percent increase in maximum longevity and approximately a twofold increase in mean life span. Five independent mutations in the same gene were reported, each of which exhibited a substantial increase in longevity in combination with a normal gene (as a heterozygote). Moreover, the heterozygote females, derived from crosses between a laboratory stock (Canton S) and mutant stocks, produced substantially more eggs than Canton S females (Rogina et al. 2000), but the heterozygotes derived from crosses to a more relevant control strain had only slightly greater fecundity than the control strain (Helfand, personal communication).

The *Indy* gene product is a protein similar to mammalian dicarboxylate co-transporters. Mammalian co-transporters are membrane proteins that transport intermediates of energy compound metabolism (Krebs cycle intermediates) into cells. The *Indy* gene is expressed at high levels in the gut, fat body, and oenocytes, which are the insect tissues/cells that play a predominant role in intermediary metabolism and storage of metabolic products. A decrease in metabolic product uptake mediated by the *Indy* mutation suggests that the mechanism of life span extension could be caloric restriction (Rogina et al. 2000), which is an environmental intervention that can increase life span in invertebrates and mammals.

For the purposes of experimental rigor, *Indy* mutations were crossed to different stocks to check longevity of the heterozygotes compared to longevity of flies from the homozygote (double mutation) stocks (Rogina et al. 2000). The heterozygotes were substantially longer lived in all but one case. When *Indy* was crossed to flies from Luckinbill selected lines (L in Table 1) there was only a 15 percent increase in longevity of the heterozygote compared to the selected line. This result suggests the intriguing possibility that the response to artificial selection for longevity includes the *"Indy* mechanism" for life span extension (Rogina et al. 2000).

Mutations in the insulin signaling pathway can also extend the life span of *D. melanogaster* (Tatar et al. 2001; Clancy et al. 2001). These results follow the pioneering work using *Caenorhabiditis elegans* in which mutations selected for extended life span were found in genes that encode interacting proteins (signaling pathway) that mediate the effects of insulin (Friedman and Johnson 1988; Kenyon et al. 1993; Dorman et al. 1995; Morris et al. 1996; Kimura et al. 1997; Paradis et al. 1998; Gill et al. 1999). The discovery of mutations in *D. melanogaster* that extend life span by reducing activity of the insulin signaling pathway suggests that a general mechanism underlies differential aging and extended longevity in animals.

D. melanogaster insulin signaling mutations that extend life span were associated with the *InR* gene that encodes the insulin receptor (Tatar et al. 2001). Increased female longevity resulted only from a specific combination of *InR* mutations. Another longevity mutation was in the *chico* gene that encodes an insulin receptor substrate protein (Clancy et al. 2001). The insulin signaling pathway plays an important role in controlling growth; loss of function associated with the *InR* and *chico* mutations results in dwarf adult flies in addition to extended longevity. *chico* has fewer and smaller cells, resulting in a body size that is 50 percent of normal (Bohni et al. 1999). *chico* mutants are known to have a relatively high proportion of lipid (Bohni et al. 1999), and perhaps correspondingly the *chico* mutant flies were found to be starvation resistant (Clancy et al. 2001). Similarly, Tatar et al. (2001) found that higher levels of energy storage fat (triglyceride lipid) were found than in a comparable fly stock. Small size, stress resistance, and high levels of lipids are to some degree associated with extended longevity of mutant *Caenorhabditis, Drosophila,* and mice (Clancy et al. 2001; Tatar et al. 2001).

Tatar et al. (2001) found that *InR* mutations conferred as much as an 85 percent extension of longevity in females and reduced late-age mortality rates in males. There was no decrease in metabolic rate of the mutant flies as measured by mass-specific oxygen consumption in a mixed sample of males and females. Tatar et al. found that the *InR* mutations have approximately 25 percent of normal juvenile hormone biosynthesis and interpreted this deficiency as a basis for increased longevity. Exogenous administration of a juvenile hormone mimic stimulated reproductive activity in *InR* mutation females and reduced life span of the mutant toward that of the wild type control. Tatar et al. argued that the longevity associated with the *InR* mutations was actually due to a deficiency in this class of insect hormone.

Clancy et al. (2001) tested mutations in various genes in the insulin signaling pathway, including the insulin receptor gene, and found that all but one was associated with normal or decreased longevity. Only the *chico*[1] mutation was found to increase male and female longevity, by up to 48 percent when the mutant was present in two copies (homozygote) and 36 percent when the mutant was combined with a normal gene (heterozygote). Resistance to elevated temperature, oxidative stress, and starvation was tested using the *chico* mutation and related normal flies. Starvation resistance was observed for heterozygotes and homozygote mutants, oxidative stress resistance was observed only for heterozygotes, and no resistance to elevated temperature was associated with the mutation. The effect of *chico* on stress resistance was not as consistent as for the long-lived mutations that affect the insulin signaling pathway of *C. elegans*. For *chico*, the homozygotes were half the size, but the moderately long-lived heterozygotes were of normal size (Clancy et al. 2001). Consequently, body size and longevity appear to be at least partially independently determined. Heterozygote *chico* females have reduced fecundity and the homozygote was sterile. Thus, the question arose whether enhanced longevity associated with *chico* was a byproduct of reduced reproduction (i.e., a reduced cost of reproduction). Clancy et al. used the *ovo*[D] mutation of normal flies that were comparable to *chico* to test the hypothesis that sterility was the underlying cause of extended longevity of *chico* females. *chico* females were found to live a substantially longer time than sterile females that were otherwise normal. These results indicate that sterility and defective insulin signaling can extend life span by different mechanisms in *D. melanogaster*. However, this result does not exclude the possibility of an interaction between the reproductive system, insulin signaling, and longevity as has been observed in *C. elegans* (Hsin and Kenyon 1999; Lin et al. 2001).

Perspectives on mutation analysis

Mutation analysis is a form of site-localized genome perturbation performed to find genes that can extend longevity. Unlike transgenic overexpression, the outcome is not constrained by a priori expectations about which genes could produce the effect. Given this lack of constraint, it is interesting that long-lived mutants have reinforced the outcome of selection experiments in identifying stress resistance as a common component of longevity. In the future, mutation analysis should be able to identify a range of genes that confer longevity in *D. melanogaster*. The potential for synergist studies using long-lived mutants and populations (lines selected for longevity) is indicated by the use of the *Indy* mutation and the Luckinbill selected lines (Rogina et al. 2000).

Gene expression

A central tenet of biology is that genes produce messenger RNA (mRNA), which in turn is used as a message to make proteins. Gene activity (expression) can be measured by the amount of message (mRNA) produced by specific genes or by reporter genes.

Helfand et al. (1995) and Rogina and Helfand (1995) monitored gene expression during aging in *D. melanogaster* using a series of transgenic lines that differed in the chromosomal location of P element reporter gene insertions in the genome. The reporter genes indicated the level of activity of nearby genes. Age-dependent differential gene expression was documented and related to longevity differences among lines.

Another approach has been to monitor the production of mRNA from specific genes as a measurement of age-dependent gene expression. For example, the abundance of mRNA corresponding to heat shock proteins has been quantified as *D. melanogaster* ages. The abundance of heat shock protein 70 (HSP70) mRNA does not increase substantially during aging (Wheeler et al. 1995), but mRNA corresponding to two smaller heat shock proteins (HSP22 and HSP23) increases markedly as a function of age (King and Tower 1999). Moreover, the level of *hsp22* gene mRNA was relatively high in the Rose lines selected for extended longevity (Kurapati et al. 2000). However, overexpression of hsp22 results in decreased longevity without affecting the level of message production by *hsp70* (Tower, personal communication). Some genes that are expressed at high levels in relatively old individuals may simply represent loss of control of proper expression as a function of age (Tower 1996; Guarente and Kenyon 2000). Alternatively, genes that are highly expressed in aging flies might provide a protective function that could be confirmed by life span extension in transgenic overexpression studies (Tower 1996, 2000).

Technology allows for simultaneous measurement of the relative amount of mRNA produced by almost all of the genes in the genome of *D. melanogaster*. In one study, gene expression was found to be much more strongly affected by sex than by age (Jin et al. 2001). In another study, the investigators measured expression of approximately 8,000 genes as a function of age and oxidative stress (Zou et al. 2000). A total of 127 genes changed expression appreciably during aging, and a third of these genes also responded to oxidative stress. This pattern of co-expression indicates a set of candidate genes that could contribute to longevity. Caloric restriction can extend the life span of *D. melanogaster*. Dietary conditions that extend life span reduce the expression of genes associated with growth, metabolism, reproduction, and stress resistance (Pletcher et al. 2002). In this study (Pletcher et al. 2002), almost a quarter of the genes changed in expression during the process of aging.

Perspectives on gene expression studies

Even at this early stage in application of genome-wide gene expression technology (microarrays), studies have identified many changes in age-related gene expression potentially relevant to extended longevity. Among the questions being explored are whether a few key genes control the expression of many other genes to extend longevity, which genes are suitable candidates for functional tests by transgene overexpression or techniques designed to suppress gene activity, and whether these genes will elucidate the relationship between delayed early-age reproduction, stress resistance, and longevity in populations.

Future approaches

A theme of this review has been that the diversity of genetic approaches used to study *D. melanogaster* is generating considerable insight into life span extension. Continuing this approach, powerful genetic technologies (mutation analysis, transgenes, genomics) combined with a rich population context (quantitative genetics, laboratory selection experiments, natural population studies such as Mitrovski and Hoffmann 2001) will be the basis for advances in understanding how genes control longevity in populations.

Given that overall fitness is requisite for individuals in outbreeding populations, the insight derived from studies on natural populations of flies may be particularly applicable to the design of interventions to extend the span of active and healthy human life.

Note

Marta Wayne kindly provided Figure 3 and Ryan Baller made major contributions to Figures 4 and 5. Discussions with Sergey Nuzhdin, Linda Partridge, Daniel Promislow, and John Tower were valuable for this review. James Curtsinger, Steven Helfand, and John Tower kindly provided information from unpublished studies. Support for this review was provided by the National Institute on Aging (AG08761).

References

Arking, Robert. 1987. "Genetic and environmental determinants of longevity in *Drosophila*," *Basic Life Sci.* 42: 1–22.

———. 1988. "Genetic analyses of aging processes in *Drosophila*," *Exp. Aging Res.* 14: 125–135.

Arking, R., S. Buck, R. A. Wells, and R. Pretzlaff. 1988. "Metabolic rates in genetically based long lived strains of *Drosophila*," *Experimental Gerontology* 23: 59–76.

Arking, R., S. Buck, A. Berrios, S. Dwyer, and G. T. Baker. 1991. "Elevated antioxidant activity can be used as a bioassay for longevity in a genetically long-lived strain of *Drosophila*," *Develop. Genetics* 12: 362–370.

Arking, R., V. Burde, K. Graves, R. Hari, E. Feldman, A. Zeevi, S. Soliman, A. Saraiya, S. Buck, J. Vettraino, K. Sathrasala, N. Wehr, and R. L. Levine. 2000a. "Forward and reverse selection for longevity in *Drosophila* is characterized by alteration of antioxidant gene expression and oxidative damage patterns," *Exp. Gerontol.* 35: 167–185.

Arking, R., V. Burde, K. Graves, R. Hari, E. Feldman, A. Zeevi, S. Soliman, A. Saraiya, S. Buck, J. Vettraino, and K. Sathrasala. 2000b. "Identical longevity phenotypes are characterized by different patterns of gene expression and oxidative damage," *Exp. Gerontol.* 35: 353–373.

Arking, Robert and S. P. Dudas. 1989. "Review of genetic investigations into the aging processes of *Drosophila*," *J. Am. Geriatr.* 37: 757–773.

Bell, Graham and V. Koufopanou. 1986. "The cost of reproduction," in R. Dawkins and M. Ridley (eds.), *Oxford Surveys of Evolutionary Biology*. Oxford: Oxford University Press, pp. 83–131.

Bohni, R., J. Riesgo-Escovar, S. Oldham, W. Brogiolo, H. Stocker, B. F. Andruss, K. Beckingham, and E. Hafen. 1999. "Autonomous control of cell and organ size by CHICO, a *Drosophila* homolog of vertebrate IRS1-4," *Cell* 97: 865–875.

Bradley, T. J., A. E. Williams, and M. R. Rose. 1999. "Physiological responses to selection for desiccation resistance in *Drosophila melanogaster*," *Amer. Zool.* 39: 337–345.

Buck, S., M. Nicholson, S. P. Dudas, G. T. Baker III, and R. Arking. 1993a. "Larval regulation of adult longevity in a genetically selected long-lived strain of *Drosophila melanogaster*," *Heredity* 71: 23–32.

Buck, S., R. A. Wells, S. P. Dudas, G. T. Baker III, and R. Arking. 1993b. "Chromosomal localization and regulation of the longevity determinant genes in a selected strain of *Drosophila melanogaster*," *Heredity* 71: 11–22.

Carey, J. R., P. Liedo, H.-G. Muller, J.-L. Wang, and J. W. Vaupel. 1998. "Dual aging modes in Mediterranean fruit fly females," *Science* 281: 396–398.

Carey, J. R., Pablo Liedo, D. Orozco, and J. W. Vaupel. 1992. "Slowing of mortality rates at older ages in large medfly cohorts," *Science* 258: 457–461.

Carlson, K. A. and L. G. Harshman. 1999. "Extended longevity lines of *Drosophila melanogaster*: Characterization of oocyte stage and ovariole numbers as a function of age and diet," *J. Gerontol.* 54: B432–440.

Carlson, K. A., T. J. Nusbaum, M. R. Rose, and L. G. Harshman. 1998. "Oocyte maturation and ovariole numbers in lines of *Drosophila melanogaster* selected for postponed senescence," *Funct. Ecol.* 52: 514–520.

Chippindale, Adam K., T. J. F. Chu, and Michael R. Rose. 1996. "Complex trade-offs and the evolution of starvation resistance in *Drosophila melanogaster*," *Evolution* 50: 753–766.

Clancy, D. J., D. Gems, L. G. Harshman, S. Oldham, H. Stocker, E. Hafen, S. J. Leevers, and L. Partridge. 2001. "Extension of life-span by loss of CHICO, a *Drosophila* insulin receptor substrate protein," *Science* 292: 104–106.

Curtsinger, J. W., H. H. Fukui, D. R. Townsend, and J. W. Vaupel. 1992. "Demography of genotypes: Failure of the limited life-span paradigm in *Drosophila melanogaster*," *Science* 258: 461–463.

Curtsinger, James W., Hidenori H. Fukui, Aziz A. Khazaeli, Andrew Kirscher, Scott D. Pletcher, Daniel E. L. Promislow, and M. Tatar. 1995. "Genetic variation and aging," *Ann. Rev. Genet.* 29: 553–575.

Curtsinger, J. W., H. H Fukui, A. S. Resler, K. Kelly, and A. A. Khazaeli. 1998. "Genetic analysis of extended life span in Drosophila melanogaster: RAPD screen for genetic divergence between selected and control lines," *Genetica* 104 : 21–32.

Djawdan, M., T. T. Sugiyama, L. K. Schlaeger, T. J. Bradley, and M. R. Rose. 1996. "Metabolic aspects of the trade-off between fecundity and longevity in *Drosophila melanogaster*," *Physiol. Zool.* 69: 1176–1195.

Djawdan, M., M. R. Rose, and T. J. Bradley. 1997. "Does selection for stress resistance lower metabolic rate?" *Ecology* 78: 828–837.

Dorman, J. B., B. Albinder, T. Shroyer, and C. Kenyon. 1995. "The *age-1* and *daf-2* genes function in a common pathway to control the lifespan of *Caenorhabditis elegans*," *Genetics* 141: 1399–1406.

Dudas, S. P. and R. A. Arking. 1995. "A coordinate upregulation of antioxidant gene activities is associated with the delayed onset of senescence in a long-lived strain of *Drosophila*," *J. Gerontol. Biol. Sci.* 50A: B117–B127.

Feder, M. E. and G. E. Hoffmann. 1999. "Heat-shock proteins, molecular chaperones and the stress response: Evolutionary and ecological physiology," *Annu. Rev. Physiol.* 61: 243–282.

Force, A. G., T. Staples, T. Soliman, and R. Arking. 1995. "A comparative biochemical and stress analysis of genetically selected *Drosophila* strains with different longevities," *Dev. Genet.* 17: 340–351.

Friedman, D. B. and T. E. Johnson. 1988. "A mutation in the *age-1* gene in *Caenorhabditis elegans* lengthens life and reduces hermaphrodite fertility," *Genetics* 118: 75–86.

Gasser, M., M. Kaiser, D. Berrigan, and S. C. Stearns. 2000. "Life-history correlates of evolution under high and low adult mortality," *Evolution* 54: 1260–1272.

Gibbs, A. G. 1999. "Laboratory selection for the comparative physiologist," *J. Exp. Biol.* 202: 2709–2718.

Gibbs, A. G., A. K. Chippindale, and M. R. Rose. 1997. "Physiological mechanisms of evolved desiccation resistance in *Drosophila melanogaster*," *J. Exp. Biol.* 200: 1821–1823.

Gill, E. B., E. M. Link, L. X. Liu, C. D. Johnson, and J. A. Lees. 1999. "Regulation of the insulin-like developmental pathway of *Caenorhabitis elegans* by a homolog of the PTEN tumor suppressor gene," *Proc. Natl. Acad. Sci. USA* 96: 2925–2930.

Graves, J. L., E. C. Toolson, C. Jeong, L. N. Vu, and M. R. Rose. 1992. "Desiccation, flight, glycogen and postponed senescence in *Drosophila melanogaster*," *Physiol. Zool.* 65: 268–286.

Guarente, L. and C. Kenyon. 2000. "Genetic pathways that regulate ageing in model organisms," *Nature* 408: 255–262.

Harman, D. 1956. "Aging: A theory based on free radical and radiation chemistry," *J. Gerontol.* 11: 298–300.

Harshman, L. G. and B. A. Haberer. 2000. "Oxidative stress resistance: A robust correlated response to selection in extended longevity lines of *Drosophila melanogaster*? *J. Gerontol. A Biol. Sci. Med. Sci.* 55: B415–417.

Harshman, Lawrence G. and Ary A. Hoffmann. 2000a. "Laboratory selection experiments using *Drosophila*: What do they really tell us?" *Trends in Ecology and Evolution* 15: 32–36.

———. 2000b. "Reply from L. G. Harshman and A. A. Hoffmann," *Trends in Ecology and Evolution* 15: 207.

Harshman, L. G., K. M. Moore, M. A. Sty, and M. M. Magwire. 1999. "Stress resistance and longevity in selected lines of *Drosophila melanogaster*," *Neurobiol. Aging* 20: 521–529.

Hartl, Daniel L. and Andrew G. Clark. 1997. *Principles of Population Genetics.* Sunderland, MA: Sinauer Associates.

Helfand, S. L., K. J. Blake, B. Rogina, M. D. Stracks, A. Centurion, and B. Naprta. 1995. "Temporal patterns of gene expression in the antenna of the adult *Drosophila melanogaster*," *Genetics* 140: 549–555.

Hoffmann, Ary A., R. Hallas, C. Sinclair, and Linda Partridge. 2001. "Rapid loss of stress resistance in *Drosophila melanogaster* under adaptation to laboratory culture," *Evolution* 52: 436–438.

Hoffmann, A. A. and P. A. Parsons. 1989. "An integrated approach to environmental stress tolerance and life history variation: Desiccation tolerance in *Drosophila*," *Biol. J. Linn. Soc.* 37: 117–136.

Hsin, H. and C. Kenyon. 1999. "Signals from the reproductive system regulate the lifespan of *C. elegans*," *Nature* 399: 362–366.

Huey, R. B. and J. G. Kingsolver. 1993. "Evolution of resistance to high temperature in ectotherms," *Am. Nat.* 142: S21–S46.

Jin, W., R. M. Riley, R. D. Wolfinger, K. P. White, G. Passador-Gurgel, and G. Gibson. 2001. "The contribution of sex, genotype and age to transcriptional variance in *Drosophila melanogaster*," *Nature Genetics* 29: 389–395.

Johnson, F. B., David A. Sinclair, and Leonard Guarente. 1999. "Molecular biology of aging," *Cell* 96: 291–302.

Kaiser, M., M. Gasser, R. Ackermann, and S. C. Stearns. 1997. "P-element inserts in transgenic lines: A cautionary tale," *Heredity* 78: 1–11.

Kenyon, C., J. Chang, A. Gensch, A. Rudner, and R. Tablang. 1993. "A *C. elegans* mutant that lives twice as long as wild type," *Nature* 366: 461–464.

Khazaeli, A. A., M. Tatar, S. D. Pletcher, and J. W. Curtsinger. 1997. "Heat-induced longevity extension in *Drosophila*. I. Heat treatment, mortality, and thermotolerance," *Journal of Gerontology, Biological Sciences 52A*, B48–B52.

Kimura, K. D., H. A. Tissenbaum, Y. Liu, and G. Ruvkun. 1997. "Daf-2, an insulin receptor-like gene that regulates longevity and diapause in *Caenorhabditis elegans*," *Science* 277: 942–946.

King, V. and J. Tower. 1999. "Aging-specific expression of *Drosophila hsp22*," *Develop. Biol.* 207: 107–118.

Krebs, R. A. and M. E. Feder. 1997. "Deleterious consequences of *Hsp70* overexpression in *Drosophila melanogaster* larvae," *Cell Stress and Chaperones* 2: 60–71.

———. 1998. "*Hsp70* and larval thermotolerance in *Drosophila melanogaster*: How much is enough and when is more too much?" *Jour. Insect Physiol.* 44: 1091–1101.

Kurapati, Raj, Hardip B. Passananti, Michael R. Rose, and John Tower. 2000. "Increased *hsp22* RNA levels in *Drosophila* lines genetically selected for increased longevity," *Jour. Gerontol. A. Biol. Sci. Med. Sci.* 55: B552–B559.

Leips, Jeff and Trudy F. C. Mackay. 2000. "Quantitative trait loci for life span in *Drosophila melanogaster*: Interactions with genetic background and larval density," *Genetics* 155: 1773–1778.

Leroi, A. M., A. K. Chippindale, and M. R. Rose. 1994. "Long-term laboratory evolution of a genetic life-history trade-off in *Drosophila melanogaster*. 1. The role of genotype-by-environment interaction," *Evolution* 48: 1244–1257.

Lin, K., H. Hsin, N. Libina, and C. Kenyon. 2001. "Regulation of the *Caenorhabditis elegans* longevity protein DAF-16 by insulin/IGF-1 and germline signaling," *Nat. Genet.* 28: 139–145.

Lin, Y.-J., L. Seroude, and S. Benzer. 1998. "Extended life-span and stress resistance in the *Drosophila* mutant *methuselah*," *Science* 282: 943–946.

Luckinbill, L. S., R. Arking, M. J. Clare, W. C. Cirocco, and S. A. Buck. 1984. "Selection for delayed senescence in *Drosophila melanogaster*," *Evolution* 38: 996–1004.

Martin, George M., Steven N. Austad, and Thomas E. Johnson. 1996. "Genetic analysis of ageing: Role of oxidative damage and environmental stresses," *Nature Genetics* 13: 25–34.

Mitrovski, P. and A. A. Hoffmann. 2001. "Postponed reproduction as an adaptation to winter conditions in *Drosophila melanogaster*: Evidence for clinical variation under semi-natural conditions.," *Proc. Roy. Soc. B* 268: 2163–2168.

Morris, J. Z., H. A. Tissenbaum, and G. A. Ruvkun. 1996. "A phosphatidylinositol-3-OH kinase family member regulating longevity and diapause in *Caenorhabditis elegans*," *Nature* 382: 536–539.

Nghiem, D., A. G. Gibbs, M. R. Rose, and T. J. Bradley. 2000. "Postponed aging and desiccation resistance in *Drosophila melanogaster*," *Exper. Gerontol.* 35: 957–969.

Nusbaum, T. J., L. D. Mueller, and M. R. Rose. 1996. "Evolutionary patterns among measures of aging," *Exp. Gerontol.* 31: 507–516.

Nuzhdin, Sergey V., Elena G. Pasyukova, C. L. Dilda, Z-B. Zeng, and Trudy F. C. Mackay. 1997. "Sex-specific quantitative trait loci affecting longevity," *Proc. Natl. Acad. Sci. USA* 94: 9734–9739.

Orr, W. C. and R. J. Sohal. 1992. "The effects of catalase gene overexpression on life span and resistance to oxidative stress in transgenic *Drosophila melanogaster*," *Arch. Biochem. Biophys.* 297: 35–41.

———. 1993. "Effects of Cu-Zn superoxide dismutase overexpression on life span and resistance to oxidative stress in transgenic *Drosophila melanogaster*," *Arch. Biochem. Biophys.* 301: 34–40.

———. 1994. "Extension of life span by overexpression of superoxide dismutase and catalase in *Drosophila melanogaster*," *Science* 263: 1128–1130.

Paradis, S. and G. Ruvkun. 1998. "*Caenorhabditis elegans* Akt/PKB tranduces insulin receptor-like signals from AGE-1 P13 kinase to the DAF-16 transcription factor," *Genes Dev.* 12: 2488–2498.

Parkes, T. L., Elia A. J. Dickinson, D. Hilliker, A. J. Phillips, and G. L. Boulianne. 1998. "Extension of *Drosophila* lifespan by overexpression of human SOD1 in motorneurons," *Nat. Genet.* 19: 103–104.

Partridge, Linda and R. Sibly. 1991. "Constraints in the evolution of life histories," *Phil. Trans. Royal Soc. Lond B.* 332: 3–13.

Partridge, L. and K. Fowler. 1992. "Direct and correlated responses to selection on age at reproduction in *Drosophila melanogaster*," *Evolution* 46: 76–91.

Partridge, Linda and N. H. Barton. 1993. "Optimality, mutation and the evolution of aging," *Nature* 362: 305–311.

Partridge, Linda, Nik Prowse, and Patricia Pignatelli. 1999. "Another set of responses and correlated responses to selection on age at reproduction in *Drosophila melanogaster*," *Proc. R. Soc. London Ser. B* 266: 255–261.

Pasyukova, Elena G., Cristina Vieira, and Trudy F. C. Mackay. 2000. "Deficiency mapping of quantitative trait loci affecting longevity in *Drosophila melanogaster*," *Genetics* 156: 1129–1146.

Pletcher, Scott D., Aziz A. Khazaeli, and James W. Curtsinger. 2000. "Why do life spans differ? Partitioning mean longevity differences in terms of age-specific mortality parameters," *Journal of Gerontology* 55A: B381–B389.

Pletcher, S. D., S. J. Macdonald, R. Marguerie, U. Certa, S. C. Stearns, D. B. Goldstein, and L. Partridge. 2002. "Genome-wide transcript profiles in aging and calorically restricted *Drosophila melanogaster*," *Current Biology* 12: 712–723.

Promislow, Daniel E. L. and Marc Tatar. 1998. "Mutation and senescence: Where genetics and demography meet," *Genetica* 102/103: 299–314.

Reiwitch, Sarah G. and Sergey V. Nuzhdin. 2002. "Quantitative trait loci of life span of mated *Drosophila melanogaster* affects both sexes," *Genetical Research* 80: 1–6.

Reveillaud, I., A. Niedzwiecki, K. G. Bench, and J. E. Fleming. 1991. "Expression of bovine superoxide dismutase in *Drosophila melanogaster* augments resistance to oxidative stress," *Mol. Cell Biol.* 11: 632–640.

Reznick, David. 1985. "Cost of reproduction: An evaluation of the empirical evidence," *Oikos* 44: 257–267.

Roff, D. A. 1992. *The Evolution of Life Histories*. New York: Chapman and Hall.

Rogina, B. and S. L. Helfand. 1995. "Regulation of gene expression is linked to life span in adult *Drosophila*," *Genetics* 141: 1043–1048.

Rogina, Blanka, Robert A. Reenan, Steven P. Nilsen, and Stephen L. Helfand. 2000. "Extended life-span conferred by cotransporter gene mutations in *Drosophila*," *Science* 290: 2137–2140.

Rose, Michael R. 1984. "Laboratory evolution of postponed senescence in *Drosophila melanogaster*," *Evolution* 38: 1004–1010.

———. 1991. *Evolutionary Biology of Aging*, Oxford: Oxford University Press.

Rose, M. R. and T. J. Bradley. 1998. "Evolutionary physiology of the cost of reproduction," *Oikos* 83: 443–451.

Rose, M. R., J. L. Graves, and E. W. Hutchinson. 1990. "The use of selection to probe patterns of pleiotropy in fitness characters," in F. Gilbert (ed.), *Insect Life Cycles*. New York: Springer-Verlag, pp. 29–42.

———. 1992. "Selection on stress resistance increases longevity in *Drosophila melanogaster*," *Exp. Gerontol.* 27: 241–250.

Rose, M. R., T. J. Nusbaum, and A. K. Chippindale. 1996. "Laboratory evolution: The experimental wonderland and the Cheshire Cat Syndrome," in M. R. Rose and G. V. Lauder (eds.), *Adaptation*. San Diego: Academic Press, pp. 221–244.

Salmon, Adam B., David B. Marx, and Lawrence G. Harshman. 2001. "A cost of reproduction in *Drosophila melanogaster*: Stress susceptibility," *Evolution* 55: 1600–1608.

Seong, K. H., T. Matsuo, Y. Fuyama, and T. Aigaki. 2001. "Neural-specific overexpression of *Drosophila* Plenty of SH3s (*DPOSH*) extends the longevity of adult flies," *Biogerontology* 2: 271–281.

Service, P. M. 1987. "Physiological mechanisms of increased stress resistance in *Drosophila melanogaster* selected for postponed senescence," *Physiol. Zool.* 60: 321–326.

———. 1989. "The effect of mating status on lifespan, egg laying and starvation resistance in *Drosophila melanogaster* in relationship to selection on longevity," *Insect Physiology* 35: 447–452.

Service, P. M. and A. J. Fales. 1993. "Evolution of delayed reproductive senescence in male fruit flies: Sperm competition," *Genetica* 91: 111–125.

Service, P. M. and R. E. Vossbrink. 1996. "Genetic variation in 'first' male effects on egg laying and remating by female *Drosophila melanogaster*," *Behav. Genet.* 26: 39–47.

Service, P. M., E. W. Hutchinson, M. D. MacKinley, and M. R. Rose. 1985. "Resistance to environmental stress in *Drosophila melanogaster* selected for postponed senescence," *Evolution* 42: 708–716.

Service, Phillip M., Charles A. Michieli, and Kirsten McGill. 1998. "Experimental evolution of senescence: An analysis using a 'heterogeneity' mortality model," *Evolution* 52: 1844–1850.

Sgro, Carla M. and Linda Partridge. 1999. "A delayed wave of death from reproduction in *Drosophila,*" *Science* 286: 2521–2524.

———. 2000. "Evolutionary responses of the life history of wild-caught *Drosophila melanogaster* to two standard methods of laboratory culture," *Am. Nat.* 156: 341–353.

Silbermann, R. and M. Tatar. 2000. "Reproductive costs of heat shock protein in transgenic *Drosophila melanogaster,*" *Evolution* 54: 2038–2045.

Simmons, F. H. and T. J. Bradley. 1997. "An analysis of resource allocation in response to dietary yeast in *Drosophila melanogaster,*" *J. Insect Physiol.* 43: 779–788.

Song, W., R. Ranjan, P. Bronk, Z. Nie, K. Dawson-Scully, Y. J. Lin, L. Seroude, H. L. Atwood, S. Benzer, and K. E. Zinsmaier. 2001. "*Methuselah,* a putative G protein-coupled receptor, regulates excitatory neurotransmitter exocytosis at the larval neuro-muscular junction of Drosophila: Abstract 51," 42nd Annual Drosophila Research Conference, Washington, DC.

Stearns, Steven C. 1989. "Trade-offs in life-history evolution," *Functional Ecology* 3: 259–268.

———. 1992. *The Evolution of Life Histories.* Oxford: Oxford University Press.

Stearns, S. C., M. Ackermann, M. Doebeli, and M. Kaiser. 2000. "Experimental evolution of aging, growth, and reproduction in fruitflies," *Proc. Natl. Acad. Sci. USA* 97: 3309–3313.

Stearns, Steven C. and Linda Partridge. 2001. "The genetics of aging in *Drosophila,*" in E. Masoro and S. Austad (eds.), *Handbook of Aging,* 5th ed. San Diego: Academic Press, pp. 345–360.

Sun J., D. Folk, T. J. Bradley, and J. Tower. 2002. "Induced overexpression of mitochondrial Mn-superoxide dismutase extends the life span of adult *Drosophila melanogaster,*" *Genetics* 161: 661–672.

Sun, J. and J. Tower. 1999. "FLP recombinase-mediated induction of Cu/Zn-superoxide dismutase transgene expression can extend the life span of adult *Drosophila melanogaster* flies," *Mol. Cell Biol.* 19: 216–228.

Tatar, Marc and J. R. Carey. 1995. "Nutrition mediates reproductive trade-offs with age-specific mortality in the beetle *Callosobruchus maculates,*" *Ecology* 76: 2066–2073.

Tatar, M., A. A. Khazaeli, and J. W. Curtsinger. 1997. "Chaperoning extended life," *Nature* 390: 30.

Tatar, Marc, A. Kopelman, D. Epstein, M.-P. Tu, C.-M. Yin, and R. S. Garofalo. 2001. "A mutant *Drosophila* insulin receptor homolog the extends life-span and impairs neuroendocrine function," *Science* 292: 107–110.

Tatar, Marc, Daniel E. L. Promislow, Aziz A. Khazaeli, and J. W. Curtsinger. 1996. "Age-specific patterns of genetic variance in *Drosophila melanogaster*. II. Fecundity and its genetic covariance with age-specific mortality," *Genetics* 143: 849–858.

Tower, J. 1996. "Aging mechanisms in fruit flies," *Bioessays* 18: 799–807.

———. 2000. "Transgenic methods for increasing *Drosophila* life span," *Mech. Aging Dev.* 118: 1–14.

Van Voorhies, W. A. and S. Ward. 1999. "Genetic and environmental conditions that increase longevity in *Caenorhabditis elegans* decrease metabolic rate," *Proc. Natl. Acad. Sci. USA* 96: 11399–11403.

Vieira, Cristina, Elena G. Pasyukova, Zhao-Bang Zeng, J. Brant Hackett, Richard F. Lyman, and Trudy F. C. Mackay. 2000. "Genotype-environment interaction for quantitative trait loci affecting life span in *Drosophila melanogaster,*" *Genetics* 154: 213–227.

Wallace, D. C. 1992. "Mitochondrial genetics: A paradigm for aging and degenerative diseases?" *Science* 250: 628–632.

Wang, Yue, Adam B. Salmon, and Lawrence G. Harshman. 2001. "A cost of reproduction:

Oxidative stress susceptibility is associated with increased egg production in *Drosophila melanogaster*," *Experimental Gerontology* 36: 1349–1359.

Watson, M. J. O. and Ary A. Hoffmann. 1996. "Acclimation, cross-generation effects, and the response to selection for increased cold resistance in *Drosophila*," *Evolution* 50: 1182–1192.

Wheeler, J. C., E. T. Bieschke, and J. Tower. 1995. "Muscle-specific expression of *Drosophila* hsp70 in response to aging and oxidative stress," *Proc. Natl. Acad. Sci. USA*. 92: 10408–10412.

Williams, George C. 1957. "Pleiotropy, natural selection, and the evolution of senescence," *Evolution* 11: 398–411.

———. 1966. "Natural selection, the costs of reproduction, and a refinement of Lack's principle," *Am. Nat.* 100: 687–690.

Zera, Anthony J. and Lawrence G. Harshman. 2001. "The physiology of life history trade-offs in animals," *Annu. Rev. of Ecol. and Systematics* 32: 95–126.

Zou, S., S. Meadows, L. Sharp, L. Y. Jan, and Y. N. Jan. 2000. "Genome-wide study of aging and oxidative stress response in *Drosophila melanogaster*," *Proc. Natl. Acad. Sci. USA*. 97: 132726–13731.

Zwann, B., R. Bijlsma, and R. F. Hoekstra. 1995. "Direct selection on life span in *Drosophila melanogaster*," *Evolution* 49: 649–659.

Interspecies Differences in the Life Span Distribution: Humans versus Invertebrates

SHIRO HORIUCHI

Life span varies considerably among species. Investigation of reasons for interspecies differences in the length of life deepens our fundamental understanding of senescence and longevity.

A statistic frequently used for interspecies comparison of life span is the longest length of life ever documented reliably for each species (Carey and Judge 2000; Comfort 1979). Measures of the typical length of life such as the mean, median, and modal ages at death are also compared among populations of different species.

Information on the maximum length and typical length of life should be used carefully, however, taking into account the *distribution* of life span in the population. In some species, life spans may concentrate in a relatively narrow age range, while in others life spans may spread widely. In some species, the maximum length may not appear too far from the typical length, while in others the longest record may be an extreme exception.

To study the life span distribution in detail, data are needed on deaths by age in a large-size cohort of several thousand or more individuals. Until around 1990, survival data on large-size cohorts were rare except for humans. Thanks to the development of biodemography, however, survival data on large cohorts were obtained for several animal species in the 1990s (Vaupel et al. 1998) and such work is continuing. These data sets provide valuable opportunities for interspecies comparison of life span distribution.

The data were collected for laboratory animals, which live in protected environments. Such animals are substantially more protected from predators, infections, food shortages, climate changes, toxic materials, and accidental injuries than animals in the natural world. Also, modern humans usually live in more protected environments than our ancestors in prehistoric eras. The majority of deaths of adult individuals in protected environments are likely to result from diseases that are related to age-associated declines in physiological conditions.[1]

Therefore, the life span distribution in a protected population should reflect two crucial aspects of longevity. First, the distribution is affected by individual differences in the level of somatic maintenance. Greater individual variations in durability of physiological systems may produce a wider dispersion in the length of life. Second, the life span distribution is affected by age-related physiological changes in individuals. In populations in which physiological conditions deteriorate more rapidly, the death rate is likely to rise more steeply, thereby making the peak of life span distribution higher. These two sources of life span distribution correspond to two major dimensions of physiological variations: physiological differences *among individuals* of the same age and physiological changes *with age* in individuals. In summary, interspecies differences in the life span distribution in protected environments permit comparison of species with respect to (1) the size of individual variation in the ability to keep the organism functioning and (2) the trajectory of age-related changes in physiological conditions.[2]

Several researchers found interspecies similarities in the age patterns of mortality among animal species. The Gompertz equation was fitted to survival data for a number of species (Finch 1990; Finch and Pike 1996). The shapes of survival curves derived from intrinsic mortality data were shown to be similar for three mammalian species: dogs, mice, and humans (Carnes, Olshansky, and Grahn 1996). The age-related mortality increase slows down or ceases at very old ages in a number of species (Vaupel et al. 1998).

In this chapter, the focus is shifted from interspecies similarities to differences. Life span distributions of six cohorts of different species are compared, their major differences are identified, and possible explanations of the differential patterns are proposed.

Data and methods

Six species were compared: humans, Mediterranean fruit flies or "medflies" (*Ceratitis capitata*), nematode worms (*Caenorhabditis elegans*), parasitoid wasps (*Diachasmimorpha longjacaudtis*), another fruit fly species (*Drosophila melanogaster*), and bean beetles (*Callosobruchus maculatus*). The five nonhuman species are invertebrates.

Life tables were obtained for 347,533 individuals born in Sweden in 1881–85; 598,118 medflies; about 550,000 nematodes; 13,358 wasps; 5,245 *Drosophila* fruit flies; and 829 beetles.[3] All of them were males except the nematodes, which were hermaphrodites.[4] The nematode cohort and *Drosophila* cohort were inbred lines, thus genetically homogeneous. (Details on the data are given in Appendix A.)

The six cohorts were compared with respect to three life table functions: the life span distribution, the age-specific death rate, and the life table aging rate. Life span differs greatly among species. Therefore, it is more reasonable to compare the life span distribution *relative to the typical life span* than on the absolute time scale (days or years). The modal life span was used as the typical length of life.[5] The modal life span is useful in studies of senescence and longevity because it is determined by adult mortality only.[6] The mean and median life spans can be strongly affected by mortality at young ages, and should not be used for comparing species that have very different patterns of birth and neonatal mortality.[7] Life table functions were rescaled using one-hundredth of the modal life span as the time unit, instead of a day or year.[8] After the rescaling, all species appear to have the same modal life span.

In addition, the data were analyzed using mathematical models called frailty models. The analysis produces estimates of (1) the size of individual variation in vulnerability to mortality risk and (2) the age pattern of mortality risk at the individual level. (See Appendix B for discussion of the statistical methods.)

Differences in the life span distribution

Figures 1, 2, and 3 display the life span distribution, age-specific death rates, and life table aging rates, respectively, for the six species. Dots indicate observed values, and solid lines show estimates by mathematical models. In each panel, the modal life span in the cohort is placed at the same location, indicated by the dotted vertical line.

Figure 1 shows two major differences between humans and invertebrates. First, the distribution is skewed to the left in humans, but is fairly symmetrical around the mode in invertebrates.[9] Second, the relative variation of life span is smaller in humans than in vertebrates. The variability differs between invertebrates as well. The dispersion is smaller in nematodes, *Drosophila*, and beetles than in medflies and wasps.

The life span distribution is uniquely determined by age-specific death rates. It has been shown for different species that the death rate tends to rise steeply with adult age, then to slow down at old ages (Vaupel et al. 1998). A series of simulations (described in Appendix C) suggests the following relationships between mortality pattern and life span distribution. A steeper age-associated increase of mortality concentrates deaths in a narrower age range, thereby raising the peak of life span distribution. Mortality deceleration weakens this effect, but to a lesser extent if the slowing down starts later. In addition, the delayed mortality deceleration shifts the peak of life span distribution to an older age, leaving fewer survivors beyond the peak age and skewing the distribution more to the left. Figures 2

FIGURE 1 The life span distribution for six species

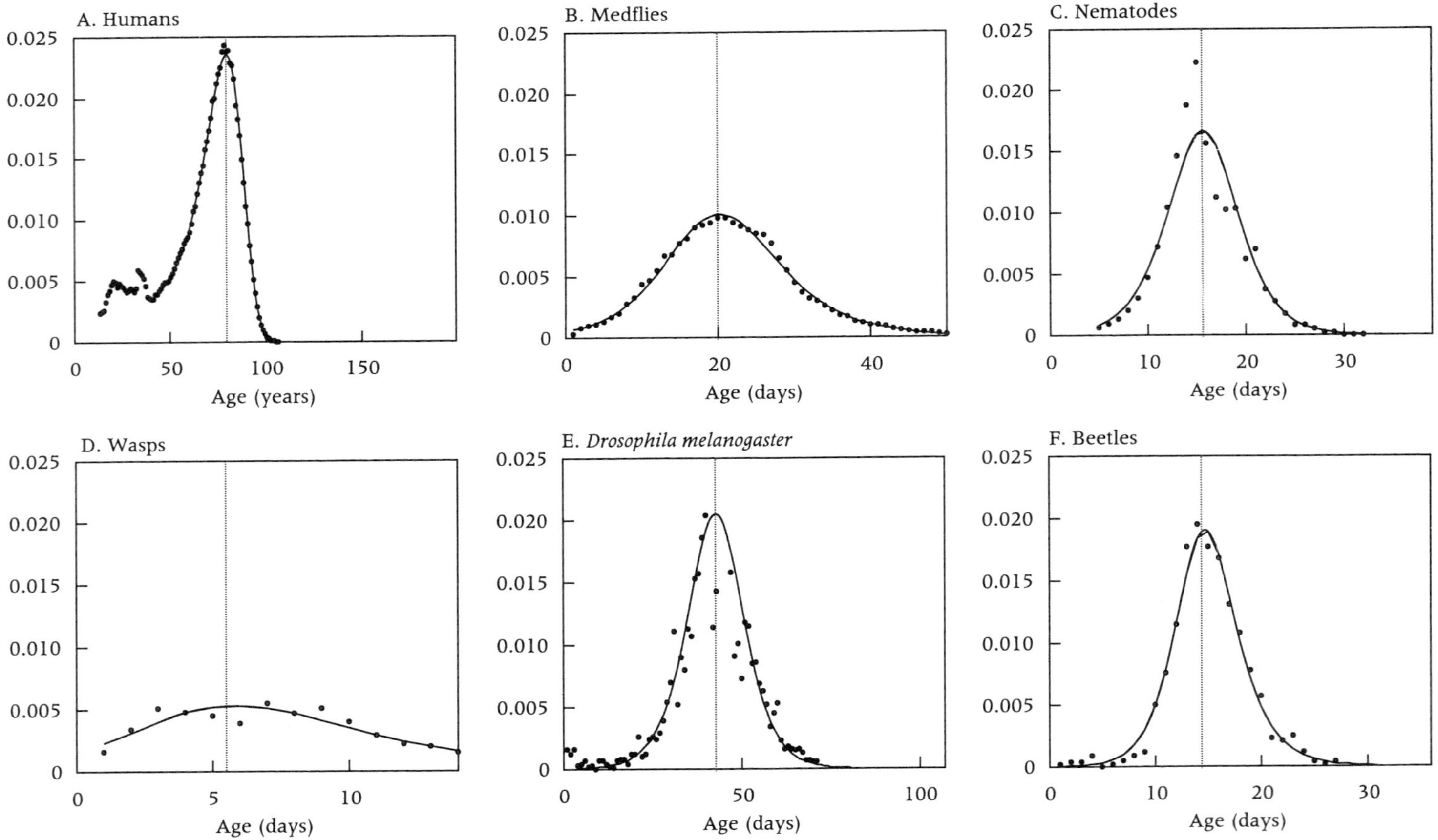

NOTE: Dots indicate values directly calculated from data, and solid lines indicate estimates using frailty models. In each panel, the vertical dotted line indicates the estimated modal life span, the vertical axis indicates the proportion of adult deaths per one-hundredth of the modal life span, and the area under the curve over the adult age range (defined in Appendix A) is unity if the age is rescaled to make the modal life span 100.

SOURCES: Humans, Swedish males born in 1881–85, from Berkeley Mortality Database (http://www.demog.berkeley.edu/wilmoth/mortality/); medflies, Carey et al. 1995, appendix; beetles, generation 1 males, Tatar and Carey 1994, Table 1; the other species, Vaupel et al. 1998, Figure 3.

FIGURE 2 Age-specific death rates for six species

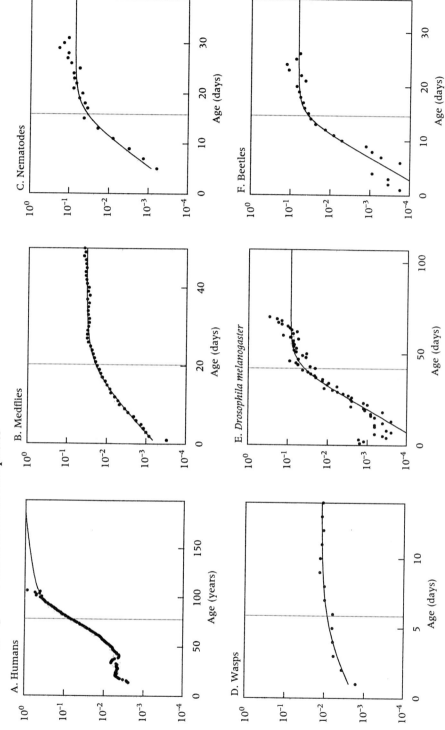

NOTE: Dots indicate values directly calculated from data, and solid lines indicate estimates using frailty models. In each panel, the vertical dotted line indicates the estimated modal life span, and the vertical axis indicates the number of deaths per individual-OHM of exposure. (OHM is one-hundredth of the modal life span.)
SOURCES: See Figure 1.

FIGURE 3 Life table aging rates for six species

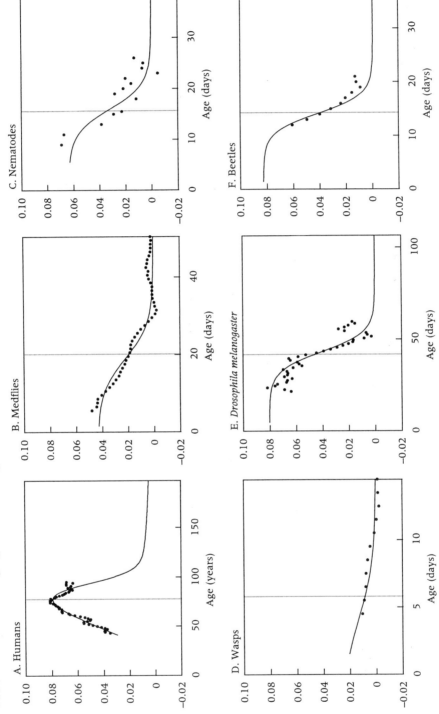

NOTE: Dots indicate values directly calculated from data, and solid lines indicate estimates using frailty models. In each panel, the vertical dotted line indicates the estimated modal life span, and the vertical axis indicates the exponential rate of mortality increase per individual-OHM. (OHM is one-hundredth of the modal life span.)
SOURCES: See Figure 1.

and 3 reveal that mortality in humans rises more steeply and slows down later than in the invertebrates.

High homogeneity of human survival: A quality control hypothesis

In invertebrates, the age-related mortality increase slows markedly around the modal age and levels off.[10] The death rate level of the plateau is in the range of 0.01 to 0.1 (per one-hundredth of the modal life span). In humans, the death rate continues to rise steeply even after having passed the death rate of 0.1 around the modal age. Thus, the life span–adjusted death rate at the modal age is considerably higher for humans than for the invertebrates (Table 1 and Figure 2).[11]

A close look at the figures reveals that mortality deceleration occurs in humans after around age 75 years, as indicated by the slight concavity of the mortality curve (Figure 2A) and the short fall of the life table aging rate (Figure 3A). However, the deceleration is less pronounced and occurs with considerably higher death rates in humans than in invertebrates. The mathematical model predicts a significant slowing of human mortality after age 100 years (solid lines in Figures 2A and 3A), but the 1881–85 Swedish cohort became almost extinct around age 100.[12]

Mortality deceleration at the population level could be produced by age-related physiological and behavioral changes in individual members. It could also be due to population heterogeneity: more frail individuals tend to die at younger ages, thereby making survivors to older ages a special group of healthy individuals. This selective survival process counteracts the age-related increase of mortality risk at the individual level. If the population is more homogeneous, the mortality deceleration starts later (Horiuchi and Wilmoth 1998: Appendix A).[13]

The frailty model analysis produced estimates of the coefficient of variation, which measures the size of individual differences in vulnerability to mortality risk. Table 1 shows that the coefficient of variation for humans is the lowest, 58 percent of the coefficient of variation for nematodes (the second lowest) and 41 percent of the coefficient of variation for wasps (the highest).[14] The results shown in Table 1 and Figures 1, 2, and 3 suggest that the human cohort is more homogeneous than the invertebrate cohorts.

Because all individuals in each invertebrate cohort were reared in essentially the same environment, their life span differences may be at least partly genetic. Not surprisingly, the coefficients of variation for genetically homogeneous populations (inbred lines of nematodes and *Drosophila*) are the first and second lowest among the invertebrates. Among the other three populations, the coefficient of variation for beetles is smaller than those of medflies and wasps. The genetic diversity of the beetle co-

TABLE 1   Estimates of the coefficient of variation (CV) in
vulnerability to mortality risk and estimates of the death
rate per OHM at the modal age at death (DRM)

Species	CV	DRM
Humans	0.544	.0984
Mediterranean fruit flies	1.152	.0188
Nematode worms	0.942	.0337
Parasitoid wasps	1.339	.0085
Drosophila melanogaster	0.964	.0417
Bean beetles	1.138	.0363

NOTE: OHM is one-hundredth of the modal life span. See discussion in the text.
SOURCE: See Figure 1.

hort may be limited, because the 829 beetles were offspring of only 19
male–female pairs.

Large differences in the coefficient of variation between humans and
two genetically heterogeneous invertebrate cohorts (medflies and wasps)
suggest that genetic variations in the level of somatic maintenance may be
greater in invertebrates than in humans. It is puzzling, however, that estimated individual differences in survivability are smaller in the genetically
heterogeneous Swedish cohort than in the genetically homogeneous nematode and *Drosophila* cohorts.

Finch and Kirkwood (2000) argue that in addition to genes and the
external environment in which individuals live after birth, "chance" is a major
source of individual differences. In particular, in embryo development from
fertilized ovum to birth or hatching, random events and fluctuations occur
at the molecular, cellular, and other levels, affecting physiological and mor-
phological characteristics of the organism. Considerable phenotypical variations have been found among individuals (even among fetuses) in genetically homogeneous populations of some species.

The wide dispersion of life span in the nematode and *Drosophila* cohorts indicates that chance may be an important source of individual differences in invertebrates. The narrow distribution of life span in the Swedish
cohort suggests that chance variations and their effects are constrained more
tightly in humans than in invertebrates. This may be explained by import-
ing the concept of "quality control" from reliability engineering.

A large number of industrial products of the same model are manu-
factured in factories. The manufacturers expect most of the products to keep
functioning properly for a certain duration of time or longer, unless the
products are damaged by accidents, inappropriate use, or unusual circumstances (i.e., mortality risks in the external environment). A certain proportion of the products, however, may not be durable enough, because of
errors (mistakes, accidents, and random fluctuations) that occurred to them

during the manufacturing processes. The purpose of quality control is to increase the proportion of final products that are satisfactorily durable. Measures taken to achieve this goal include reduction of the frequency and extent of errors, detection and correction of errors, control of adverse consequences of uncorrected errors, and identification and elimination of defective products.

It is possible to think of a biological analogue of quality control. Chance variations in embryo development and their potentially detrimental effects on somatic durability may be more tightly controlled, thereby decreasing the proportion of newborn individuals that are not capable of surviving long enough to achieve reproductive success.

It seems useful to distinguish between two types of defective embryos. First, some embryos will not be able to survive to the adult form. Many of them may suffer from errors in gametogenesis and meiosis and result in natural abortions, hatching failures, and larval deaths, without any direct effects on the distribution of adult life span. Second, some embryos are "substandard." They are viable in the adult form but their prospective fitness is considerably lower than expected from their genotypes.[15] Reduced production of these embryos and their elimination before birth or during early development could affect the distribution of adult life span if their low fitness comes from their low survival abilities.

The timing and effectiveness of quality control may be related to such factors as the cost of pregnancy, parental care, and number of offspring. It would be advantageous for defective embryos to be discarded as early as possible if the cost of pregnancy (in terms of length, energy, etc.) is high.[16] Quality control increases the cost-effectiveness of parental investment, particularly if offspring require parental care. A smaller number of offspring may make it more important to have no or few children who have "substandard" fitness.

Quality control in manufacturing shifts the peak of product life distribution to the right by reducing the number of products that are not durable enough. Without changes in the fundamental design of the product, however, the maximum length of product life remains unchanged, that is, the end of the right tail of the distribution cannot move further to the right. Therefore, quality control compresses the failure time distribution to the right and rectangularizes the product's survival curve. The variation in the product's life span is thereby reduced, and the distribution is skewed to the left. It seems reasonable to expect that quality control in embryo development should have similar effects on the life span distribution of living organisms.

In summary, the differences between humans and invertebrates in the relative timing of mortality deceleration and the shape of the life span distribution may be explained by the following "quality control" hypothesis. Chance variations in embryo development and their potentially adverse effects are more tightly controlled in humans than in invertebrates, which makes human populations more homogeneous than invertebrate populations with respect to the level of somatic maintenance.

Mortality acceleration in humans: A repair senescence hypothesis

A steeper mortality increase tends to make the life span distribution more narrowly concentrated around the peak. In humans, the death rate starts to ascend around age 40 and continues to rise steeply (Figure 2A), as implied by high values of the life table aging rate in Figure 3A.[17] The rapid age-associated increase of human mortality makes the death rate (per one-hundredth of the modal life span) at the modal age greater, the peak of life span distribution higher, and the relative life span variation smaller in humans than in invertebrates.

Figure 3 reveals, however, that the mortality increase is not necessarily faster (i.e., the life table aging rate is not necessarily greater) for humans in their 40s than for invertebrates in the corresponding age ranges (around the midpoint between age 0 and the modal age). The mortality increase around the modal age is the steepest in humans because the slope of the logarithmic mortality curve becomes progressively steeper from about age 40 to about age 75. The acceleration of mortality increase is substantial, from nearly 4 percent per one-hundredth of the modal life span (5 percent per year) around age 40 to about 8 percent per one-hundredth of the modal life span (10 percent per year) around age 75 (Figure 3A). Such a distinct phase of mortality acceleration is absent in all five invertebrate cohorts. The age-associated relative increase in mortality accelerates significantly at younger old ages in humans, but not in invertebrates.[18]

Thus, the overall life table aging rate patterns are strikingly different between humans and invertebrates. In humans, it rises first, then declines. The bell-shaped pattern, indicating a combination of mortality acceleration at younger old ages and deceleration at older old ages, has been observed widely among human populations of various countries in different periods, in both cohort and period life tables (Coale and Guo 1989; Horiuchi and Coale 1990; Horiuchi 1997; Horiuchi and Wilmoth 1998).[19] In contrast, the life table aging rate declines with age (i.e., the age-associated relative increase of mortality decelerates) fairly consistently in all of the invertebrate species.[20]

The frailty model analysis suggests that the cohort mortality acceleration reflects underlying mortality patterns of individuals. Figure 4 displays estimated mortality risks for the standard human and standard medfly.[21] The human log-mortality curve is notably convex, implying a significant acceleration of mortality risk at the individual level.[22]

**FIGURE 4   Estimated age trajectory of mortality risk at the
individual level, for the standard human and standard medfly**

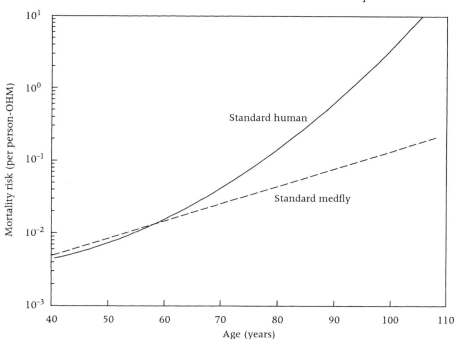

NOTE: The horizontal axis shows human age. Medfly ages were converted to corresponding human ages by
equalizing the modal life spans of the two cohorts. The vertical axis indicates the number of deaths per
individual-OHM of exposure. (OHM is one-hundredth of the modal life span.) In this analysis, the standard
human is a person whose mortality risk at age 40.5 years is equal to the cohort average, and the standard
medfly is a medfly whose mortality risk at age 1.0 day is equal to the cohort average.
SOURCES: See Figure 1.

Mortality increase may be related to damage accumulation.[23] Mechanisms of somatic maintenance are not perfect, allowing damage to accumulate (Kirkwood 1997). The mathematical model of senescence developed
by Strehler and Mildvan (1960) suggests that if damage accumulates at a
constant rate, the trajectory of logarithmic mortality risk is linear
(Gompertzian). However, a modified version of the Strehler–Mildvan model
suggests that if the mechanisms for preventing and repairing damage deteriorate with age, damage accumulates at an increasing rate and the trajectory of logarithmic mortality risk is quadratic and convex. (See Appendix B
for the mathematical derivation.)

It is reasonable to assume that functions of damage prevention and
repair decline with age. Various immune functions deteriorate with age
(Miller 1995). The rate of protein degradation declines with age, and this is
likely to allow abnormal proteins to accumulate (Van Remmen et al. 1995).
Wound healing is slower at older ages (Chuttani and Gilchrest 1995). The
ability to mount acute host defenses to oxidant injury decreases with age in

rats (Ikeyama et al. 2002). The apoptotic response to genotoxic stress declines with age in mice (Suh et al. 2002).

Damage prevention and repair are probably important for most species. In particular, long-lived species are expected to have highly efficient
mechanisms of damage prevention and repair that can keep the organism
functioning for a few decades or longer. Therefore, it seems possible that
senescence in long-lived species is strongly affected by the deterioration of
prevention and repair functions later in life, whereas senescence in short-
lived species is mainly attributable to relatively low effectiveness of prevention and repair mechanisms, which makes the rate of damage accumulation already high early in life.

The preceding discussion can be summarized as follows. In long-lived
species, highly effective mechanisms of damage prevention and repair make
their senescent processes slower than in short-lived species. However, the
age-related deterioration of prevention and repair mechanisms may make
the senescent processes of long-lived species more accelerated than those
of short-lived species. This "repair senescence" hypothesis is consistent with
the accelerated relative increase of the human death rate at younger old
ages.[24]

Conclusions

This study compared variations and differences in life table functions relative to the modal life span. Relative and absolute variations should be distinguished strictly and carefully.

In addition, the interspecies comparison has at least three major limitations. First, because of data availability, comparison was made between
two extremes: humans versus invertebrates. It would be desirable to include, for example, some mammalian species that live for a few to several
years. To study shapes of life table functions, however, requires a large number of observations that produces fairly smooth age trajectories up to very
old ages. Although many studies have been conducted on mice and rats,
rodent survival data on cohorts that are sufficiently large for analysis of life
table aging rates, to the knowledge of this author, do not exist. (Appendix
D shows survival data on 223 mice and 205 dogs just for crude comparison
with humans and invertebrates, but the method of analysis and reliability
of the results are very limited because of the small cohort sizes.)

Second, animal populations, after being exposed to the force of gene
selection (or the lack of it) in the laboratory for many generations, may
differ significantly in the size of genetic variations from their ancestors in
the natural world. This applies to modern humans as well. Life table patterns in protected populations may not sensitively reflect the effects of natural
selection on the abilities of somatic maintenance in the wild.

Third, although both humans in modern societies and animals in laboratories can be considered to live in protected environments, their conditions may not be highly comparable. The laboratory animals were more strictly protected from epidemics and food shortage and probably from violence than were Europeans born in the late nineteenth century.[25] On the other hand, contemporary humans (and, to a lesser extent, pet animals) benefit greatly from modern medical care, while insects and worms in laboratories do so much less, if at all.

These limitations strongly suggest that the results of this study should be interpreted with caution. Nevertheless, the major differences between human and invertebrate life table patterns are clear enough to make us think that they may reflect some important underlying factors. Interspecies differences in the life span distribution are worthy of further investigation.

Appendix A: Data and measurement

The human data (the male Swedish cohort born in 1881–85) were downloaded from the Berkeley Mortality Database (http://www.demog. berkeley.edu/wilmoth/ mortality/), which later evolved to the Human Mortality Database (http://www. mortality.org). The medfly and beetle data were drawn from Carey et al. (1995: Appendix) and Tatar and Carey (1994: Table 1), respectively, and data on the other species were provided by the authors of Vaupel et al. (1998). See these references and Johnson et al. (2001) for more information on the data.

Males and females may exhibit different mortality patterns. In this study, male data were used, except for the hermaphrodite nematodes, mainly because male medflies were preferred to female medflies. In the original study (Carey et al. 1992), the male and female medflies were maintained on a sugar-only (protein-deprived) diet. Protein deprivation alters female mortality patterns markedly but affects male patterns only mildly (Müller et al. 1997). Physiologically, female medfly mortality is more sensitive than male medfly mortality to environmental variations (Carey et al. 2001). The extent to which these sex differentials hold for other invertebrate species remains to be investigated.

The Swedish 1881–85 cohort was selected for several reasons. First, selective survival processes can be followed only in cohort mortality, and cohort data were used for the other species. Second, a cohort born more than 110 years ago was preferred, to ensure that virtually all of its members had died. Third, Sweden is one of the few countries whose demographic data during the nineteenth century were considered already highly accurate. Finally, a previous study has shown that this particular cohort followed the typical pattern of old-age mortality that was widely observed in human male and female populations (Horiuchi and Wilmoth 1998).

The life span distribution was obtained over the entire age range (except the hatching day) for medflies, wasps, and beetles, and for ages 14 years and older for humans, 4 days and older for *Drosophila*, and 5 days and older for nematodes. The death rate declined with age for the first 13 years in humans and for the first 3 days in *Drosophila*. These periods were not included in the data analysis because the focus of this study was on longevity and senescence, not on development. Nematode mortality data during the first 4 days were not collected in the original study. The age ranges for which the life span distribution was computed can be considered to approximately represent "adult life," including both reproductive and postreproductive stages, but excluding early stages in which the death rate declines with age.

For five out of the six cohorts, deaths of all cohort members were recorded; for nematodes, samples were drawn repeatedly for measuring mortality at different ages. The nematode data set comprises results of five studies that used the same wild-type (i.e., not mutant) inbred line, TJ1060 (Johnson et al. 2001). Because dead nematodes cannot be identified immediately, it takes considerably more time to count them than dead beetles, flies, or wasps. Therefore, in each of the studies, samples were drawn every other day, and death rates for the samples during the following 24-hour period were obtained. After the cohort size had become relatively small, the entire cohort was followed.

Because of this procedure, nematode mortality data were not obtained for the 6th, 8th, 10th, 12th, 14th, and 16th days. Without death rates for these dates, it is impossible to estimate the life span distribution for the 5th day and later. Thus death rates for these dates were geometrically interpolated from death rates for the adjacent days. These interpolated death rates are neither included in Figure 2C nor used for computing life table aging rates in Figure 3C, but are used for deriving the life span distribution in Figure 1C (dots).

Because the invertebrates experience metamorphosis, their life span in this study was defined to be the length of life in the adult form. Larval survival data were not collected for those cohorts.

Ages of the invertebrate animals were counted as follows. Let "hatching day" be the day of emergence into the adult form. When an entire invertebrate cohort is followed, usually a hatching day is set up and newborn individuals are counted at the end of the day. Deaths between hatching and day's end are not counted. Ages of the cohort members were counted assuming that they were 0.5 days old at the end of the hatching day.

Four life table functions were obtained for each of the cohorts: the proportion surviving (not shown), life span distribution, age-specific death rate, and life table aging rate. Following the convention, they are denoted by $l(x)$, $d(x)$, $m(x)$, and $k(x)$. Their relationships, assuming the continuity and differentiability of these functions with respect to age x, are given by: $d(x) = -dl(x)/dx$, $m(x) = d(x)/l(x)$, and $k(x) = dm(x)/m(x)dx$.

The $d(x)$ function was scaled such that its total over the adult age range (described earlier in this appendix) is unity. For medflies, wasps, *Drosophila*, and beetles, the death rate for the age range $[x,x+1)$ was calculated by $-log(N(x+1)/N(x))$, where $N(x)$ is the number of survivors at age x. The life table aging rate at age x was obtained as the weighted least squares estimate of the slope parameter of the Gompertz model, by fitting the Gompertz equation to logarithmic death rates for an age interval centered at x.

Appendix B: Gamma Gompertz model and gamma log-quadratic model

Assumptions and estimation

In this study, frailty models were used for summarizing and interpreting the observed survival patterns. In frailty models, the mortality risk of each individual is assumed to be the product of two terms: (1) a hypothetical variable called frailty, denoted by z, and (2) a function of age x, denoted by $\mu_s(x)$ (Vaupel et al. 1979):

$$\mu(z,x) = z\mu_s(x). \tag{B.1}$$

Frailty z may be considered as the combined age-adjusted effects of genetic and environmental characteristics of the individual on his/her mortality risk, and $\mu_s(x)$ represents the combined effects of age-related physiological and behavioral changes on the mortality risk. $\mu_s(x)$ is called the "standard" mortality schedule, because it is the trajectory of mortality risk for individuals with $z = 1$, which is the mean value of z at the lowest end of the age range in which mortality is assumed to follow the model.

As in many previous studies, z is assumed to be gamma-distributed. One of the two parameters of the gamma distribution uniquely determines the coefficient of variation of z. Because of selective survival, the mean and variance of z change with age. However, z remains gamma-distributed and the coefficient of variation is constant over age (Vaupel et al. 1979).

The most widely used form for $\mu_s(x)$ is the exponential equation (Gompertz model):

$$\mu_s(x) = \mu_0 \exp(\theta x). \tag{B.2}$$

The combination of the gamma-distributed frailty in the cohort and the Gompertzian mortality risk at the individual level is called the gamma Gompertz model. The combination of the two assumptions makes the cohort death rate at age x, denoted by $m(x)$, a three-parameter logistic function of age (Beard 1959):

$$m(x) = \frac{Be^{\theta x}}{1 + Ce^{\theta x}}. \tag{B.3}$$

For the five invertebrate cohorts, the gamma Gompertz model fits $d(x)$, $m(x)$, and $k(x)$ patterns well (Figures 1, 2, and 3). The model is fundamentally incompatible with the typical old-age pattern of human mortality, however. It is mathematically impossible for the gamma Gompertz model to produce a bell-shaped $k(x)$ curve, because equation B.3 implies that $k(x)$ changes monotonically with age. Thus a modified model was adopted for humans, assuming that the age trajectory of mortality risk at the individual level follows the form:

$$\mu_s(x) = \exp(ax^2 + bx + c). \tag{B.4}$$

This pattern can be called "log-quadratic," because the logarithm of the mortality risk is a quadratic function of age. (The Gompertz equation is log-linear.) By combining this equation with the gamma-distributed z, the cohort death rate at age x is given by

$$m(x) = \bar{z}(x)\exp(ax^2 + bx + c), \tag{B.5}$$

where the mean of z at age x is given by

$$\bar{z}(x) = \cfrac{1}{1 + \cfrac{1}{\alpha}\displaystyle\int_{x_0}^{x}\exp(at^2 + bt + c)dt}.$$

x_0 is the age at which the mortality risk starts to follow the quadratic equation, and α is the inverse of the squared coefficient of variation of the gamma distribution. The mean of z at age x_0 is one. Equation (B.5) is called the gamma log-quadratic model.

The gamma Gompertz model and the gamma log-quadratic model were fitted to age-specific death rates by the method of weighted least squares. The parameters were estimated by minimizing

$$L = \sum_{i} D_i(\ln M_i - \ln m_i)^2, \tag{B.6}$$

where D_i and M_i are the number of deaths and the death rate, respectively, for the i-th age group; m_i is calculated by equation B.3 or B.5. The weight, D_i, is approximately equal to the inverse of the variance of $\ln M_i$.

The $d(x)$-based method of maximum likelihood estimation (Pletcher 1999; Promislow et al. 1999) was not used, partly because it cannot be applied to the sample-based data on nematode mortality, and partly because the $m(x)$-based method seemed more sensitive to $k(x)$ variations. Estimates by weighted least squares are approximately maximum likelihood estimates if, for each age group, the number of living individuals and the number of deaths are large and the death rate is not extremely high.

In model estimation, some consecutive youngest ages and/or oldest ages were excluded in order to avoid ages with fewer than ten deaths. In humans, the model was fitted to ages 40 and older because the death rate follows a smooth upward trajectory only after age 40 (i.e., $x_0 = 40.5$).

Gerontological interpretation

Equation B.4 is consistent with the idea that mechanisms for preventing and repairing somatic damage deteriorate with age. This can be shown using the model by Strehler and Mildvan (1960), in which the "vitality" of a living organism is assumed to decline linearly with age. This assumption is supported by empirical evidence that various physiological indicators change with age approximately linearly. The linear decrease of vitality is expressed as:

$$v(x) = v_0 - gx, \tag{B.7}$$

where v_0 is the initial vitality level and g is the rate of physiological decline. For simplicity, vitality is assumed to decline from age 0. If physiological conditions deteriorate with age because of accumulation of damage, g can be interpreted as the rate of damage accumulation. Strehler and Mildvan derived the Gompertz equation from the assumption of linear vitality decline.

Living organisms have mechanisms for preventing and repairing the damage. (Hereafter, only "repair" will be mentioned for simplicity; "prevention" can be treated mathematically in the same way.) Thus, the model may be modified as

$$v(x) = v_0 - (g - r)x, \tag{B.8}$$

where r is the repair rate and $0 < r < g$. However, the repair mechanisms may deteriorate with age. A simple model of this process is a linear decline of repair rate with age:

$$r(x) = r_0 - ax, \tag{B.9}$$

where r_0 is the initial repair rate and a is the rate of deterioration of repair mechanisms.

Substituting B.9 into B.8 and using the vitality–mortality relationship in the Strehler–Mildvan model, we get:

$$\mu(x) = \exp(ax^2 + bx + c),$$

where $b = g - r_0 > 0$ and $c = -v_0 < 0$. (If this process does not start at age 0 but at a certain age x_0, then $b = g - r_0 - 2ax_0$ and $c = -v_0 - gx_0 + r_0x_0 + ax_0^2$.)

The log-quadratic equation is consistent with results of three previous studies on human mortality. First, in a study on longevity of Danish twins, the underlying pattern of mortality risk of individuals was estimated, using a semi-parametric frailty model. The derived mortality trajectory was noticeably log-convex (Yashin and Iachine 1997). Second, a multivariate hazard model analysis was applied to genealogical records on British noblemen. The estimated age pattern of mortality beyond age 50 years at the individual level was also log-convex (Doblhammer and Oeppen 2002).

Third, it has been shown that major medical causes of death have significantly different age trajectories of mortality at younger old ages. The age pattern is associated with the selectiveness of the cause. Diseases with well-known strong effects of risk factors (such as cancers, liver cirrhosis, and hemorrhagic stroke) can be considered highly selective, because individuals who have the risk factors are particularly vulnerable. Mortality curves of those diseases tend to be log-concave, and patterns of diseases that are considered less selective (such as pneumonia, gastroenteritis, and congestive heart failure) tend to be log-convex (Horiuchi and Wilmoth 1997, 1998). Greater selectivity, whether it is due to environmental or genetic differences, makes the mortality curve more log-concave. Therefore, the differential trajectories of cause-specific mortality suggest that age patterns of total mortality risk of individuals, which are not distorted by selection, might be even more log-convex than those of aggregate-level death rates from less selective diseases.

Appendix C: Simulation of relationships between mortality pattern and life span distribution

A simulation study was conducted to investigate relationships between the age pattern of mortality and the shape of life span distribution. The three-parameter logistic equation (gamma Gompertz model), which was fitted to death rates for the five invertebrate cohorts (but not those for the human cohort), was used for generating hypothetical mortality schedules.

For this simulation, the widely used form (equation B.3) was converted into:

$$m(x) = \frac{\lambda e^{\theta x}}{1 + \dfrac{\lambda}{\alpha\theta}(e^{\theta x} - 1)},$$

(C.1)

because the parameters in this form can be more easily interpreted than those in equation B.3 (Vaupel 1990). λ is the death rate at age zero (i.e., $\lambda = m(0)$), θ affects the steepness of age-related mortality increase, and α is negatively associated with the timing of mortality deceleration. By combining different values of these parameters, 150 life tables were produced. In all of the life tables, the modal life span was fixed at 100 years. The ranges of λ, θ, and α are [0.00002, 0.00230], [0.0202, 0.0838], and [0.444, 3.400]. These ranges cover λ, θ, and α values estimated for the five invertebrate cohorts as well as α for the Swedish cohort. (Note that α values in the gamma Gompertz and gamma log-quadratic models are comparable.)

Relationships between these three parameters and two measures of life span distribution were examined. The two variables are (1) the density at the modal age (i.e., the height of the peak), which measures the dispersion/concentration of deaths around the modal age, and (2) the proportion of all deaths between ages 0 and 200 that occur under age 100, which indicates the extent of skewness to the left.

Major results of the simulation are shown in Figure 5. Panel A shows that if the modal life span is fixed, λ and θ are negatively associated. That is, the same modal life span could be derived from a combination of low initial mortality and steep mortality rise and a combination of high initial mortality and slow mortality rise. (Note that although θ is not a life table aging rate, θ and the life table aging rate are on the same scale. Thus, the horizontal axes of the three panels of Figure 5 are directly comparable to the vertical axes of Figure 3. In particular, if α is infinitely large, equation C.1 becomes the Gompertz model, in which θ is the age-independent life table aging rate.)

Because of the strong correlation between λ and θ, the other two panels of Figure 5 use θ, but not λ. Panel B indicates that both a steeper rise of mortality (necessarily combined with lower initial mortality) and a later start of mortality deceleration make the life span variation smaller and the peak of distribution higher. Note that the vertical axis of this panel is directly comparable to the vertical axes of Figure 1.

Panel C reveals that greater delay of mortality deceleration skews the life span distribution more to the left. The life span distribution is exactly symmetric if α is unity (i.e., if frailty is exponentially distributed). The effect of steepness of mortal-

FIGURE 5 Simulated relationships between mortality pattern and life span distribution

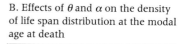

B. Effects of θ and α on the density of life span distribution at the modal age at death

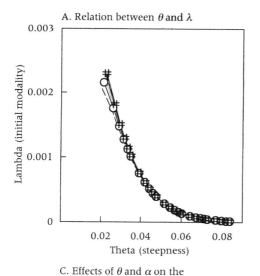

A. Relation between θ and λ

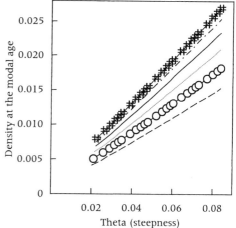

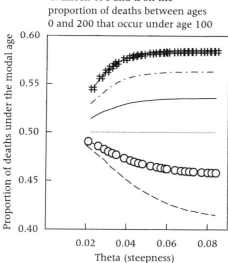

C. Effects of θ and α on the proportion of deaths between ages 0 and 200 that occur under age 100

NOTE: Summary of 150 simulated logistic mortality schedules in which the modal life span is fixed at 100. In the majority of the mortality schedules, almost 100 percent of deaths occur under age 200. In all three panels, the six curves indicate different α values: 4/9 (dashed), 2/3 (circle), 1 (dotted), 1.5 (solid), 2.25 (dash-dot) and 3.4 (asterisk). For definitions of λ, θ, and α, see equation C.1.

ity rise on the skewness depends on the timing of mortality deceleration, and does not appear substantial for left-skewed distributions, as indicated by the upper three curves. Although these results are based on the particular mathematical model, it seems reasonable to expect the same directions of relationship for typical patterns of adult mortality in empirical life tables.

Appendix D: Life span distribution in mice and dogs

Figure 6 displays life span distributions and mortality patterns for an inbred line (B6CF$_1$) of 223 mice and a cohort of 205 beagle dogs. The data have been drawn from a previous study of interspecies mortality comparison (Carnes, Olshansky, and Grahn 1996). Figure 6 was produced in the same way as Figures 1 and 2. The solid lines indicate the gamma Gompertz model fitted to the data. The estimated modal life span, indicated by the vertical dotted line, is 952 days for mice and 4,704 days for beagle dogs. Because of the small size of each cohort, it is impossible to obtain a smooth sequence of reliable life table aging rates. Thus, we cannot examine whether the relative mortality increase with age accelerates at younger old ages in these cohorts.

FIGURE 6   Life span distribution (A and B) and age-specific death rates (C and D) for mice and dogs

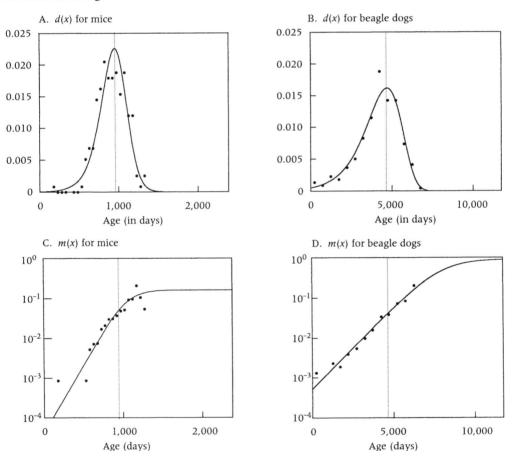

NOTE: Produced in the same way as Figures 1 and 2.
SOURCE: Carnes, Olshansky, and Grahn 1996.

The skewness of the life span distribution splits vertebrates and invertebrates clearly. The three mammalian curves are more skewed than the five invertebrate curves. Thus, estimated coefficients of variation for mice and dogs (0.680 and 0.221) are smaller than those for invertebrates. Moreover, their coefficients of variation are closer to the human figure (0.544) than to the figure for the lowest invertebrate (0.942). If we expect quality control in embryogenesis to be tighter among vertebrates than invertebrates, this result is consistent with the conjecture.

It appears puzzling that in terms of the estimated coefficient of variation, humans are placed between mice and dogs. However, it is possible that dogs of the same breed living in the same laboratory are far more homogeneous in mortality risk than people in an entire country. In addition, the very low coefficient of variation of the dogs may simply be a statistical artifact of the cohort size, which might have become too small before the starting age of noticeable mortality deceleration.

The height of the peak of the life span distribution fails to distinguish vertebrates and invertebrates clearly. Although the estimated peak for mice is the second highest, the peak for dogs is lower than peaks for *Drosophila* and beetles. This seems due to the slope of the mortality curve for dogs before the modal age, which is apparently less steep than the slopes for *Drosophila* and beetles. Therefore, of the two major features of human life span distribution in Figure 1A, the great skewness may be more biologically essential than the high peak.

Because of the small cohort sizes, however, these results should not be taken as highly reliable. For this type of analysis, it is desirable to collect survival data for considerably larger mammalian cohorts.

Notes

I thank James Carey, James Curtsinger, Thomas Johnson, S. Jay Olshansky, Vladimir Shkolnikov, and James Vaupel for providing data. I am also grateful to Caleb Finch, Thomas Johnson, Ryuichi Kaneko, S. Jay Olshansky, and James Vaupel for useful comments on earlier versions. This research was supported by grants K02-AG00778 and R01-AG14698 from the US National Institute on Aging, National Institutes of Health.

1 However, a higher proportion of human deaths at younger adult ages than at older ages is due to injuries (including accidents, suicide, and homicide) and infectious diseases (see, for example, Horiuchi 1999: Figure 2). Probably this is why both the proportion of the life span and the death rate for the Swedish cohort rise smoothly and consistently after age 40 years, but not under age 40 (Figures 1A and 2A).

2 These two effects cannot be separated in a mechanistic manner. In order to investigate these two effects on the life span distribution, additional information and/or assumptions are needed.

3 The exact size of the nematode cohort is not known, because samples were drawn from the cohort.

4 A very small proportion of the nematodes were males. They were not included in the mortality measurement.

5 The modal life span is 79.5 years for humans, 20.2 days for medflies, 15.7 days for nematodes, 5.8 days for wasps, 42.8 days for *Drosophila*, and 14.7 days for beetles. Because the life span distribution was erratic in some of the cohorts, modal ages estimated by mathematical models were adopted.

6 Although the modal life span is not used as widely as the mean life span (life expectancy) and median life span, the usefulness of the modal life span in aging research has been increasingly recognized (e.g., Kannisto 2001).

7 For example, neonatal deaths can be distinguished from perinatal miscarriages and stillbirths fairly accurately in humans and large-sized mammalian species. This distinction is a necessary condition for measuring infant mortality and life expectancy at birth in human demography. Usually the distinction is difficult or meaningless, however, when invertebrate eggs hatch and go through metamorphosis.

8 The vertical axes of Figures 1, 2, and 3 are the proportion of deaths per one-hundredth of the modal life span (OHM), the number of deaths per individual-OHM, and the exponential rate of mortality increase per OHM, respectively. These OHM-based life table functions can be obtained simply by calculating them in terms of the conventional time unit (year or day), multiplying them by the modal life span (in the same conventional time unit), and dividing by 100.

9 In this context, "being skewed to the left" means that the left arm of the bell-shaped curve is less steep than the right arm.

10 Although some old-age death rates in Figure 2 (in particular, those for nematodes, *Drosophila*, and beetles) appear to be far above the estimated plateau, most of those points are based on small numbers of deaths and have large stochastic variations. In medflies and wasps, the death rate not only reaches a plateau, but decreases noticeably at very old ages (Vaupel et al. 1998). This occurs outside the age range of Figure 2B. The three-parameter logistic model cannot capture this part of the mortality trajectory.

11 Also note that the death rate and the life table aging rate are exactly equal at the modal age, which is a mathematical characteristic of the life table.

12 An international data set that includes a relatively large number of centenarians and super-centenarians (ages above 110 years) has revealed that the mortality increase slows down considerably after age 100 (Kannisto 1994; Thatcher, Kannisto, and Vaupel 1998). This finding is consistent with the extrapolated mortality deceleration in Figures 2A and 3A.

13 The heterogeneity hypothesis explains not only the occurrence of mortality deceleration but also various related phenomena (reviewed in Wilmoth and Horiuchi 1999; also

Ewbank 2001). A major issue raised about the heterogeneity hypothesis is that individual differences estimated by frailty models tend to greatly exceed variations of quantitative physiological traits that are usually observed among individuals of the same species. It is important, however, to strictly distinguish between physiological variations and risk variations. Small physiological differences may lead to large differences in mortality risk.

At least two kinds of data suggest substantial individual variations in mortality risk. First, age variations in mortality are considerably larger than age variations in physiological characteristics. In the Swedish male cohort born in 1881–85, the death rate increased by more than ten times from age 40 to age 73. It seems difficult, on the other hand, to find quantitative physiological traits with tenfold differences between 40-year-old and 73-year-old persons. Second, many epidemiological studies indicate substantial individual differences in mortality risk. According to a multivariate hazard model analysis of data from National Health Interview Surveys in the United States, the mortality risk (adjusted for age and income) in 1990–95 for "divorced/separated and unemployed black males with less than 12 years of education" is 9.38 times higher (= 1.36 x 1.78 x 1.25 x 2.00 x 1.55) than for "married and employed non-black females with 16 or more years of education" (Rogers, Hummer, and Nam 2000: Table 3.2). These two are the highest and lowest risk profiles that have been produced by only five basic demographic and socioeconomic factors (sex, race, education, marital status, and employment status). This risk difference corresponds to the 94.5 percent median-centered range (97.25 percentile versus 2.75 percentile) in the frailty distribution estimated for the Swedish cohort and to the 57.0 percent range for the medfly cohort. Additional inclusion of four health-behavior factors (smoking, alcohol consumption, exercise, and obesity) increases the ratio further to 32.85, corresponding to the 99.6 percent range for the Swedish cohort and to the 76.9 percent range for medflies (ibid.: Table 3.3). Thus, individual differences estimated by frailty model analyses do not seem far greater than those estimated from epidemiological data.

14 This is not a statistical artifact of using different models (the gamma log-quadratic

model for humans and the gamma Gompertz model for the others), because the latter model produced a lower estimate of the coefficient of variation for humans than the former model.

15 Gavrilov and Gavrilova (2001) have developed reliability-theoretical models of mortality and senescence, in which living organisms are assumed to be born with stochastically varying numbers of nonfatal defects.

16 It is believed that a high proportion of human embryos are aborted before diagnosis of the pregnancies and even before implantation of the fertilized eggs, though it has not been feasible to estimate the proportion involved (Ellison 2001: chapter 2).

17 The life table aging rate (LAR) is the rate of relative mortality increase with age. A rise of LAR indicates an acceleration of age-related proportional increase in mortality (a convex curvature of the logarithmic mortality curve); a decline of LAR indicates a deceleration (a concave curvature); and a constant LAR implies an exponential mortality increase (a straight line). Figure 3 reveals that with respect to the assumed constancy of the life table aging rate, all of the cohorts deviate considerably from the Gompertz model. The striking difference between human and invertebrate patterns suggests that the life table aging rate is a powerful statistical tool for biodemography. Detailed descriptions of the rate are given in Horiuchi and Wilmoth (1997, 1998).

18 It is necessary to distinguish relative (proportional, geometric) and absolute increases. In humans, the absolute mortality increase usually continues to accelerate throughout adult ages up to around 100 years or more, but the relative mortality increase starts to slow down around age 75.

19 Male populations in some industrialized countries deviate from this typical bell-shaped pattern of the life table aging rate (Horiuchi 1997). The deviation may be related to the diffusion of smoking after World War I. Historical time series of mortality data for Sweden have revealed that male mortality in the nineteenth century followed the bell-shaped pattern, but the deviation emerged and became noticeable during the twentieth century (Horiuchi and Wilmoth 1998). My analysis of the time series of cause-specific mortality in France since 1925

has shown that smoking-related diseases such as lung cancer contributed significantly to the deviation (unpublished results).

20 The difference between humans and invertebrates can be seen, though less clearly, in curvatures of death rate functions as well (Figure 2). The logarithmic mortality curves for invertebrates appear consistently concave, as implied by the declines in the life table aging rate. However, the logarithmic mortality curve for humans is convex from around age 40 to around age 76, and concave thereafter. It may not be easy to visually detect these curvatures in the death rate function, but life table aging rates in Figure 3A show clearly that the accelerations and decelerations are significant.

21 The standard individual is a person whose mortality risk at a given age is equal to the cohort average. In this analysis, age 40.5 years for humans and age 1.0 day for medflies were used as the reference ages. These are the ages at which death rates of the respective cohorts are assumed to follow the adopted mathematical models.

22 The model estimates that the mortality risk at the individual level continues to accelerate, but the mortality rate at the aggregate level ceases to accelerate and starts to decelerate around age 76 years. The divergence between these two levels occurs at older old ages because the effect of selective survival on the age pattern of mortality is stronger at ages characterized by a higher death rate.

23 In particular, accumulation of DNA damage seems to play an essential role in senescence (de Boer et al. 2002; Stern et al. 2002).

24 The human mortality acceleration at younger old ages contributed to the reduction of life span variation. In general, however, a high concentration of deaths around the modal age does not necessarily imply a mortality acceleration. More essential for death concentration is a high level of mortality around the modal age, which could be reached not only by an accelerated mortality increase but also by a steep constant increase.

25 Deaths from accidents, suicide, homicide, and infectious diseases at younger adult ages might have contributed to the asymmetry of life span distribution for the Swedish cohort.

References

Beard, R. E. 1959. "Note on some mathematical mortality models," in G. E. W. Wolstenholme and M. O'Connor (eds.), *CIBA Foundation Colloquia on Ageing, The Life Span of Animals* 5. Boston: Little, Brown and Company, pp. 302–311.

Carey, James R. and Debra S. Judge. 2000. *Longevity Records: Life Spans of Mammals, Birds, Amphibians, Reptiles and Fish*. Odense: Odense University Press.

Carey, J. R., P. Liedo, H. G. Müller, J. L. Wang, B. Love, L. Harshman, and L. Partridge. 2001. "Female sensitivity to diet and irradiation treatments underlies sex-mortality differentials in the Mediterranean fruit fly," *Journal of Gerontology: Biological Sciences* 56(2): B89–93.

Carey, J. R., P. Leido, D. Orozco, M. Tatar, and J. W. Vaupel. 1995. "A male-female longevity paradox in medfly cohorts," *Journal of Animal Ecology* 64: 107–116.

Carey, J. R., P. Liedo, D. Orozco, and J. W. Vaupel. 1992. "Slowing of mortality rates at older ages in large medfly cohorts," *Science* 258: 458–461.

Carnes, B. A., S. J. Olshansky, and D. Grahn. 1996. "Continuing the search for a fundamental law of mortality," *Population and Development Review* 22: 231–264.

Chuttani, A. and B. A. Gilchrest. 1995. "Chapter 11. Skin," in Edward J. Masoro (ed.), *Handbook of Physiology: Section 11: Aging*. New York: Oxford University Press, pp. 309–324.

Coale, A. J. and G. Guo. 1989. "Revised regional model life tables at very low levels of mortality," *Population Index* 55: 613–643.

Comfort, A. 1979. *The Biology of Senescence*. New York: Elsevier.

de Boer, Jan, Jaan Olle Andressoo, Jan de Wit, Jan Huijmans, Rudolph B. Beems, Harry van Steeg, Geert Weeda, Gijsbertus T. J. van der Horst, Wibeke van Leeuwen, Axel P. N. Themmen, Morteza Meradji, and Jan H. J. Hoeijmakers. 2002. "Premature aging in mice deficient in DNA repair and transcription," *Science* 296: 1276–1279.

Doblhammer, Gabriele and Jim Oeppen. 2002. "Reproduction and longevity among the British peerage: The effect of frailty and health selection," unpublished paper presented at the annual meeting of the Population Association of America, 9–11 May, Atlanta.

Ellison, Peter T. 2001. *On Fertile Ground: A Natural History of Human Reproduction*. Cambridge: Harvard University Press.

Ewbank, Douglas C. 2001. "Demography in the age of genomics: A first look at the prospects," in C. E. Finch, J. W. Vaupel and K. Kinsella (eds.), *Cells and Surveys*. Washington, DC: National Academy Press.

Finch, Caleb E. 1990. *Longevity, Senescence, and the Genome*. Chicago: The University of Chicago Press.

Finch, Caleb E. and Thomas B. L. Kirkwood. 2000. *Chance, Development, and Aging*. New York: Oxford University Press.

Finch, C. E. and M. C. Pike. 1996. "Maximum life span predictions from the Gompertz mortality model," *Journal of Gerontology: Biological Sciences* 51(3): B183–B194.

Gavrilov, Leonid A. and Natalia S. Gavrilova. 2001. "The reliability theory of aging and longevity," *Journal of Theoretical Biology* 213: 527–545

Horiuchi, S. 1997. "Postmenopausal acceleration of age-related mortality increase," *Journal of Gerontology: Biological Sciences* 52(1): B78–92.

Horiuchi, S. 1999. "Epidemiological transitions in human history," in *Health and Mortality: Issues of Global Concern*. New York: United Nations, pp. 54–71.

Horiuchi, S. and A. J. Coale. 1990. "Age patterns of mortality for older women: An analysis using the age-specific rate of mortality change with age," *Mathematical Population Studies* 2: 245–267.

Horiuchi, S. and J. R. Wilmoth. 1997. "Age patterns of the life-table aging rate for major causes of death in Japan, 1951–1990," *Journal of Gerontology: Biological Sciences* 52A: B67–B77.

Horiuchi, S. and J. R. Wilmoth. 1998. "Deceleration in the age pattern of mortality at older ages," *Demography* 35(4): 391–412.

Ikeyama, S., G. Kokkonen, S. Shack, and X. T. Wang. 2002. "Loss in oxidative stress tolerance with aging linked to reduced extracellular signal-regulated kinase and Akt kinase activities," *FASEB Journal* 16(1): 114–116.

Johnson, Thomas E., Deqing Wu, Patricia Tedesco, Shale Dames, and James W. Vaupel. 2001. "Age-specific demographic profiles of longevity mutants in *Caenorhabditis elegans* show segmental effects," *Journal of Gerontology: Biological Sciences* 56: B331–B339.

Kannisto, V. 1994. *Development of Oldest-Old Mortality, 1950–1990: Evidence from 28 Developed Countries*. Odense: Odense University Press.

Kannisto, V. 2001. "Mode and dispersion of the length of life," *Population: An English Selection* 13(1): 159–172.

Kirkwood, T. B. 1997. "The origins of human ageing," *Philosophical Transactions of the Royal Society of London—Series B: Biological Sciences* 352(1363): 1765–1772.

Miller, Richard A. 1995. "Immune system," in Edward J. Masoro (ed.), *Handbook of Physiology: Section 11: Aging*. New York: Oxford University Press, pp. 555–590.

Müller, H. G., J. L. Wang, W. B. Capra, P. Liedo, and J. R. Carey. 1997. "Early mortality surge in protein-deprived females causes reversal of sex differential of life expectancy in Mediterranean fruit flies," *Proceedings of the National Academy of Sciences* 94: 2762–2765.

Pletcher, S. D. 1999. "Model fitting and hypothesis testing for age-specific mortality data," *Journal of Evolutionary Biology* 12(3): 430–439.

Promislow, D. E. L., M. Tatar, S. D. Pletcher, and J. Carey. 1999. "Below threshold mortality: Implications for studies in evolution, ecology and demography," *Journal of Evolutionary Biology* 12(2): 314–328.

Rogers, Richard G., Robert A. Hummer, and Charles B. Nam. 2000. *Living and Dying in the USA: Behavioral, Health, and Social Differentials of Adult Mortality*. San Diego: Academic Press.

Stern, N., A. Hochman, N. Zemach, N. Weizman, I. Hammel, Y. Shiloh, G. Rotman, and A. Barzilai. 2002. "Accumulation of DNA damage and reduced levels of nicotine adenine dinucleotide in the brains of Atm-deficient mice," *Journal of Biological Chemistry* 277: 602–608.

Strehler, B. L. and A. S. Mildvan. 1960. "General theory of mortality and aging," *Science* 132: 14–21.

Suh, Y., K.-A. Lee, W.-H. Kim, B.-G. Han, J. Vijg, and S. C. Park. 2002. "Aging alters the apoptotic response to genotoxic stress," *Nature Medicine* 8: 3–4.

Tatar, M. and J. R. Carey. 1994. "Sex mortality differentials in the bean beetle: Reframing the questions," *American Naturalist* 144: 165–175.

Thatcher, A. R., V. Kannisto, and J. W. Vaupel. 1998. *The Force of Mortality at Ages 80 to 120*. Odense: Odense University Press.

Van Remmen, Holly, Walter Ward, Robert V. Sabia, and Arlan Richardson. 1995. "Gene expression and protein degradation," in Edward J. Masoro (ed.), *Handbook of Physiology: Section 11: Aging*. New York: Oxford University Press, pp. 171–234.

Vaupel, J. W. 1990. "Relative risks: Frailty models of life history data," *Theoretical Population Biology* 37: 220–234.

Vaupel, J. W., J. R. Carey, K. Christensen, T. E. Johnson, A. I. Yashin, N. V. Holm, I. A. Iachine, V. Kannisto, A. A. Khazaeli, P. Liedo, V. D. Longo, Y. Zeng, K. G. Manton, and J. W. Curtsinger. 1998. "Biodemographic trajectories of longevity," *Science* 280: 855–860.

Vaupel, J. W., K. G. Manton, and E. Stallard. 1979. "The impact of heterogeneity in individual frailty on the dynamics of mortality," *Demography* 16: 439–454.

Wilmoth, John R. and Shiro Horiuchi. 1999. "Do the oldest old grow old more slowly?," in B. Forette, C. Franceschi, J. M. Robine, and M. Allard (eds.), *The Paradoxes of Longevity*. Heidelberg: Springer-Verlag, pp. 35–60.

Yashin, A. I. and I. A. Iachine. 1997. "How frailty models can be used for evaluating longevity limits: Taking advantage of an interdisciplinary approach," *Demography* 34(1): 31–48.

Embodied Capital and the Evolutionary Economics of the Human Life Span

HILLARD KAPLAN

JANE LANCASTER

ARTHUR ROBSON

A fundamental question concerning aging is whether the life spans of organisms evolve and, if so, what forces govern their evolution. In this chapter we argue that life spans do evolve and present a general theory of life spans, with a particular focus on humans. We employ a qualitative definition of life span: the amount of time between birth and the age at which the likelihood of death becomes high, relative to the likelihoods at younger ages.[1] Most multicellular organisms exhibit a phase in which mortality decreases with age and then a second phase in which mortality increases with age.[2] Our definition focuses on this second phase, on the age at which death becomes imminent because of physiological deterioration or some environmental condition (such as winter). We chose this approach over the "maximum life span" concept, because it is more biologically meaningful. It focuses on the length of time that organismic function is adequate to sustain life. Since it is concerned with likelihoods rather than actual events, it assumes that many individuals do not live their full life span. For many organisms, including humans, this qualitative definition corresponds to a more precise quantitative definition: the modal age at death, conditional on reaching adulthood. The principal argument we develop is that life spans evolve as part of an integrated life-history program and that the program for development and reproduction is fundamentally related to the age of death.

Our first section outlines an evolutionary economic framework for understanding the effects of natural selection on life histories, previously referred to as "embodied capital theory." It combines the basic structure of life-history theory as developed in biology with the formal analytical approach developed in the analysis of capital in economics. We next discuss specialization and flexibility in life histories, with special emphases on the fast–slow continuum and on the relationship between brain evolution and life-history evolution. This is followed by a graphical presentation of analytical models of life-history evolution, based on embodied capital theory.

Our second section focuses on the special features of the human life course. This section briefly reviews a theory of human life-history evolution, developed and partially tested in earlier work. The theory posits that large brains and slow life histories result from a dietary specialization that has characterized the last 2 million years of human evolution. Empirical findings suggest that humans have a particular life course with characteristic schedules of growth, development, fertility, mortality, and aging. The approach here does not assume that those schedules are fixed and unresponsive to environmental variation. Rather it implies structured flexibility based upon the variation experienced in human evolutionary history and a set of specialized anatomical, physiological, and psychological adaptations to the niche humans occupied during that history. Together, those adaptations result in a life span for the species that can vary within a limited range.

We conclude with a discussion of two themes: short- and long-term flexibility in the human life span and the building blocks for a more adequate theory of senescence and life span.

Embodied capital and life-history theory

Fundamental tradeoffs in life-history theory

Life-history theory in biology grew out of the recognition that all organisms face two fundamental reproductive tradeoffs (see Charnov 1993; Lessells 1991; Roff 1992; Sibly 1991; Stearns 1992, for general reviews and Hill and Kaplan 1999, for a review of the application of life-history theory to humans). The first tradeoff is between current and future reproduction. The second is between quantity and quality of offspring. With respect to the first (the principal focus here), early reproduction is favored by natural selection, holding all else constant. This is the result of two factors. First, earlier reproduction tends to increase the length of the reproductive period. Second, shortening generation length by early reproduction usually increases the growth rate of the lineage.

The forces favoring early reproduction are balanced by benefits derived from investments in future reproduction. Those investments, often referred to as "somatic effort," include growth and maintenance. The allocation of energy to growth has three potential benefits. It can increase a) the length of the life span, by lowering size-dependent mortality, b) the efficiency of energy capture, thus allowing for a higher rate of offspring production, and c) the rate of success in intrasexual competition for mates. For this reason, organisms typically have a juvenile phase in which fertility is zero until they reach a size at which some allocation to reproduction increases fitness more than it increases growth. Similarly, for organisms that engage in repeated bouts of reproduction (humans included), some energy during the reproductive phase should be diverted from reproduction and allocated to maintenance so that organisms can live to reproduce again. Natural selection is expected to optimize the allocation of energy to current reproduction and to future reproduction (via investments in growth and maintenance) at each point in the life course so that genetic descendents are maximized (Gadgil and Bossert 1970; Sibly et al. 1985; see Hill and Hurtado 1996 for an application of those models to humans).

Specialization and flexibility in life histories and the fast–slow continuum

Variation across taxa and across conditions in optimal energy allocations and optimal life histories is shaped by ecological factors, such as food supply, disease, and predation rates. It is generally recognized that there are species-level specializations that result in bundles of life-history characteristics, which, in turn, can be arrayed on a fast–slow continuum (Promislow and Harvey 1990). For example, among mammals, species on the fast end exhibit short gestation times, early reproduction, small body size, large litters, and high mortality rates, with species on the slow end having opposite characteristics (ibid.). Similarly, among plants, some species that specialize in secondary growth are successful at rapidly colonizing newly available habitats, but their rapid life history means that they invest little in chemical defense and structural cells that would promote longevity. On the other end of the continuum are trees, such as the bristle cone pine, that are slow to mature but suffer very low mortality rates and are very long-lived (Finch 1998).

It is also recognized that many, if not most, organisms are capable of slowing or accelerating their life histories, depending upon environmental conditions such as temperature, rainfall, food availability, density of conspecifics, and mortality hazards. Within-species variation in life-history characteristics can operate over several time scales. For example, there is abundant evidence that allocations to reproduction, as measured by fecundity and fertility, vary over the short term among plants, birds, and humans in response to the balance between food supply and energy output (see, for example, Hurtado and Hill 1990; Lack 1968). The impacts of the environment may extend over longer time intervals through developmental effects. For example, calorie restriction in young rats tends to slow growth rates and leads to reduced adult stature, even when food becomes abundant in the later juvenile period (Shanley and Kirkwood 2000). Some intraspecific variation arises at even longer time scales, where this involves differential selection on genetic variants in different habitats. For example, rates of senescence vary across populations of grasshoppers, with those at higher altitudes and experiencing earlier winters senescing faster than those at lower altitudes (Tatar, Grey, and Carey 1997).

A central thesis of this chapter is that both specialization and flexibility are fundamental to understanding the human life span. On the one hand, the large human brain supports the ability to respond flexibly to environmental variation and to learn culturally, facilitating short-term flexibility. On the other hand, the commitment to a large brain and the long period of development necessary to make it fully functional constrains the human life course by requiring specializations for a slow life history.

Embodied capital and life-history theory

The embodied capital theory integrates life-history theory with capital investment theory in economics (Becker 1975; Mincer 1974) by treating the processes of growth, development, and maintenance as investments in stocks of somatic or embodied capital. In a physical sense, embodied capital is organized somatic tissue: muscles, digestive organs, brains, and so on. In a functional sense, embodied capital includes strength, speed, immune function, skill, knowledge, and other abilities. Since such stocks tend to depreciate with time, allocations to maintenance can also be seen as investments in embodied capital. Thus, the present–future reproductive tradeoff becomes a tradeoff between investments in own embodied capital and reproduction, and the quantity–quality tradeoff becomes a tradeoff between the embodied capital of offspring and their number.

The embodied capital theory allows us to treat problems that have not been addressed with standard life-history models. For example, physical growth is only one form of investment. The brain is another form of embodied capital, with special qualities. On the one hand, neural tissue monitors the organism's internal and external environment and induces physiological and behavioral responses to stimuli (Jerison 1973, 1976). On the other hand, the brain has the capacity to transform present experiences into future performance. This is particularly true of the cerebral cortex, which specializes in the storage, retrieval, and processing of experiences. The expansion of the cerebral cortex among higher primates represents an increased investment in this capacity (Armstrong and Falk 1982; Fleagle 1999; Parker and McKinney 1999). Among humans, the brain supports learning and knowledge acquisition during both the juvenile and adult periods, well after the brain has reached its adult mass. This growth in the stock of knowledge and functional abilities is another form of investment.

The action of natural selection on the neural tissue involved in learning, memory, and the processing of stored information depends on the costs and benefits realized over the organism's lifetime. There are substantial energetic costs of growing the brain early in life and of maintaining neural tissue throughout life. Among humans, for example, it has been estimated that about 65 percent of all resting energetic expenditure is used to support the maintenance and growth of the brain in the first year of life (Holliday 1978). Another potential cost of the brain may be decreased performance early in life. The ability to learn may entail reductions in "preprogrammed" behavioral routines, thereby decreasing early performance. The incompetence of human infants, even children, in many motor tasks is an example.

Taking these costs into account, the net benefits from the brain tissue involved in learning are only fully realized as the organism ages (see Figure 1). In a niche where there is little to learn, a large brain might have higher costs early in life and a relatively small influence on productivity late in life. Natural selection may then tend to favor the small brain. In a more challenging niche, however, although a small brain might be slightly better early in life, because of its lower cost, it would be much worse later, and the large brain might be favored instead.

The brain is not the only system that learns and becomes more functional through time. Another example is the immune system, which re-

FIGURE 1   Age-specific effects of brains on net production: Easy and difficult foraging niches

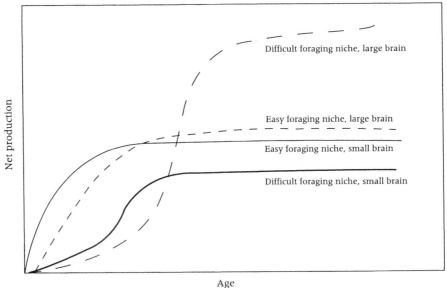

SOURCE: Kaplan et al. (2000).

quires exposure to antigens in order to become fully functional. Presumably, indeed, the maturation of the immune system is a primary factor in the decrease in mortality with age from birth until the end of the juvenile period.

Furthermore, a positive relationship between brain size and life span (controlling for body size) is found in empirical studies of mammals (Sacher 1959) and primates (Allman, McLaughlin, and Hakeem 1993; Hakeem et al. 1996; Judge and Carey 2000; Kaplan and Robson 2002). Such considerations led us to propose that brain size and longevity coevolve for the following reasons. Since the returns to a large brain lie in the future, ecological conditions favoring large brains also favor greater expenditure on survival. Conversely, exogenous ecological conditions that lower mortality favor increased expenditure on survival and hence also much greater investment in brain capital (Kaplan et al. 2000; Kaplan and Robson 2002; see Carey and Judge 2001 for an alternative coevolutionary model of human life spans).

This logic suggested an alternative approach to standard treatments of life-history theory. Standard treatments generally define two types of mortality: 1) extrinsic, which is imposed by the environment and is outside the control of the organisms (e.g., predation or weather), and 2) intrinsic, over which the organism can exert some control in the short run or which is subject to selective control over longer periods. In most models of growth and development, mortality is treated as extrinsic and therefore not subject to selection (Charnov 1993; Kozlowski and Wiegert 1986; for exceptions see Janson and Van Schaik 1993; Charnov 2001). Models of aging and senescence (Promislow 1991; Shanley and Kirkwood 2000) frequently treat aging as affecting intrinsic mortality, with extrinsic mortality, in turn, selecting for rates of aging. For example, in the Gompertz–Makeham mortality function where the mortality rate, μ, equals $A + Be^{\mu x}$ (with A, B, and μ being parameters and x referring to age), this entails treating the first term on the right-hand side of the equation, A, as the extrinsic component and the second term as the intrinsic component.

In our view, this distinction between types of mortality is unproductive and generates confusion. Organisms can exert control over virtually all causes of mortality in the short or long run. Susceptibility to predation can be affected by vigilance, choice of foraging zones, travel patterns, and anatomical adaptations, such as shells, cryptic coloration, and muscles that facilitate flight. Each of these behavioral and anatomical adaptations has energetic costs that reduce energy available for growth and reproduction. Similar observations can be made regarding endogenous responses to disease and temperature. The extrinsic mortality concept has been convenient, because it provided a reason for other life-history traits, such as age of first reproduction and rates of aging. However, this has prevented the examination of how mortality rates themselves evolve by natural selection.

Since all mortality is, to some extent, intrinsic or endogenous, a more useful approach is to examine the functional relationship between mortality and effort allocated to reducing it (see Figure 2). Exogenous variation can be thought of in terms of varying "assault types" and varying "assault rates" of mortality hazards. For example, warm, humid climates favor the evolution of disease organisms and therefore increase the assault rate and diversity of diseases affecting organisms living in those climates. Exogenous variation also may affect the functional relationship between mortality hazards and endogenous effort allocated to reducing them.

The recognition that all mortality is partially endogenous and therefore subject to selection complicates life-history theory because it requires multivariate models, but it also generates insights about evolutionary co-adaptation or coevolution among life-history traits. One of the benefits of modeling life-history evolution formally in terms of capital investments is that the analysis of such investments is well developed in economics with many well-established results. The next section summarizes some formal results of applying capital investment theory to life-history evolution.

FIGURE 2   Mortality as a function of investments

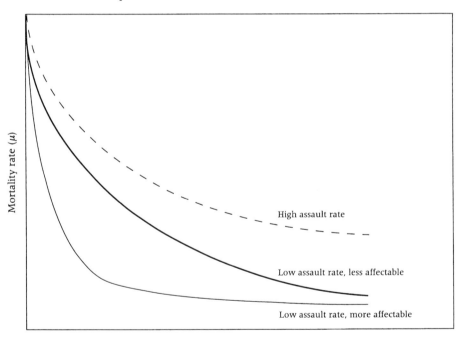

Investment in mortality reduction (*s*)

Capital investments and endogenous mortality

As a first step, it is useful to think of capital as the bundle of functional abilities of the soma. Organisms generally receive some energy from their parents, represented as an initial stock of capital, say K_0. Net energy acquired from the environment, F, at each point in time, t, is a positive function of the capital stock, with diminishing returns to capital. This energy can be used in three ways, which are endogenous and subject to selection. It can be reinvested in increasing the capital stock, that is, in growth. Define $v(t)$ as flow of investment at time t, so that dK/dt equals $v(t)$. Since growth and development take time, it is useful to impose a maximal investment rate, $\bar{v}$. Some energy, s, may also be allocated to reducing mortality, μ, for example via increased immune function, as illustrated in Figure 2. The probability of reaching any age, $p(t)$, is then a function of mortality rates at each earlier age, so that

$$p(t) = e^{-\int_0^t \mu(t)dt}.$$

Finally, energy can be used for reproduction, which is the net excess energy available after allocations to capital investments and mortality reduction, y; so $y(t) = F(K) - v(t) - s(t)$.

The dynamic optimization program is to find the largest solution r of

$$\int_0^\infty p(t)y(t)e^{-rt}dt = C_0,$$

where C_0 is the cost of producing a newborn. This equation is an economic extension of the continuous-time Euler–Lotka equation for the long-run growth rate in a species without parental investment after birth. Under most conditions (for example, for most of human evolutionary history), the average r must be close to zero. It can then be shown that an optimal life history would choose capital investment and mortality reduction so as to maximize total expected surplus energy over the life course. The results of the analysis have been presented and proven formally (Robson and Kaplan 2002). At each point in time, the marginal gain from investments in capital and the marginal gain from increased expenditure on survival must equal their marginal costs. During the capital investment period, where v is greater than zero, the value of life, J, which is equal to total expected future net energy, is increasing with age, since productivity is growing with increased capital. The optimal value of s also then increases. At some age, a steady state is reached where capital is at its optimum level and both capital and mortality rates remain constant.

Two important comparative results emerge from this analysis. An environmental change that increases the productivity of capital has two reinforcing effects: it increases the optimal level of capital investment (and hence the length of the investment period) and it decreases mortality through increases in s. A reduction in mortality rates has two similar effects: it increases the optimal capital stock and produces a reinforcing increase in s.

We note that the model does not result in senescence, as defined by increasing mortality rates with age. Even if capital were to depreciate over time (say, if $dK/dt = (1 - \lambda)K(t) + v(t)$, with λ being the proportional depreciation rate), a steady state still would be achieved where depreciation would be exactly offset by investment (Arrow and Kurz 1970: 85; Intriligator 1971). We address this issue in the final section.

Embodied capital and the evolution of human life histories

There has been a series of radiations within the primate order toward increased brain size, relative to body size, and toward increased longevity. These involve a transition from primitive prosimian primates to monkeys, then from monkeys to apes, and finally from apes to humans. For example, a human has a brain that is roughly three times as big as that of a chimpanzee and lives about twice as long. Can the theory illustrated above explain those radiations resulting in the long lives and large brains characteristic of the genus *Homo* and, particularly, of modern *Homo sapiens*? We posit that this extreme brain size and extreme longevity are coevolved responses to learning-intensive foraging strategies and a dietary shift toward high-quality, nutrient-dense, and difficult-to-acquire food resources. The following logic underlies our proposal. First, high levels of knowledge, skill coordination, and strength are required to exploit the suite of resources humans consume. The attainment of those abilities requires time and a significant commitment to development. This extended learning phase during which productivity is low is compensated for by higher productivity during the adult period, with an intergenerational flow of food from old to young. Since productivity increases with age, the time investment in skill acquisition and knowledge leads to selection for lowered mortality rates and greater longevity, because the returns on the investments in development occur at older ages.

Second, we believe that the feeding niche specializing in large, valuable food packages, particularly hunting, promotes cooperation between men and women and high levels of male parental investment, because it favors sexual specialization in somatic investments and thus generates a complementarity between male and female inputs. The economic and reproductive cooperation between men and women facilitates provisioning of juveniles, which both bankrolls their somatic investments and allows lower mortality during the juvenile and early adult periods. Cooperation between males and females also allows women to allocate more time to childcare, increasing both survival and reproductive rates. Finally, large packages also appear to promote interfamilial food sharing. Food sharing reduces the risk of food shortfalls due to the vagaries of foraging, providing insurance against illness and against variance in family size resulting from stochastic mortality and fertility. These buffers favor a longer juvenile period and higher investment in other mechanisms to increase life span.

Thus, we propose that the long human life span coevolved with the lengthening of the juvenile period, with increased brain capacities for information processing and storage, and with intergenerational resource flows— all as a result of a significant dietary shift. Humans are specialists in that they consume only the highest-quality plant and animal resources in their local environment and rely on creative, skill-intensive techniques to exploit them. Yet, the capacity to develop new techniques for extractive foraging and hunting allows them to exploit a wide variety of foods and to colonize all of the Earth's terrestrial and coastal ecosystems. In the following sections we review the specialized adaptations associated with this life history.

Digestion and diet

There is mounting evidence from various sources, including digestive anatomy, digestive biochemistry, bone isotope ratios, archeology, and observations of hunter-gatherers, that humans are specialized toward the consumption of calorie-dense, low-fiber foods that are rich in protein and fat. Contrary to early generalizations based on incomplete analysis and limited evidence (Lee 1979; Lee and DeVore 1968), more than half of the calories in hunter-gatherer diets are derived, on average, from meat. There are ten foraging societies and five chimpanzee communities for which caloric production or time spent feeding has been monitored systematically (Kaplan et al. 2000). All modern foragers differ considerably in diet from chimpanzees. Measured in calories, the major component of forager diets is vertebrate meat. Meat accounts for between 30 percent and 80 percent of the diet in the sampled societies, with most diets being more than 50 percent vertebrate meat, whereas chimpanzees obtain about 2 percent of their food energy from hunted foods. Similarly, using all 229 hunter-gatherer societies described in the Ethnographic Atlas (Murdock 1967) and Murdock's estimates based upon qualitative ethnographies, Cordain et al. (2000) found median dependence on animal foods in the range of 66 to 75 percent.

The next most important food category in the ten-society sample is extracted resources, such as most invertebrate animal products, roots, nuts, seeds and difficult-to-extract plant parts such as palm fiber or growing shoots. These are mostly nonmobile resources embedded in a protective context such as underground, in hard shells, or bearing toxins that must be removed before they can be consumed. In the ten-forager sample, extracted foods accounted for about 32 percent of the diet, as opposed to 3 percent among chimpanzees.

In contrast to hunted and extracted resources, which are difficult to acquire, collected resources form the bulk of the chimpanzee diet. Collected resources, such as fruits, leaves, flowers, and other easily accessible plant parts, are simply gathered and consumed. They account for 95 percent of the chimpanzee diet, on average, but only 8 percent of the human forager diet. The data suggest that humans specialize in rare but nutrient-dense resource packages or patches (meat, roots, nuts) whereas chimpanzees specialize in ripe fruit and plant parts with low nutrient density.

Comparative data on digestive anatomy confirm that these contemporary differences reflect long-term adaptations. Gorillas, chimpanzees, and humans can be arrayed along a continuum in terms of their digestive anatomy (Schoeninger et al. 2001). The gorilla has a very long large intestine and caecum in order to use bacterial fermentation for the breakdown of plant cellulose in leaves and other structural plant parts as a source of dietary protein. Although gorillas eat significant quantities of fruit, they derive a large proportion of their calories and most of their protein from leaves and other nonreproductive plant parts. Chimpanzee caeca are somewhat smaller. Chimpanzees supplement leaf consumption with hunted foods, insects, and nuts for fat and protein.

Human digestive anatomy is specialized for a very different diet. Humans have very small large intestines and are incapable of digesting cellulose in large quantities as a source of protein, and they have very long small intestines for the digestion of lipids (ibid.). Moreover, humans are very inefficient at chain elongating and desaturating various carbon fatty acids to produce the fatty acids that are essential cellular lipids (Emken et al. 1992; cited in Cordain et al. 2002, upon which this discussion is based). Since humans share this trait with other obligate carnivores and since those essential fatty acids are found only in animal foods, it appears that human digestion is specialized toward meat consumption and low-fiber diets. If chimpanzees consumed as much meat as humans, the nitrogen would destroy their foregut bacteria; and if they consumed a diet as low in fiber, they would suffer from colonic twisting (Schoeninger et al. 2001). Humans, on the other hand, must reduce dietary fiber. When they acquire foods that are high in fiber, such as roots and palm fiber, they remove the fiber before ingestion (ibid.).

Although the data are still scarce, it appears that this dietary shift occurred at the origin of the genus *Homo* about 2 million years ago. Compared to chimpanzees and australopithecines, early *Homo* appears to have had a reduced gut (Aiello and Wheeler 1995); and radio-isotope data from fossils also suggest a transition from a plant-based diet to greater reliance on meat (see Schoeninger et al. 2001 for a review). There is significant archeological evidence of meat eating by *Homo* in the early Pleistocene (Bunn 2001). Finally, radio-isotope evidence from Neanderthal specimens (Richards et al. 2000) and from anatomically modern humans in Europe (Richards and Hedges 2000) during the late Pleistocene shows levels of meat eating that are indistinguishable from carnivores. It is interesting that this dietary transition occurs at about the same time as the hominid brain expanded beyond the size of the ape's brain (Aiello and Wheeler 1995).

The brain and cognitive development

Although it has long been recognized that intelligence is the most distinctive human trait, it is now becoming increasingly clear that our larger brains and greater intellectual capacities depend upon the stretching out of development at every stage. The production of cortical neurons in mammals is limited to early fetal development, and, compared to monkeys and apes, human embryos spend an additional 25 days in this phase (Deacon 1997; Parker and McKinney 1999). The greater original proliferation of neurons in early fetal development has cascading effects in greatly extending other phases of brain development, ultimately resulting in a larger, more complex, and more effective brain. For example, in monkeys, such as macaques, myelination of the brain begins prenatally and is largely complete in 3.5 years, whereas in humans this process continues for at least 12 years (Gibson 1986). Dendritic development is similarly extended to age 20 or later in humans.

The timing of cognitive development is extended in chimpanzees relative to monkeys, and in humans relative to apes (see Parker and McKinney 1999, upon which this discussion is based and references therein for reviews of comparative cognitive development in monkeys, apes, and humans). In terms of Piagetian stages, macaque monkeys traverse only two subperiods of cognitive development regarding physical phenomena by 6 months of age and peak in their logical abilities at around 3 years of age; however, they can never represent objects symbolically, classify objects hierarchically, or recognize themselves in a mirror. Chimpanzees traverse three to four subperiods of cognitive development by about 8 years of age.[3] They can recognize themselves in a mirror and are much better skilled at classification than macaques, but can never construct reversible hierarchical classes or engage in abstract, logical reasoning. Human children traverse eight subperiods of cognitive development over the first 18 to 20 years.

Although humans take about 2.5 times as long to complete cognitive development as do chimpanzees, humans actually learn faster. In most cognitive spheres, especially language, a 2-year-old child has the abilities of a 4-year-old chimpanzee. Humans have much more to learn and their brains require more environmental input to complete development. Formal abstract logical reasoning does not emerge until ages 16 to 18. This is the time when productivity begins to increase dramatically among modern hunter-gatherers. The ability to construct abstract scenarios and deduce logical relationships appears to induce a growth in knowledge that results in peak productivity in the mid-30s.

Elongated development in humans appears to be associated with slowed aging of the brain. Macaques exhibit physiological signs of cognitive impairment, as evidenced by Alzheimer-like neuropathology and cerebral atrophy by ages 22 to 25, and chimpanzees exhibit this by age 30. This contrasts with humans, for whom such changes are rare until age 60 (<1 percent) and only common (>30 percent) in their 80s (see Finch 2002; Finch and Sapolsky 1999 for reviews; however, the evidence on chimpanzees is mixed).

Physical growth

Physical growth in humans differs from that in chimpanzees and gorillas. It is first faster, then slower, and finally faster. Human neonates weigh about 3,000 grams (Kuzawa 1998), whereas mean birthweights for gorillas and chimpanzees are 2,327 and 1,766 grams (Leigh and Shea 1996). The differences are due not only to the longer gestation length among humans, but also to weight gain per day (ibid.). The comparison with gorillas is especially pronounced since adult female gorillas weigh about 60 percent more than average women among contemporary hunter-gatherers. Body composition also differs, with human neonates being much fatter (3.75 times more fat than that found in mammals of comparable weight), suggesting an even greater difference in the calories stored in human newborns (Kuzawa 1998).

It seems likely that these differences in neonatal body size are associated with the differences in brain growth rates. A human brain is twice as big at birth as a chimpanzee neonate's; indeed, it is about the same size as a chimpanzee adult's, despite the 15-fold greater total body weight of the latter. The bigger body and greater stores of energy in the form of fat are probably necessary to support the human brain and its rapid postnatal growth.

Following infancy and early childhood, humans grow absolutely more slowly and proportionally much more slowly than do chimpanzees. Growth is almost arrested for human children during middle and late childhood (see Figure 3). By age 10, chimpanzees have caught up with and surpassed human children in body size. Only with the adolescent growth spurt do humans achieve their final larger body size. Children in the foraging societies for which data are available do not acquire enough calories to feed themselves until they have completed growth. Growth is supported through within- and between-family food sharing (Kaplan et al. in press). Middle childhood is generally a time of low appetite, which is then followed by the voracious appetite associated with adolescence.

We propose that human infants grow quickly until both their body and gut are of sufficient size that they can comfortably support the large brain.

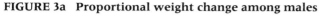

FIGURE 3a Proportional weight change among males

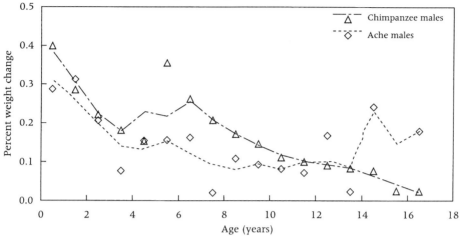

FIGURE 3b Proportional weight change among females

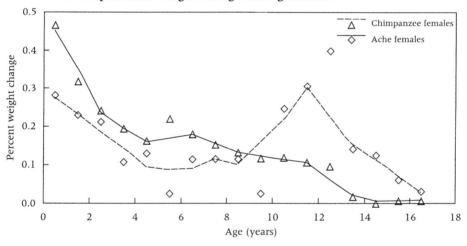

SOURCE: Ache data collected by Hill, Hurtado, and Kaplan. Chimpanzee data collected by Pusey and Williams, personal communication.

Growth rates are then slow, because children do not need large bodies since they do very little work. Instead they learn through observation and through play. When their brains are almost ready for large bodies, growth rates increase rapidly and adult body size is achieved relatively quickly.

Production, reproduction, and energy flows

Figure 4 compares age profiles of production for humans and chimpanzees. The chimpanzee net production curve shows three distinct phases. The first phase, lasting to about age 5, is the period of complete and then partial dependence upon mother's milk. Net production during this phase is nega-

FIGURE 4 **Net food production and survival: Human foragers and chimpanzees**

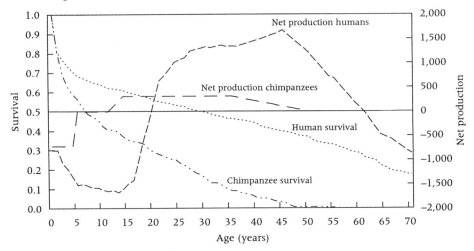

SOURCE: Adapted from Kaplan and Robson (2002).

tive. The second phase, during which net production is zero, is one of independent juvenile growth, lasting until adulthood, about age 13 for females. The third phase is reproductive, during which females, but not males, produce a surplus of calories used for nursing.

Humans, in contrast, produce less than they consume for close to 20 years. Net production is negative and falling until about age 14, with the growth in consumption due to increased body size outstripping the growth in production, and only then begins to climb. Net production in adulthood among humans is much higher than among chimpanzees and peaks at a much older age. Peak net production among humans reflects the payoffs to the long dependency period. It is about 1,750 calories per day, but it is not reached until about age 45. Among chimpanzee females, peak net production is only about 250 calories per day, and since fertility decreases with age, net productivity probably decreases during adulthood.

This great increase in net production among humans during adulthood is a consequence of the difficulty of acquiring foods, as shown by the age profiles of production for collected, extracted, and hunted resources. In most environments, fruits are the easiest resources that people collect. Daily production data among Ache foragers show that both males and females reach their peak daily fruit production by their mid to late teens. Some fruits that are simply picked from the ground are collected by 2- to 3-year-olds at 30 percent of the adult maximum rate. Ache children acquire five times as many calories per day during the fruit season as during other seasons of the year (Kaplan 1997). Similarly, among the Hadza, teenage girls acquired 1,650 calories per day during the wet season when fruits were available but only 610 calories per day during the dry season when they were not. If we weight the wet and dry season data equally, Hadza teenage girls acquire 53 percent of their calories from fruits, compared to 37 percent and 19 percent for reproductive-aged and post-reproductive women (Hawkes, O'Connell, and Blurton Jones 1989).

In contrast, the acquisition rate of extracted resources often increases through early adulthood as foragers acquire the necessary skills. Data on Hiwi women show that root acquisition rates do not peak until about ages 35 to 45 (Kaplan et al. 2000), and the rate for 10-year-old girls is only 15 percent of the adult maximum. Hadza women appear to attain maximum root digging rates by early adulthood (Hawkes, O'Connell, and Blurton Jones 1989). Hiwi honey extraction rates by males peak at about age 25. Again the extraction rate of 10-year-olds is less than 10 percent of the adult maximum. Experiments among Ache women and girls clearly show that young adult girls are not capable of extracting palm products at the rate attained by older women (Kaplan et al. 2000). Ache women do not reach peak return rates until their early 20s. !Kung (Ju/'hoansi) children crack mongongo nuts at a much slower rate than adults (Blurton Jones, Hawkes, and Draper 1994b), and nut cracking rates among the neighboring Hambukushu do not peak until about age 35 (Bock 1995). Finally, even chimpanzee juveniles focus on more easily acquired resources than adult chimpanzees. Difficult-to-extract resources such as termites and ants and activities such as nut cracking are practiced less by chimpanzee juveniles than adults (Boesch and Boesch 1999; Hiraiwa-Hasegawa 1990; Silk 1978).

The skill-intensive nature of human hunting and the long learning process involved are demonstrated by data on hunting return rates by age (see Kaplan et al. 2001 for details on why hunting is so cognitively demanding). Hunting return rates among the Hiwi do not peak until ages 30 to 35, with the acquisition rate of 10-year-old and 20-year-old boys reaching only 16 percent and 50 percent of the adult maximum. The hourly return rate for Ache men peaks in the mid-30s. The return rate of 10-year-old boys is a mere 1 percent of the adult maximum, and the return rate of 20-year-old juvenile males is still only 25 percent of the adult maximum. Marlowe (unpublished data, personal communication) obtains similar results for the Hadza. Also, boys switch from easier tasks, such as fruit collection, shallow tuber extraction, and baobab processing, to honey extraction and hunting in their mid to late teens among the Hadza, Ache, and Hiwi (Blurton Jones, Hawkes, and O'Connell 1989, 1997; Kaplan et al. 2000). Even among chimpanzees, hunting is strictly an adult or near-adult activity (Boesch and Boesch 1999; Stanford 1998; Teleki 1973).

A complex web of intrafamilial and interfamilial food flows and other services supports this age profile of energy production. First, there is the sexual division of labor. Men and women specialize in different kinds of skill acquisition and then share the fruits of their labor. The specialization generates two forms of complementarity. Hunted foods acquired by men complement gathered foods acquired by women, because protein, fat, and carbohydrates complement one another with respect to their nutritional functions (Hill 1988). The fact that male specialization in hunting produces high delivery rates of large, shareable packages of food leads to another form of complementarity. The meat inputs of men shift the optimal mix of activities for women, increasing time spent in childcare and decreasing time spent in food acquisition (Hurtado et al. 1992). They also shift women's time to foraging and productive activities that are compatible with childcare and away from dangerous ones. In the ten-group sample, men, on average, acquired about twice as many calories and seven times as much protein as women (68 percent vs. 32 percent of the calories and 88 percent vs. 12 percent of the protein) (Kaplan et al. 2000). We estimate that on average 31 percent, 39 percent, and 30 percent of those calories support adult female, adult male, and offspring consumption (Kaplan et al. 2001). This implies that after taking into account own consumption, women supply only 3 percent of the calories to offspring and men provide the remaining 97 percent. Men supply not only all of the protein and fat to offspring, but also the bulk of the protein and fat consumed by women. This contrasts sharply with most (>97 percent) mammalian species, among which the female supports all of the energetic needs of an offspring until it begins eating solid foods (Clutton-Brock 1991) and males provide little or no investment. The high productivity of men has probably allowed for the evolution of physiological adaptations among women, such as fat storage at puberty and again during pregnancy—adaptations not found in apes.

Mortality

Figure 5 shows the mortality rates of chimpanzees (synthesized from five chimpanzee sites; Hill et al. 2001) and two foraging groups (Ache: Hill and Hurtado 1996: Table 6.1; Hadza: Blurton Jones et al. 2002: Table 2). Although there are differences among chimpanzee and human foraging populations, the contrast between the two species is clear. Before the age of 5, mortality rates are not very different between human foragers and chimpanzees. The average mortality rate during the adult period differs greatly between the two species, as does the age at which mortality rates rise steeply. Mortality within the two foraging groups remains quite low from adolescence to ages 35 to 40 and rises abruptly after about age 65. Among chimpanzees, mortality rates begin to rise quickly after their lowest point prior to reproduction.

The low adult mortality rates among humans are necessary components of delayed production. Only about 30 percent of chimpanzees ever born reach 20, the age when humans produce as much as they consume, and less than 5 percent of chimpanzees reach 45, when human net production peaks. The relationship between survival rates and age profiles of production is made

FIGURE 5 Age-specific mortality rates: Chimpanzees and human foragers

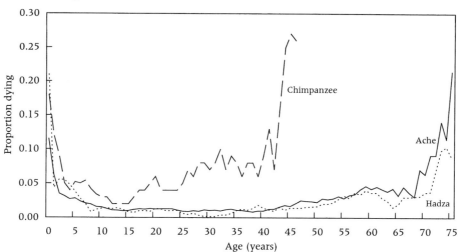

even clearer in Figure 6 (adapted from Kaplan and Robson 2002). This plots net expected *cumulative* productivity by age, multiplying the probability of being alive at each age times the net productivity at that age and then summing over all ages up to the present. The unbroken and dotted lines show *cumulative* productivity by age for chimpanzees and humans, respectively. The longer human training period is evident when the troughs in the human and chimpanzee curves are compared. The dashed line is a hypothetical cross of

FIGURE 6 Cumulative expected net caloric production by age: Humans and chimpanzees

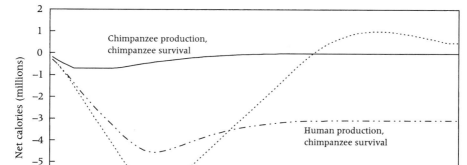

SOURCE: Adapted from Kaplan and Robson (2002).

human production profiles with chimpanzee survival rates. It shows that the human production profile would not be viable with chimpanzee survival rates, because expected lifetime net production would be negative.

Finally, although the mortality data for the Ache and the Hadza are based on small sample sizes and thus subject to considerable sample error, there is some evidence that proportional rates of change in mortality are not constant during the adult period. Figure 7 shows proportional rates of change in mortality. Mean yearly mortality rates were determined for the following age classes (10–14, 15–19, 20–29, 30–39, 40–49, 50–59, 60–69, 70–79) and then transformed into natural logarithms. Differences in those values between an age class and the preceding one were divided by the number of years in the age class to determine an average yearly rate of increase or decrease. For example, the first set of bars compares mortality rates of 10–14-year-olds with those of 15–19-year-olds and shows that rates decreased by 1 percent per year and 6 percent per year among the Ache and Hadza, respectively. For the Ache, the data show virtually no change in mortality rates from age 10 to age 39. From age 40 to age 69, rates of increase vary between 2 and 6 percent per year, with the lowest rate of increase occurring during the 60s. However, mortality rates show a steep rise to about 14 percent per year after age 70. For the Hadza, mortality rates actually decrease during the teens and 20s and stay essentially the same during the 30s. From age 40 to age 69, rates of increase vary between 0 and 9 percent per year, also with the lowest rate of increase during the 60s. Again, mortality rates show a steep rise to about 12 percent per year after age 70.

FIGURE 7   Yearly proportional change in mortality rates

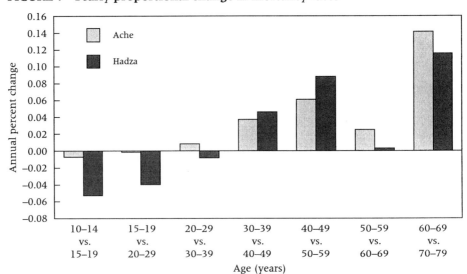

FIGURE 8 Adult mortality density

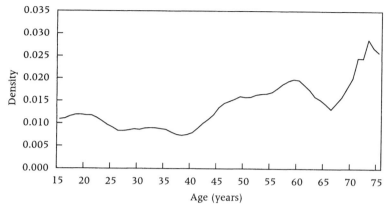

The relative constancy of mortality rates during prime adulthood is striking in both the Ache and the Hadza, as is the steep increase during the 70s. This suggests that "rectangularization" of the mortality profile among humans (Fries 1980, 1989) predates modernization. It may be a fundamental feature of our species, reflecting both strict genetic quality control before implantation of the fetus (as discussed in detail by Ellison 2001) and increased effort at slowing the rate of senescence during adulthood. Figure 8 shows the adult mortality density function, conditional on surviving to age 15, for the Ache and Hadza (averaging the two groups and then smoothing with a five-year running average). Although samples are small at older ages, there is an apparent mode at 73 years of age. The pattern also might suggest that a good candidate for the natural human life span is about 65 to 75 years, defined in terms of the imminence of death. This corresponds well with our impressions regarding physical deterioration. While some individuals show marked decline in their late 50s to early 60s, others remain vigorous to age 70 or so. After this physical decline, death often follows within a few years. Unfortunately, however, very little is known about the timing and population distribution of physical decline among hunter-gatherers or other traditional subsistence-level peoples.

The life span

The human life course and human life span

Our fundamental thesis in this chapter is that the human life course is an integrated adaptation to a specialized niche. Digestive physiology and anatomy; nutritional biochemistry; brain growth and cognitive development; tempo of body mass increases and appetite; age profiles of productivity, reproduction, parental investment, and mortality; and, ultimately, the life span are coadapted to a learning-intensive feeding niche, giving humans access to the most nutrient-dense and highest-quality food resources. The joint examination of these domains suggests a highly structured life course, in which six distinct stages can be recognized.

Brain growth occurs from the early fetal stage to about 5 years of age, with 90 percent occurring by age 3.0 to 3.5. Human mothers and their babies maintain large fat reserves to support this brain growth (Ellison 2001; Kuzawa 1998). While cognitive development unfolds over many more years, a great deal of linguistic competence, especially comprehension, is achieved during this first stage of life. Thus, it would seem that the human specialization evident in the first stage of life is building the "physical plant" (i.e., the brain) and the knowledge acquisition pathway (i.e., language ability) to support a long period of learning.

The second stage, childhood, is characterized by very slow physical growth, a large energetic allocation to building the immune system (McDade and Worthman 1999; Worthman 1999a), several important phases of cognitive development facilitated by play and other forms of practice, very low productivity, and very low mortality. Parents insist that children remain in safe places and encourage them to produce food only when it is easily and safely acquired (Blurton Jones, Hawkes, and Draper 1994a). The unique feature of human childhood is that it is fully supported by familial energy inputs, reducing exposure to mortality hazards and allowing time for learning. Faster physical growth would only make children more expensive before their brains were ready for food production.

Adolescence follows, during which physical growth is accomplished rapidly, the reproductive system matures, and the final phases of cognitive development occur. It is during this phase that the brain and the rest of the body become ready for adult productivity. While productivity increases during adolescence, it is also largely supported by familial food inputs.

For males, "on-the-job training" characterizes the period from early adulthood to prime adulthood in the mid to late 30s. Both physical strength and information-processing speed (fluid intelligence) peak in early adulthood, but knowledge-based abilities (crystallized intelligence) continue to rise (Horn 1968). As a result, productivity increases many-fold during this period. Mortality rates remain virtually constant and low. Resource production at this stage in the life course of women is very different. It is characterized by a reduction in productivity in the interests of fertility and parental investment. Women face a tradeoff between resource acquisition and childcare because the burden of children reduces foraging efficiency and because foraging often exposes children to environmental hazards. As a result, women reduce their work or shift their activities toward food processing and other, less dangerous, efforts.

Middle age is a period of simultaneous parenthood and grandparent-hood. Dependency loads on parents peak around age 40, just before grand-parenthood begins. Although productivity is also at a peak level, net pro-ductivity including dependency loads is negative, supported by resource transfers from other families. Through middle age, dependency loads di-minish, as does productivity.

Old age commences around age 60, and during this seventh decade of life physical deterioration proceeds rapidly and brain aging becomes evident, followed by a sharp increase in mortality rates. Parenting is finished and work effort decreases along with productivity. This is not to say that there are no positive contributions to fitness during this phase. Older adults attempt to be as productive as possible, reallocating their time to skill-intensive but less en-ergy-intensive activities (e.g., craft production and childcare; Gurven and Kaplan 2001). They may also affect the productivity of the younger popula-tion through their knowledge of the habitat and through their political skills.

If this set of stages characterized the average life course of the last 100,000 to 200,000 years, it is likely that the processes of maintenance and repair from the intracellular level to the whole-organism level were shaped by selection to achieve a life span including all six stages. One possibility is that the sixth stage is actually an artifact of selection intended to maintain the body and mind in good condition through the first five stages of life. The sixth stage would then be the result of the impossibility of nature's designing a body that collapses the moment high functionality is no longer needed. Another possibility is that this sixth stage has been positively se-lected because of the fitness benefits produced during it.

Some (e.g., Hawkes et al. 1998; Williams 1957) have hypothesized that the benefits provided by grandmothers to grandchildren have selected for menopause and the long human life span. The data on child dependency burdens suggest an alternative hypothesis: the life span is the result of se-lection on adults, allowing them to support all of their children through adolescence, and this requires high productivity to about age 60 (Lancaster and King 1985; see Peccei 2001 for a recent statement and review). The post-reproductive period might then not begin until children are fully reared. Grandchildren are only half as related to grandparents as children are to parents. Therefore, at the margin, the benefits to grandchildren would have to be twice as great as those to children to favor the same investment in continued life.

This hypothesis of course depends upon menopause, which must be explained as well. It is still unclear whether menopause is the result of tradeoffs between egg quality and reproductive life span, tradeoffs between supporting multiple dependent young and production of additional offspring, tradeoffs between grandparental investment and reproduction, or something else (see Kaplan 1997 for a discussion).

Flexibility and variation in the human life span

The human brain is the physical medium through which culture is maintained and transmitted. As such, it is generally thought to have greatly expanded the behavioral flexibility of our species relative to other animals. However, the commitment to building and programming the brain requires a highly structured life history that places constraints on the timing of life events. Our species is committed to long-term neural and cognitive capital accrual and to a long life span. The characteristic life history of our ancestors has shaped age profiles of growth, tissue repair, and physical decline.

Nevertheless, human life histories show evidence of systematic variation in response to environmental variation. Those effects appear to be the result of the interaction between changes in environmental conditions and human physiology and behavior. Perhaps the most striking example of that interaction is the pattern of changes accompanying the secular trend toward modernization. Increased nutrition and decreased work and disease loads have systematic effects on human developmental physiology. Physical growth rates have increased and maturation now begins earlier, resulting in greater stature, higher body weight, and earlier age of menarche in girls (Eveleth 1986; Lancaster 1986; Worthman 1999b). This response is very likely the result of adaptive flexibility in growth and maturation in the face of variation in food supply and in disease assault rates that was experienced during human evolutionary history.

In contrast to this increase in the rate of physical development, aging may be slowed in response to better nutrition and decreased work and disease loads. Although it is possible that humans would also show slowed aging in response to radical reductions in caloric intake (as do rats in feeding experiments—see Shanley and Kirkwood 2000 for a review), it is also possible that within the usual range of variation, rates of aging are slowed and life spans are lengthened when nutrition is better and disease loads are lower (Fogel and Costa 1997). This outcome would also be adaptive. The changing mortality rates among older people accompanying modernization and the fact that some chronic diseases occurred at earlier ages in the nineteenth-century United States than today (Costa 2000) are consistent with this possibility.

On the other hand, increased risk of heart disease, diabetes, and cancer from overweight and lack of exercise may also be the result of evolved responses. Given the common activity regimes in our past and the variability in food supply, human appetites and nutritional biochemistry may be designed to store fat and increase blood lipid levels when food is abundant. Those adaptations might reduce the life span in the context of modern patterns of activity levels and food access and consumption.

In addition to these physiological adaptations, there are also behavioral responses to modernization. The models outlined in our first section above are as applicable to short-term behavioral variation as they are to long-term life-history adaptations. Two noteworthy effects of modernization are increased economic payoffs to educational capital and decreased mortality owing to improvements in public health. The present models predict reinforcing endogenous behavioral responses to such changes. Increased payoffs to education should promote increased investment both in educational capital and in staying alive. Improvements in public health should also promote reinforcing increases in capital investment and staying alive. It is an open question whether we are reaching the upper limit of our flexibility in the life span. It appears that with respect to stature and perhaps age of menarche, we have reached the limit. There appears to be more scope for variation in the life span, given investments in medical technology designed to reduce disease and the effects of aging. In any case, knowledge about the human genome is likely to lead to manipulations of genes and gene products, resulting in life span increases of very large magnitude.

Building blocks for an adequate theory of senescence and the life span

Given the definition of life span guiding this discussion (the span from birth to the age when death is imminent as a result of physical deterioration), an adequate theory of life span will require a theory of senescence. Each of the principal theories suffers from important weaknesses, rendering the theory incomplete. This section first reviews those weaknesses and then offers a framework intended to remedy them.

Senescence is generally defined as an increasing mortality rate with age. The theories proposed by Medawar (1952), Williams (1957), and Hamilton (1966) share the common premise that senescence is the manifestation of the decreasing force of selection with age, resulting from extrinsically imposed hazards of mortality. In essence, their argument is that sources of mortality that strike when the individual is old are less stringently selected against than those striking when young, simply because an individual has a greater probability of being young than old.

According to Williams (1957), antagonistic pleiotropy accounts for the increase in mortality with age. Senescence is due to the presence of genes with opposing pleiotropic effects at different ages, increasing survival at younger ages but decreasing survival at older ages or increasing fertility at younger ages at the expense of reduced survival at older ages. Since extrinsically imposed mortality decreases the force of selection with age, genes with such positive effects at younger ages and negative effects at older ages will accumulate through selection, producing increasing mortality rates with age. He speculated that selection on mortality at each age would vary with reproductive value (i.e., expected future reproduction conditional on being alive at that age). Hamilton (1966) formalized this argument and showed that sensitivities of fitness to changes in mortality rates decrease with the age of action. However, his results suggested that reproductive value is not the critical determinant, because the sensitivity of fitness to changes in mortality rates depends both on the probability of reaching that age and on expected future reproduction at that age.

According to the formal model discussed above, the fact that mortality discounts the future does not, in itself, select for increasing mortality rates with age. Our approach allows both capital investments and expenditure on mortality reduction to be subject to natural selection. If aging, in the sense of explicit time dependence, is not built-in, then positive but constant mortality rates and a constant capital stock are optimal. This is so because our model assumes that current mortality can be reduced by current energy expenditure. From an initial viewpoint, reducing future mortality is indeed less important than reducing present mortality, just because survival is not certain. However, future expenditures on mortality reduction should be discounted for the same reason, and in the long run these two effects are precisely offsetting. If mortality is avoided during a time interval, the organism faces the same tradeoff between reproduction and survival as earlier.

This result does not depend on the assumption that current mortality is reduced by current expenditures. An alternative model might assume the existence of a stock of somatic capital for determining mortality, as seems a plausible explicit interpretation of Kirkwood's (1990) "disposable soma theory." However, it can be shown that the optimal life history may again entail a steady state in which this stock, and hence the mortality rate, are constant. From a current viewpoint, the tradeoff between the present and the future, as reflected in the decision about the size of the stock, is stationary. Even when decisions are made at the beginning of life, both the costs and benefits of future mortality control are deflated in the same way. Finally, even if somatic capital depreciates over time, results from capital investment theory show that the stock will be maintained at an optimal level (Arrow and Kurz 1970; Intriligator 1971). Optimal investment in the capital stock will continue to precisely offset depreciation. As long as production is constant with a constant capital stock, optimal mortality rates remain constant as well.[4]

What then causes senescence? Suppose there are cost functions for building embodied capital and for repairing and maintaining embodied capital. Imagine that the embodied capital stock is described in terms of two state variables, the quantity of the stock and its efficiency. From its inception, the organism builds somatic capital, adding to its quantity until the optimal quantity is reached. However, because of its own metabolic activity and assaults from outside agents, the efficiency of the capital stock is subject to decay. For example, free radicals and other harmful molecules may accumulate in cells, accidents may cause tissue damage, and pathogens may disrupt physiological function and damage cells. In the same sense that the costs of producing new car seats may be different from the costs of repairing tears in those seats, the costs of producing new cells and adding to the quantity of embodied capital may be different from the costs of repairing damage to them. Thus, during development, the optimal life-history program will have to equalize marginal fitness returns from three different investments: adding to new capital, repairing existing capital, and reducing current mortality. When the capital stock reaches the level at which some allocation to reproduction is also optimal, marginal returns from investments in producing new capital in the form of descendants must also be equalized with the marginal returns from the other three investments.

It is possible that for many capital stocks, quantity increases with age while efficiency decreases because, as suggested by Kirkwood (1990), optimal repair is not complete. Such a process appears to characterize cognitive development and aging. For example, verbal knowledge tends to increase, at least through middle age, but information-processing speed and memory peak during the third decade and decline thereafter (Schaie 1996). During the first phase of life, those increases in quantity have larger effects on the value of life (i.e., the expected future contribution to fitness) than do the decreases in efficiency. Thus productivity increases and optimal mortality decreases. However, at some age, the effects of disrepair overwhelm the effects of growth and learning, and overall productivity and the ability to fend off mortality hazards are lessened. This would lead to increasing optimal mortality with age.

This framework, with four kinds of investment, may provide a more adequate basis for a general theory of life-history evolution, and of life span in particular. Without making the unrealistic assumption that any life-history component, such as mortality, is extrinsic, selection can act to optimize and coadapt each of those components. Given the results discussed above, it is plausible that exogenous factors increasing the productivity of capital should select not only for increased capital investments and reduced mortality rates, but also for higher optimal allocations to repair, because of the higher probability of reaching old age. This would then lead to a longer life span. Similarly, exogenous factors that affect mortality rates may select not only for greater capital investment and reinforcing increases in optimal allocations to reducing mortality rates, but also for slower senescence. Given niche differentiation in the payoffs to body size and learning, in the payoffs to investment in offspring, in the payoffs to reducing mortality rates, and possibly in the payoffs to repair, the wide array of life histories found in nature is not surprising.

Whether such a framework will be useful awaits the formal development of models designed to analyze the effects of selection on those investment functions. It does, however, direct empirical attention to measuring the shapes of those functions. What are the costs and benefits of reducing the quantities of harmful molecules in cells, of DNA repair, and of differing numbers and kinds of immune cells? How do those costs and benefits compare to those associated with increases in muscle mass, brain mass, and learning? And, how do each of those functional relationships compare with those characterizing the production of offspring of different sizes and different functional abilities? Regardless of the ultimate productivity of this evolutionary economic life-history framework, understanding those relationships is likely to be illuminating.

Notes

This chapter was written with support from the National Institute on Aging, grant number AG15906. The authors acknowledge the contributions of Kim Hill to the data sets and their analyses on the comparative diets and demography of chimpanzees and foragers published previously (Kaplan et al. 2000). We also thank Kim Hill and Magdalena Hurtado for their data on resource acquisition by age and sex among the Hiwi and Ache (Kaplan et al. 2000). These data sets and analyses form a critical base for the second part of this chapter.

1 Of course, a quantitative definition would require specification of that ratio.

2 In some cases, there may be some additional mortality increases and drops due to phase transitions such as weaning.

3 The fourth subperiod, covering abilities such as recognition of conservation of quantities of liquids under container transformations, seems to require tutelage and symbolic training.

4 Medawar's (1952) model, in which deleterious mutations with age-specific effects late in life accumulate owing to the weaker force of selection at older ages, does not suffer from this problem. However, it can only account for specific diseases with late onset, not for the general and progressive deterioration of the soma with age, nor the physiological processes underlying aging.

References

Aiello, L. and P. Wheeler. 1995. "The expensive-tissue hypothesis: The brain and the digestive system in human and primate evolution," *Current Anthropology* 36: 199–221.

Allman, J., T. McLaughlin, and A. Hakeem. 1993. "Brain weight and life-span in primate species," *Proceedings of the National Academy of Sciences* 90: 118–122.

Armstrong, E. and D. Falk (eds.). 1982. *Primate Brain Evolution*. New York: Plenum Press.

Arrow, K. J. and M. Kurz. 1970. *Public Investment, the Rate of Return, and Optimal Fiscal Policy*. Baltimore, MD: Johns Hopkins Press.

Becker, G. S. 1975. *Human Capital*. New York: Columbia University Press.

Blurton Jones, N. G. 2002. "Antiquity of postreproductive life: Are there modern impacts on hunter-gatherer postreproductive life spans?" *American Journal of Human Biology* 14: 184–205.

Blurton Jones, N. G., K. Hawkes, and P. Draper. 1994a. "Differences between Hadza and !Kung children's work: Original affluence or practical reason?," in E. S. Burch and L. Ellana (eds.), *Key Issues in Hunter Gatherer Research*. Oxford: Berg, pp. 189–215.

———. 1994b. "Foraging returns of !Kung adults and children: Why didn't !Kung children forage?," *Journal of Anthropological Research* 50: 217–248.

Blurton Jones, N., K. Hawkes, and J. O'Connell. 1989. "Modeling and measuring the costs of children in two foraging societies," in V. Standen and R. A. Foley (eds.), *Comparative Socioecology of Humans and Other Mammals*. London: Basil Blackwell, pp. 367–390.

————. 1997. "Why do Hadza children forage?," in N. L. Segal, G. E. Weisfeld, and C. C. Weisfield (eds.), *Uniting Psychology and Biology: Integrative Perspectives on Human Development*. New York: American Psychological Association, pp. 297–331.

Bock, J. A. 1995. "The determinants of variation in children's activities in a southern African community," unpublished Ph.D. dissertation, Department of Anthropology University of New Mexico, Albuquerque.

Boesch, C. and H. Boesch. 1999. *The Chimpanzees of the Tai Forest: Behavioural Ecology and Evolution*. Oxford: Oxford University Press.

Bunn, H. T. 2001. "Hunting, power scavenging, and butchering among Hadza foragers and Plio-Pleistocene *Homo*," in C. B. Stanford and H. T. Bunn (eds.), *Meat-eating and Human Evolution*. Oxford: Oxford University Press, pp. 199–218.

Carey, J. R. and D. S. Judge. 2001. "Life span extension in humans is self-reinforcing: A general theory of longevity," *Population and Development Review* 27: 411–436.

Charnov, E. L. 1993. *Life History Invariants: Some Explanations of Symmetry in Evolutionary Ecology*. Oxford: Oxford University Press.

Clutton-Brock, T. H. 1991. *The Evolution of Parental Care*. Princeton: Princeton University Press.

Cordain, L., J. Brand Miller, S. B. Eaton, N. Mann, S. H. A. Holt, and J. D. Speth. 2000. "Plant-animal subsistence ratios and macronutrient energy estimations in hunter-gatherer diets," *American Journal of Clinical Nutrition* 71: 682–692.

Cordain, L., S. B. Eaton, J. Brand Miller, N. Mann, and K. Hill. 2002. "The paradoxical nature of hunter-gatherer diets: Meat based, yet non-atherogenic," *European Journal of Clinical Nutrition* 56(supp): 542–552.

Costa, D. L. 2000. "Understanding the twentieth century decline in chronic conditions among older men," *Demography* 37: 53–72.

Deacon, T. 1997. *The Symbolic Species*. New York: W. W. Norton.

Ellison, P. T. 2001. *On Fertile Ground: A Natural History of Human Reproduction*. Cambridge, MA: Harvard University Press.

Emken, R. A., R. O. Adlof, W. K. Rohwedder, and R. M. Gulley. 1992. "Comparison of linollenic and linoleic acid metabolism in man: Influence of linoleic acid," in A. Sinclair and R. Gibson (eds.), *Essential Fatty Acids and Eicosanoids: Invited Papers from the Third International Conference*. Champaign, IL: AOCS Press, pp. 23–25.

Eveleth, P. B. 1986. "Timing of menarche: Secular trend and population differences," in J. B. Lancaster and B. A. Hamburg (eds.), *School-Age Pregnancy and Parenthood*. Hawthorne, NY: Aldine de Gruyter, pp. 39–53.

Finch, C. E. 1998. "Variations in senescence and longevity include the possibility of negligible senescence," *Journal of Gerontology: Biological Sciences* 53A: B235–B239.

————. 2002. "Evolution and the plasticity of aging in the reproductive schedules in long-lived animals: The importance of genetic variation in neuroendocrine mechanisms," in D. Pfaff, A. Arnold, A. Etgen, S. Fahrback, and R. Rubin (eds.), *Hormones, Brain and Behavior*. San Diego: Academic Press.

Finch, C. E. and R. M. Sapolsky. 1999. "The evolution of Alzheimer disease, the reproductive schedule and the apoE isoforms," *Neurobiology of Aging* 20: 407–428.

Fleagle, J. G. 1999. *Primate Adaptation and Evolution*. New York: Academic Press.

Fogel, R. W. and D. L. Costa. 1997. "A theory of technophysio evolution, with some implications for forecasting population, health care costs and pension costs," *Demography* 34: 49–66.

Fries, J. F. 1980. "Ageing, natural death and the compression of morbidity," *New England Journal of Medicine* 303: 130–136.

————. 1989. "The compression of morbidity: Near or far?," *Milbank Quarterly* 67: 208–232.

Gadgil, M. and W. H. Bossert. 1970. "Life historical consequences of natural selection," *American Naturalist* 104: 1–24.

Gibson, K. R. 1986. "Cognition, brain size and the extraction of embedded food resources," in J. G. Else and P. C. Lee (eds.), *Primate Ontogeny, Cognition, and Social Behavior*. Cambridge: Cambridge University Press, pp. 93–105.

Gurven, M. and H. Kaplan. 2001. "Determinants of time allocation to production across the lifespan among the Machiguenga and Piro Indians of Peru," Albuquerque, NM: Department of Anthropology, University of New Mexico.

Hakeem, A., G. R. Sandoval, M. Jones, and J. Allman. 1996. "Brain and life span in primates," in R. P. Abeles, M. Catz, and T. T. Salthouse (eds.), *Handbook of the Psychology of Aging*. San Diego: Academic Press, pp. 78–104.

Hamilton, W. D. 1966. "The molding of senescence by natural selection," *Journal of Theoretical Biology* 12: 12–45.

Hawkes, K., J. F. O'Connell, and N. Blurton Jones. 1989. "Hardworking Hadza grandmothers," in V. Standen and R. A. Foley (eds.), *Comparative Socioecology of Humans and Other Mammals*. London: Basil Blackwell, pp. 341–366.

Hawkes, K., J. F. O'Connell, N. G. Blurton Jones, H. Alvarez, and E. L. Charnov. 1998. "Grandmothering, menopause, and the evolution of human life histories," *Proceedings of the National Academy of Science* 95: 1336–1339.

Hill, K. 1988. "Macronutrient modifications of optimal foraging theory: An approach using indifference curves applied to some modern foragers," *Human Ecology* 16: 157–197.

Hill, K., C. Boesch, J. Goodall, A. Pusey, J. Williams, and R. Wrangham. 2001. "Mortality rates among wild chimpanzees," *Journal of Human Evolution* 39: 1–14.

Hill, K. and A. M. Hurtado. 1996. *Ache Life History: The Ecology and Demography of a Foraging People*. Hawthorne, NY: Aldine.

Hill, K. and H. Kaplan. 1999. "Life history traits in humans: Theory and empirical studies," *Annual Review of Anthropology* 28: 397–430.

Hiraiwa-Hasegawa, M. 1990. "The role of food sharing between mother and infant in the ontogeny of feeding behavior," in T. Nishida (ed.), *The Chimpanzees of the Mahale Mountains: Sexual and Life History Strategies*. Tokyo: Tokyo University Press, pp. 267–276.

Holliday, M. A. 1978. "Body composition and energy needs during growth," in F. Falker and J. M. Tanner (eds.), *Human Growth*. New York: Plenum Press, pp. 117–139.

Horn, J. L. 1968. "Organization of abilities and the development of intelligence," *Psychological Review* 75: 242–259.

Hurtado, A. M. and K. Hill. 1990. "Seasonality in a foraging society: Variation in diet, work effort, fertility, and the sexual division of labor among the Hiwi of Venezuela," *Journal of Anthropological Research* 46: 293–345.

Intriligator, M. D. 1971. *Mathematical Optimization and Economic Theory*. Englewood Cliffs, NJ: Prentice-Hall.

Jerison, H. J. 1973. *Evolution of the Brain and Intelligence*. New York: Academic Press.

———. 1976. "Paleoneurology and the evolution of mind," *Scientific American* 234: 90–101.

Judge, D. S. and J. R. Carey. 2000. "Postreproductive life predicted by primate patterns," *Journal of Gerontology: Biological Sciences* 55A: B201–B209.

Kaplan, H. S. 1997. "The evolution of the human life course," in K. Wachter and C. Finch (eds.), *Between Zeus and the Salmon: The Biodemography of Longevity*. Washington, DC: National Academy of Sciences, pp. 175–211.

Kaplan, H., M. Gurven, K. R. Hill, and A. M. Hurtado. In press. "The natural history of human food sharing and cooperation: A review and a new multi-individual approach to the negotiation of norms," in H. Gintis, S. Bowles, R. Boyd, and E. Fehr (eds.), *Moral Sentiments: Theory, Evidence and Policy*. Cambridge: Cambridge University Press.

Kaplan, H. S., K. Hill, A. M. Hurtado, and J. B. Lancaster. 2001. "The embodied capital theory of human evolution," in P. T. Ellison (ed.), *Reproductive Ecology and Human Evolution*. Hawthorne, NY: Aldine de Gruyter.

Kaplan, H. S., K. Hill, J. B. Lancaster, and A. M. Hurtado. 2000. "A theory of human life history evolution: Diet, intelligence, and longevity," *Evolutionary Anthropology* 9: 156–185.

Kaplan, H. S. and A. Robson. 2002. "The emergence of humans: The coevolution of intelligence and longevity with intergenerational transfers," *Proceedings of the National Academy of Sciences* 99: 10221–10226.

Kirkwood, T. B. L. 1990. "The disposable soma theory of aging," in D. E. Harrison (ed.), *Genetic Effects on Aging II*. Caldwell, NJ: Telford Press, pp. 9–19.

Kozlowski, J. and R. G. Wiegert. 1986. "Optimal allocation to growth and reproduction," *Theoretical Population Biology* 29: 16–37.

Kuzawa, C. W. 1998. "Adipose tissue in human infancy and childhood: An evolutionary perspective," *Yearbook of Physical Anthropology* 41: 177–209.

Lack, D. 1968. *Ecological Adaptations for Breeding in Birds*. London: Methuen.

Lancaster, J. B. 1986. "Human adolescence and reproduction: An evolutionary perspective," in J. B. Lancaster and B. A. Hamburg (eds.), *School-Age Pregnancy and Parenthood*. Hawthorne, NY: Aldine de Gruyter, pp. 17–39.

Lancaster, J. B. and B. J. King. 1985. "An evolutionary perspective on menopause," in V. Kerns and J. K. Brown (eds.), *In Her Prime: A View of Middle Aged Women*. Garden City, NJ: Bergen and Garvey, pp. 13–20.

Lee, R. B. 1979. *The !Kung San: Men, Women, and Work in a Foraging Society*. Cambridge: Cambridge University Press.

Lee, R. B. and I. DeVore (eds.). 1968. *Man the Hunter*. Chicago: Aldine.

Leigh, S. R. and B. T. Shea. 1996. "Ontogeny of body size variation in African apes," *American Journal of Physical Anthropology* 99: 43–65.

Lessells, C. M. 1991. "The evolution of life histories," in J. R. Krebs and N. B. Davies (eds.), *Behavioural Ecology*. Oxford: Blackwell, pp. 32–65.

McDade, T. W. and C. M. Worthman. 1999. "Evolutionary process and the ecology of human immune function," *American Journal of Human Biology* 11: 705–717.

Medawar, P. B. 1952. *An Unsolved Problem in Biology*. London: Lewis.

Mincer, J. 1974. *Schooling, Experience, and Earnings*. Chicago: National Bureau of Economic Research.

Murdock, G. P. 1967. "Ethnographic atlas: A summary," *Ethnology* 6: 109–236.

Parker, S. T. and M. L. McKinney. 1999. *Origins of Intelligence: The Evolution of Cognitive Development in Monkeys, Apes and Humans*. Baltimore: Johns Hopkins Press.

Promislow, D. E. L. 1991. "Senescence in natural populations of mammals: A comparative study," *Evolution* 45: 1869–1887.

Promislow, D. E. L. and P. H. Harvey. 1990. "Living fast and dying young: A comparative analysis of life history variation among mammals," *Journal of Zoology* 220: 417–437.

Richards, M. P. and R. M. Hedges. 2000. "Focus: Gough's Cave and Sun Hole Cave human stable isotope values indicated a high animal protein diet in the British Upper Paleolithic," *Journal of Archeological Science* 27: 1–3.

Richards, M. P., P. B. Pettitt, E. Trinkaus, F. H. Smith, M. Paunovic, and I. Karavanic. 2000. "Neanderthal diet at Vindija and Neanderthal predation: The evidence from stable isotopes," *Proceedings of the National Academy of Sciences* 97: 7663–7666.

Robson, A. and H. Kaplan. 2002. "The coevolution of intelligence and longevity in hunter-gatherer economies." London, Ontario: Department of Economics, University of Western Ontario.

Roff, D. A. 1992. *The Evolution of Life Histories*. London: Chapman and Hall.

Sacher, G. A. 1959. "Relation of lifespan to brain weight and body weight in mammals," in G. E. W. Wolstenhome and M. O'Connor (eds.), *Ciba Foundation Colloquia on Ageing*. London: Churchill, pp. 115–133.

Schaie, K. W. 1996. *Intellectual Development in Adulthood: The Seattle Longitudinal Study*. New York: Cambridge University Press.

Schoeninger, M., H. T. Bunn, S. Murray, T. Pickering, and J. Moore. 2001. "Meat-eating by the fourth African ape," in C. B. Stanford and H. T. Bunn (eds.), *Meat-eating and Human Evolution*. Oxford: Oxford University Press, pp. 179–195.

Shanley, D. P. and T. B. L. Kirkwood. 2000. "Calorie restriction and aging: A life-history analysis," *Evolution* 54: 740–750.

Sibly, R. M. 1991. "The life-history approach to physiological ecology," *Functional Ecology* 5: 184–191.

Sibly, R., P. Calow, and N. Nichols. 1985. "Are patterns of growth adaptive?," *Journal of Theoretical Biology* 112: 553–574.

Silk, J. B. 1978. "Patterns of food-sharing among mother and infant chimpanzees at Gombe National Park, Tanzania," *Folia Primatologica* 29: 129–141.

Stanford, C. B. 1998. *Chimpanzee and Red Colobus: The Ecology of Predator and Prey.* Cambridge, MA: Harvard University Press.

Stearns, S. C. 1992. *The Evolution of Life Histories.* Oxford: Oxford University Press.

Tatar, M., D. W. Grey, and J. R. Carey. 1997. "Altitudinal variation in senescence in a Melanoplus grasshopper species complex," *Oecologia* 111: 357–364.

Teleki, G. 1973. *The Predatory Behavior of Wild Chimpanzees.* Lewisburg, PA: Bucknell University Press.

Williams, G. C. 1957. "Pleiotropy, natural selection and the evolution of senescence," *Evolution* 11: 398–411.

Worthman, C. M. 1999a. "Epidemiology of human development," in C. Panter-Brick and C. M. Worthman (eds.), *Hormones, Health and Behavior: A Socioecological and Lifespan Perspective.* Cambridge, UK: Cambridge University Press, pp. 47–105

———. 1999b. "Evolutionary perspectives on the onset of puberty," in W. R. Trevathan, E. O. Smith, and J. J. McKenna (eds.), *Evolutionary Medicine.* Oxford: Oxford University Press, pp. 135–163.

Rescaling the Life Cycle: Longevity and Proportionality

RONALD LEE
JOSHUA R. GOLDSTEIN

Since 1900 life expectancy in the United States has increased by about 30 years, from 48 to 77 years. Most projections foresee continued increase at a slower rate, reaching 85 or 90 years by 2100 (Lee and Carter 1992; Tuljapurkar et al. 2000). Some analysts foresee the possibility of far greater life expectancy increases—to 100, 150 years, or more (Manton et al. 1991; Ahlburg and Vaupel 1990; Schneider et al. 1990). Individuals welcome longer life, but for populations, increases of this magnitude could impose heavy costs on the working-age groups and could have other substantial but uncertain economic and social consequences. These consequences will depend in large part on how the additional expected years of life are distributed across the various social and economic stages of the life cycle. This chapter speculates about how the gains will be used and how the life cycle will be modified.

Proportional rescaling of the life cycle, in which every life cycle stage and boundary simply expands in proportion to increased life expectancy, provides a convenient benchmark. Under proportional rescaling, if longevity doubles, then so would childhood, the length of work and retirement, the span of childbearing, and all other stages of the life cycle. Proportional rescaling of the life cycle would appear to be neutral in some sense, so that life, society, and economy could continue as before, except that there would be proportionately more time spent in every stage. However, we will find that while this is (or could be) true for some aspects of life, other aspects would vary with the square or some other power of longevity.

Although proportional rescaling may seem natural, and indeed does occur in nature as we discuss, powerful forces impede its application to humans. First are biological constraints related to human development, such as menarche and menopause, even though our vigor and health may advance with longevity. Second are institutional constraints, for example on schooling and retirement. Although these may adjust in the long run, over shorter horizons they may block proportional rescaling. Third, stock-flow inconsistencies may cause human and physical capital to rise with longevity more rapidly than labor force size, causing wages and incomes to rise and the interest rate and returns to human capital to fall, triggering further adjustments. If time in retirement is a luxury, then it may rise rapidly as incomes rise, so that age at retirement does not rise in proportion to longevity—as in fact it has not. For these and other reasons, we should not actually expect to observe proportional rescaling in connection with increased life span in the past, nor should we anticipate it for the future. Nonetheless, proportionality is a useful benchmark against which to compare past changes and the hypothetical future.

What would proportional rescaling look like?

A perfectly proportional rescaling of the life cycle accompanying increased life expectancy would appear indistinguishable from the effect of a simple change in the units of measurement of age/time, as if we had changed from dollars to pesos or from inches to centimeters. For example, suppose that an original 75-year life cycle were instead measured in units of six lunar months, or half-years. The new life cycle would be 150 units long instead of 75, but of course nothing would be different. (We use doubling as a convenient example, but we believe that a 15 percent increase in life cycle is more likely for the twenty-first century.) Every life cycle stage would last twice as long as before, measured in the new units. Every rate would be only half as great, since only half as many events would take place in 6 months as in 12 months.

Proportional life cycle changes can occur in both strong and weak forms. In the strongest form, the change affects not only the average timing of life cycle transitions but also the whole distribution of timing within a population. Thus, for example, the spread around the mean age of death or mean age of childbirth would also double if the mean age of death doubled. We call this "strong proportionality," which can apply both to the distribution of an event among the individuals in a population and to the timing or level of an event in the individual life cycle. If the change in life cycle timing is only proportional with respect to the mean ages of each life cycle transition or stage, then we call this "weak proportionality." Weak proportionality would occur if, for example, as longevity increased, the mean age of reproduction increased proportionately but the variance of the net maternity function stayed constant.

It is important to distinguish between "flow" or "rate" variables, which are measured per unit of time, and "stock" variables, which are not. Completed fertility, accumulated wealth, the probability of ever marrying, or of ever having a first birth, are all examples of stocks. Birth rates, death rates, wage rates, income, and rate of knowledge acquisition are flows, all measured per unit of time. Under the perfectly proportional rescaling of the life cycle discussed above, all stocks are unchanged, provided they are assessed at the same proportional stage of the life cycle (for example at the age equal to 60 percent of life expectancy). In a population with a life expectancy twice as great, the net reproduction rate (NRR) would not be twice as high; it would be unchanged. All flow or rate variables, however, would be reduced in inverse proportion to the increase in longevity. Adding these reduced rates across twice as long an age span of reproduction would then yield an unchanged completed fertility.[1] We might say that perfectly proportional rescaling of the life cycle is always "stock constrained," in the sense that the magnitude of stocks is preserved under rescaling.

In some cases, this kind of stock-constrained rescaling is not an interesting scenario to contemplate. For example, if people were to live twice as long, we would not expect that at each age they would earn only half as much, or that they would learn only half as much during each year in school. In these cases, it is more natural to assume that the rate of productivity, or earning, or acquisition of knowledge, remains constant, so that the stock of lifetime earnings, or completed education, would double. Such cases we call "flow constrained."

Figure 1 illustrates some hypothetical examples of what proportional rescaling would look like under different assumptions, all of which assume the convenient, yet implausible doubling of longevity. The top row illustrates strong proportionality with stock constraints, such as might be the case with fertility. Here the ages at which flows occur are doubled, but at the same time the flow itself is halved. The result is that lifetime completed fertility remains unchanged, although the age at which a given cumulative fertility is achieved is doubled. The middle row shows strong proportionality with flow constraints, such as might be the case with earnings. The age at which a given flow occurs in the rescaled age profile is twice that of the original profile. Keeping flows constant over a longer period of time, however, produces a doubling of the stock, for example cumulative lifetime earnings (and perhaps derived stocks, such as lifetime savings, which we discuss later). The bottom row illustrates a particular case of weak proportionality. Here the mean age of the flow doubles from 25 to 50, but the width of the age profile does not expand proportionally.

Proportional rescaling in nature

The object of this chapter is to explore the feasibility of proportional rescaling within a single species, namely humans. A large body of work already exists looking at proportional rescaling between biological species. In biology this falls under the study of "biological invariants" (Charnov 1993). Figure 2 shows

FIGURE 1   Illustrative examples of proportional rescaling under a hypothetical doubling of longevity

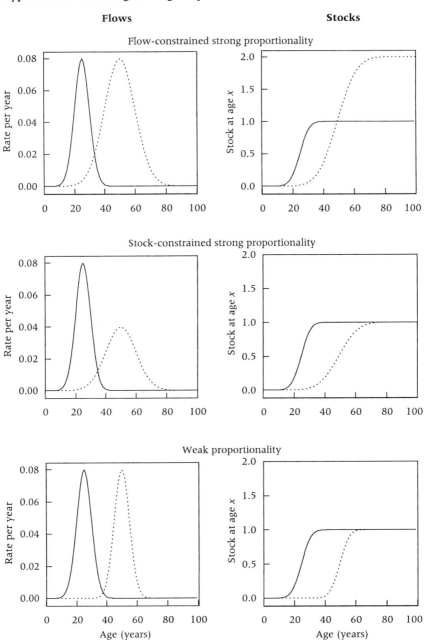

NOTES: Original age profile is shown as a solid line, rescaled profile as dashed line. Stocks at age *x* are the integral of flows from age zero to *x*. In this figure, flow-constrained strong proportionality is obtained by doubling ages at which a given rate is originally observed, resulting in a doubling of eventual stocks; stock-constrained strong proportionality is obtained by simultaneously doubling ages and halving flows; and weak proportionality is obtained by doubling the mean of the flows while keeping the standard deviation of flows unchanged.

FIGURE 2   An example of proportional rescaling in nature: Age at maturity versus expected life as a mature adult in 23 mammalian species (logarithmic scales)

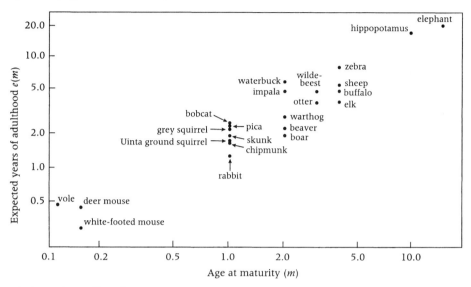

NOTE: Data from Millar and Zammuto 1983.

an example of such an invariant across mammalian species: the relationship between age at physical maturity (m) and expected years of adulthood $e(m)$. Goldstein and Schlag (1999) show a similar figure for the relationship between mean age of reproduction and life expectancy at birth. Other examples of biological invariants include the ratio of body mass to life expectancy at birth; the product of the number of offspring and the chance of survival to the mean age of reproduction; and metabolic invariants (Austad 1997). "Invariant" is used by biologists not as a rule to which there are no exceptions but rather to describe general tendencies, with variation around them. In many cases, the rescaling involves a power transformation and is therefore not proportional. For example, body mass rises approximately with the cube of body length, so if body mass is proportional to life expectancy, then body length will not be. Note, however, that in equilibrium every animal population will have a net reproduction rate of unity, so this stock measure will be invariant across species, consistent with proportional rescaling.

Biological invariants result from evolutionary forces, and explanations for them are given in terms of evolutionary theory and maximization of reproductive fitness. The increased longevity of modern humans has not resulted from natural selection. It is the result of scientific advances, changes in life style, social organization, nutrition, and manmade environments. For this reason, we would not expect the biological invariants in nature to apply to human life cycle timing as we live longer. Still, the analogy is useful, because the optimizing principle still obtains even if what is being optimized is not reproductive fitness, but rather a more general notion of utility that might include economic as well as reproductive success and hedonic consumption as well as productive investments.

Furthermore, it appears that the evolution of proportionally related life cycles across species is not a result of independent adaptations across a range of traits but is rather linked to some underlying biological mechanisms that control the rate of metabolism and other features of the life cycle clock (e.g., Carey et al. 1998; Finch 1990; Biddle et al. 1997; Lin et al. 1997). Evidence for this includes the ability to breed longer-living animals simply by the selective breeding of individuals that reproduce late in life. The result is that both the onset of reproduction and the age at death are proportionally delayed. Restricted diets in the laboratory appear to have a similar effect, delaying mortality but also delaying physical maturity. Finally, some genetic modifications appear to have proportional effects on longevity and the timing of reproduction.

Life table stretching and the proportional life cycle

We can divide the human life cycle into stages of childhood, working ages, and old age, marked approximately by the boundaries of age 15 or 20 and 60 or 65. When life expectancy increases in a population, person-years of life are added to the life table within each of these three life cycle stages, not only at the end of life in old age. This is not a natural way to think about the lives of individuals; for individuals, alterations in length of life seem always to come at the end. But that is only so ex post; ex ante, individuals are subject to risks of death at every age, and therefore their expected years of life at each age throughout the potential life cycle are subject to modification when these risks change.

Historically, we can observe at what ages person-years of survival are added when life expectancy at birth has risen. When life expectancy has risen by one year from a low level such as 20, this one year has been distributed as follows: 0.7 years between 15 and 65; 0.2 years between 0 and 15; and only 0.1 year after 65. As life expectancy rises further, the incremental gains in childhood and the working ages decline, and the gains in old age rise. Further increases from the current life expectancy level of 77 years in the United States will be concentrated in old age, with 0.7 years coming after age 65 and hardly any coming before age 15 (Lee 1994; Lee and Tuljapurkar 1997).

Increasing survival does not affect all stages of life equally. A mechanical reason for this is that in the above calculations we have not rescaled the age boundaries of youth and old age as longevity increased. But even if these boundaries were rescaled, historical improvements of mortality have been inconsistent with proportional rescaling. Mortality has declined faster in infancy than over the rest of the life cycle, and this has resulted in nonproportional changes in the survival curve.

As with all other aspects of the life cycle, we can visualize strong proportional change by asking what would happen if we simply changed the units of measuring age, say to 6-month units. Then for the proportion of survivors, or for life expectancy, values would be attained at age x that were previously attained at age $x/2$. The higher the age in the initial life table, the farther out we would have to move on the new age scale to reach a corresponding level. For age $x=3$, the new age would be 6; and for age $x=60$, the new age would be 120. (Death rates would be shifted in the same way, but also proportionately reduced, as discussed earlier.)

To state this is to see that historical shifts have not corresponded to this simple assumption. Fries (1980) emphasized that the actual pattern of mortality change is of the opposite sort, and called it "compression of morbidity." When mortality declines, the upper end of the $l(x)$ curve shifts relatively little, rather than relatively more as required by strong proportionality. Indeed, some observers suggest that the maximal length of life (first x such that $l(x) = 0$) has hardly changed at all in recent centuries. Wilmoth et al. (2000), however, show that the age of the oldest death in Sweden has been rising at least since 1861, accelerating in recent decades; since 1970, it has been rising at about the same speed as life expectancy (Wilmoth and Robine, in this volume.)

One useful diagnostic is to look at the $l(x)$ value for some age x in 1900, for example, and find the corresponding age at which that $l(x)$ value is reached in 1995.[2] Calculation shows that comparing US mortality in 1900 and 1995, the drop in survivorship reached at age 1 in 1900 was not matched until age 59 in 1995, an age increase of 5,800 percent. The corresponding increases at ages 30, 60, and 90 were 137 percent, 33 percent, and 10 percent. Under strong proportionality, these percent increases would be equal at all ages. We have made similar calculations for the projected survivorship changes between 1995 and 2080, based on the mortality projections of the Social Security Administration. The drop in survivorship reached at age 1 in 1995 will not be matched until age 23 in 2080, an age increase of 2,200 percent. The corresponding increases at ages 30, 60, and 90 are 50 percent, 13 percent, and 5 percent.

We can get some analytic insight into the likely pattern of mortality change in the future by drawing on a simple result from Vaupel (1986). Suppose that mortality after age 50 follows Gompertz's Law, with death rates rising across age at a constant exponential rate $\theta = 10$ percent per year. Suppose further that death rates at each age over 50 decline over time at $\rho = 1$ percent per year. With these assumptions, mortality at ages over 50 declines in such a way that the mortality curve shifts ρ/θ years to the right every year.[3] With values of $\rho = .01$ and $\theta = 0.1$, the mortality curve shifts 0.1 years to the right every year; and every decade, the death rate previously experienced at age x will now be experienced at age $x+1$. Once mortality decline has proceeded to the point where survival to age 50 is close to unity,[4] then the survival curve and life expectancy will be displaced to the right by one year each decade, at every year of age. Because this shift to the right is equal across all ages over 50, rather than increasing in proportion to age, it is not strictly consistent with proportional stretching.

Under proportional stretching of the life cycle, the time spent disabled or in ill health would rise in proportion to longevity, as would the time spent free of disability. Recent research on disability, chronic illness, and functional status reveals patterns that are broadly consistent with such proportional changes, at least for the last two decades in the United States (Costa 2000; Crimmins et al. 1997; Manton et al. 1997; Freedman and Martin 1999; Manton and Gu 2001).[5] For example, Crimmins et al. (1999) conclude that persons in their late 60s in 1993 are functionally like those in their early 60s in 1982.[6] Research (Lubitz and Prihoba 1984; Lubitz et al. 1995; Miller 2001) also shows that health care costs in old age are more closely related to time until death than to chronological age, so that as life expectancy rises and fewer persons at any age are near death, health care costs would fall, other things being equal. All these findings are qualitatively consistent with proportional rescaling of the life cycle for health, functional status, and disability; but because of imprecision of measure they are also consistent with compression of morbidity, with disabled years shrinking as a proportion of the life cycle.

Individual demographic aspects of rescaling

Mortality, survival, health, and longevity are the bare bones of the life cycle. Now we enrich the story by discussing fertility and other social behavior.

Transitions to adulthood: Education, marriage, and the onset of childbearing

In animals, the life cycle stage is often divided into infancy and maturity and is closely tied to physical growth and sexual maturity (Kaplan 1997). In humans, the transition to adulthood is typically seen as being much more complicated, involving a change in social, economic, and familial roles (Modell et al. 1976; Marini 1987). Sexual maturity is only the beginning of the transition to adulthood, which may be completed when children are economically and residentially independent of their parents, get married, and begin families of their own.

Some easily measured indicators of the transition to adulthood include the age of educational completion, first marriage, and first birth. Table 1 shows the pace of change of the mean ages of these indicators for several industrial countries between 1975 and 1995. Education, marriage, and child-bearing are all being postponed at a rapid pace in the United States, Japan, and Sweden.[7] The rates of rescaling vary by indicator, but all are faster than the pace of longevity increase. Whereas life expectancy at birth is increasing roughly 0.2 percent per year, mean age at first birth is increasing about twice as fast, and mean age at first marriage perhaps four times as fast.

The differential rates of rescaling of education, marriage, and fertility mean not only that the transition to adulthood as a whole is being shifted to later ages but also that the interrelations between the various components of the transition are changing. For example, the faster pace of marriage postponement than of first births is evidence of the increase in premarital births and the rise of cohabitation in the United States and Sweden. The synchrony in marriage and birth postponement in Japan is due in part to low levels of premarital childbearing in that country.

The changing age profile of educational enrollment over the last half-century in the United States shows dramatic increases at both older and younger ages. Figure 3 shows the percent enrolled in school by age group among the civilian noninstitutionalized population. We see massive increases in enrollments in both the youngest age groups (3–4 and 5–6 years) and the older age groups (above age 21). For older ages, the increases in enrollment correspond nicely to the proportional stretching scenario. The increases in enrollment at younger ages, however, are not at all consistent with the proportional stretching argument, according to which the first ages of enrollment should be increasingly postponed over time. One explanation of increased enrollment at younger ages is that it represents not so much a

TABLE 1   The pace of rescaling of selected life cycle indicators in the United States, Japan, and Sweden: Annual rates of change (in percent), 1975 to 1995

Country	Life expectancy at birth (females)	Mean age at first birth (women)	Mean age at first marriage (women)	Mean age at end of school enrollment (both sexes)
United States	0.1	0.4	0.9	0.5
Japan	0.4	0.3	0.3	0.3
Sweden	0.2	0.5	0.7	NA

NOTES: (1) US ages at first birth and first marriage are period medians. (2) Mean age at end of school enrollment is calculated from period enrollment rates. The mean age is estimated as $m = \text{sum}(nEx * n) + x0$, where nEx is the enrollment rate for the age group x to $x+n$ (e.g., 31.9 percent for 20- and 21-year-olds in 1970), n is the width of the age group, and $x0$ is the youngest age group of full enrollment (age 6 in the US data). For Japan, entrance rates to high school, junior college, and university were available. The mean age at end of school enrollment was estimated as $m = 14 + (4 * \text{HS entry}) + (2 * \text{HS entry} * \text{Jr. college entry}) + (4 * \text{HS entry} * \text{university entry})$.
SOURCES: Japan (1999, 2000); Sweden (1995); NCHS (2000).

FIGURE 3 Educational enrollment rates in percent of civilian noninstitutionalized population by age: United States 1947, 1964, 1980, and 1998

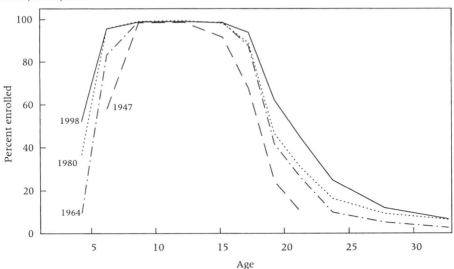

SOURCE: US Census Bureau, Current Population Survey.
(http://www.census.gov/population/socdemo/school/taba-2.txt)

change in the timing of the educational stage of the life cycle as an institutional shift from the private sphere of home education of infants to the public sphere of day care and kindergarten. Still, if entry into a socializing environment (being surrounded by nonfamily members) is thought to be part of the transition from infancy to childhood, early enrollments represent an acceleration in life cycle timing.

While some of the transitions that signify the completion of entry into adulthood, such as marriage and childbearing, are being delayed, many are being advanced to younger ages. Biologically, the long-term trend, until recently, has been earlier ages at menarche and physical maturity (Eveleth and Tanner 1990). Legally, the trend worldwide is for voting rights to be extended to younger ages.[8] Adult criminal penalties in both the United States and Japan are increasingly being extended to minors. Socially, precociousness appears to be the rule rather than the exception, with children allowed various forms of independence at increasingly younger ages.

Whereas half a century ago transitions to maturity were compressed into a narrow age span, the transition appears to be becoming more diffuse. Children who move away from their parents increasingly return home (Goldscheider and Goldscheider 1999). Education, labor force participation, and the establishment of new families are often less clear-cut stages than they once were. People in their 20s and 30s may be working, going to school, receiving support from their parents, and starting a family of their own—all simultaneously rather than in a series of ordered steps.

What are the implications of an expansion of early adulthood? One result is a mismatch, at least temporarily, between certain life cycle–linked institutions and the life cycle timing of individuals. As an example of this, in the United States young adults can find themselves without health insurance because they are too old to be covered by their parents' plans but are not yet economically secure enough to have their own plans.

A second consequence of a longer transition to adulthood is more time for career and partner searches. Social and sexual interactions with potential spouses may now last a decade or more. Likewise, career experimentation and repeated exit from and entry to education are possible thanks to less time pressure to support a family and achieve economic independence. Because the efficiency of searches probably remains constant per unit of calendar time (i.e., searches are flow constrained), the quality of searches should, all other things equal, improve.

A third consequence is the inversion of traditional sequences (Rindfuss et al. 1987). A traditional sequence in the first half of the twentieth century might have been: educational completion, departure from parents' home, entry into labor force, marriage, and childbearing. Now, with the extended time of the transition and the moving back and forth between transitions, childbearing may precede marriage; entry into the work force may precede leaving the parental home; divorce may be followed by moving back to the parents' home. The extended time over which the various transitions to adulthood occur allows greater opportunity to reverse transitions and to change their order.

From the point of view of a rational life cycle planner, an extended period of quasi-adulthood probably makes sense for those who can expect to live a long time. A long investment horizon makes it worthwhile to invest more in one's own human capital and stay in school longer. Likewise, it makes experimentation less costly and potentially more rewarding. It is not clear whether time spent by 20- and 30-year-olds who have not yet committed themselves to careers or to families is a productive human capital investment or leisure (a kind of pre-career retirement). In some sense, it may be both. The prolonged period of transition to adulthood may be akin to the wrestling of young chimpanzees, who look to us as if they are just playing but are actually learning skills that will be useful and necessary to them as adults.

Fertility

Over the course of the last century, increased longevity has been accompanied by declines in total fertility, without a consistent change in the mean age of childbearing.[9] Delays in the timing of the first birth have been counterbalanced by an earlier end of childbearing as the level of fertility has declined. In recent decades, however, both the onset and end of childbearing have been slowly shifting to older ages. A continuation of this trend is consistent with proportional stretching of the life cycle, but from a theoretical perspective it is difficult to make firm predictions about either the timing or the amount of fertility that will be associated with longer life.

As an example of the pattern of fertility change over the last century, Figure 4 shows age-specific fertility rates for women in the United States in 1920, 1980, and 1998. The upper panels show the unadjusted age-specific fertility rates and the lower panels show the same age pattern, but normalized so that all of the curves have the same fertility level over the life cycle. The figure indicates that as fertility declined from 1920 to 1980, the timing of childbearing became more concentrated in the 20 to 25 year age group.

FIGURE 4   Period age-specific fertility profiles of US women: 1920, 1980, and 1998

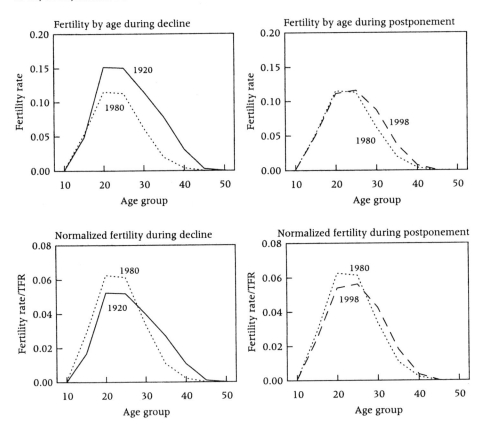

NOTE: Normalized profiles are obtained by dividing the age-specific rates in a given year by the total fertility rate in that year.

This was a result of declining fertility at both younger and older ages. Since 1980, however, the whole distribution of fertility has shifted slightly to older ages. This recent shift is consistent with proportional change, in that birth rates at older ages are increasing faster than at younger ages.

Some believe that further increases in age at first birth will not be accompanied by a proportional increase in the age at the end of childbearing. As a result the distribution of childbearing will be compressed to a narrower band of the life cycle, increasingly between the ages of 30 and 40 years. One reason for believing such an assessment is the biology of female reproduction and the onset of menopause, in which there has been little if any recorded change in the last century. A counterargument is that if biological advances can extend life, then advances in reproductive technology can extend childbearing. Already, the ability to store eggs and sperm, as well as frozen embryos, suggests that parents can have children at arbitrarily late ages with the present technology. In the future, it may well be possible for women to conceive, gestate, and give birth to children at older ages. It does not seem a priori a more difficult problem to increase the age of women's maximum reproduction than to increase the age of maximum longevity.

In the near term, proportional increases in the mean age of childbearing—weak proportionality—are possible even without advances in reproductive technology. For example, if female life expectancy were to increase 25 percent to 100 years, the same proportional increase would require a mean age of childbearing of 35, well within the realm of current biology.

Would later childbearing reduce total fertility? Timing may have a direct effect on the level of fertility, particularly if delays in the onset of fertility are not accompanied, because of biological limits, by higher fertility at older ages. But later childbearing may also influence the demand for children.

Proportional rescaling could increase the costs of childbearing, both the direct costs like children's education and the indirect costs like the earnings and promotion opportunities forgone (Willis 1973; Calhoun and Espenshade 1988). Parents have higher potential earnings at older ages, so later births have a higher opportunity cost as participation in the labor force is curtailed. Opportunity costs of childbearing will depend in part on how the age-earnings curve is rescaled. But in some careers (e.g., academia and law firms) job security increases with age as a result of institutional practices such as tenure. It may actually involve less sacrifice for some people to have children at older ages, depending on institutional arrangements. Also, discounting will reduce the opportunity costs of later-born children.

Higher forgone earnings may be more than offset by higher lifetime earnings, and the relative costs of childbearing may decrease. Furthermore, if the spacing of children does not change, there might be a substantial decline in the relative costs of bearing, say, two children. Consider a woman who lives 75 years and has two children 3 years apart. She will be the mother of children less than 3 years old for a total of 6 years out of her 75-year life span. On the other hand, a woman who lived to be 100 and had the same number of children with the same spacing would still only spend 6 years of her life with infants. In this case the nonproportional change of the length of infancy resulting from the flow-constrained nature of human growth may change the relative cost of childbearing.

Population-level implications of rescaling

The overlap of generations

Under perfect proportionality, population size and structure remain unchanged (Goldstein and Schlag 1999). However, if the mean age of reproduction does not change at the same pace as longevity, then population size, the overlap of generations, and dependency ratios will be affected.

Consider a stylized life cycle in which childhood lasts until age 30, retirement begins at age 60, and everyone dies at age 90. Let reproduction occur at age 30 and the population be stationary. In this case, three generations will be alive at once, individuals will spend one-third of their life working, and the total dependency ratio for the population as a whole will be 2:1. Under proportional rescaling, say a doubling, none of these population characteristics would change; population size would also remain constant.

If, on the other hand, longevity increased without changing generational length, more generations would be alive at once and the total population size would increase. Such a nonproportional change would increase the share of life spent between childbearing and retirement and would also change the total dependency ratio of the whole population. If we doubled the length of life and the age of retirement as above but without changing the age of reproduction, the number of generations alive at once would increase from three to six, the population size would double, and the total dependency ratio would shrink from 2:1 to 1:1. Nonproportional changes in generation length could also occur in the reverse direction. If generation length increases faster than longevity, the result is a decline in population size and, if childbearing still signals the entry into adulthood, an increase in the dependency ratio.

Is rescaling a solution for subreplacement fertility?

While the above discussion has assumed replacement-level fertility, it is perhaps of greater practical interest to consider how rescaling of the life cycle might offset population aging that accompanies below-replacement fertility. For example, age 65 in a stationary population might be mapped to age 70 in a shrinking population, in order to keep retirees a constant proportion of the entire population.

The nonlinearity of the age pyramid means that changes in population growth rates cannot be offset by proportional rescaling of the life cycle. Taking the United States life table of 1992 as an example, Table 2 shows the stable age pyramid for two cases: a population growth rate of zero and a negative growth rate of 1 percent (equivalent to a total fertility rate of about 1.5). By redefining ages, we can distort the subreplacement fertility age pyramid so that it has the same shape as the replacement-fertility age pyramid.

This table indicates the ages at which each decile of the population age distribution is reached. Thus, in the stationary population 10 percent of the population is under age 6.7 years, while in the subreplacement population all children under age 10.2 years are needed to fill the first decile. The table also shows that the kind of restructuring of the life cycle that would be needed to offset a shift to subreplacement fertility is not proportional. The column labeled "ratio of ages" shows that more rescaling would be needed at younger ages than at older ages. The age below which 90 percent of the population will find itself shifts from 73.4 years to 77.9 years, a change of only 6 percent as opposed to the more than 50 percent rescaling that would be needed at younger ages. The table also shows the absolute difference in ages that would be required by rescaling. Here we see that the greatest changes would be needed in the middle of the life cycle, shifts of about 8 years, or about twice the magnitude of the shifts at the two extremes of the life cycle.

A striking and counterintuitive result is that less proportional rescaling would be needed at older ages than at younger ages. We might interpret this optimistically, since it is presumably hardest to make large changes at older ages.

TABLE 2   Rescaling implied by a change in population growth rates. Ages corresponding to cumulative deciles in a stationary population (r = 0) and in a declining population (r = –.01). Proportional and absolute rescaling implied by change in growth rate

Percentile	Age r = 0	Age r = –.01	Ratio of ages	Difference of ages
0.1	6.7	10.2	1.52	3.51
0.2	14.4	20.3	1.41	5.91
0.3	22.2	29.6	1.34	7.43
0.4	30.0	38.2	1.27	8.20
0.5	37.9	46.2	1.22	8.34
0.6	45.9	53.9	1.17	7.95
0.7	54.2	61.4	1.13	7.15
0.8	63.0	69.1	1.10	6.02
0.9	73.4	77.9	1.06	4.50

NOTE: Stable populations based on 1992 combined-sex life table for the United States (Berkeley Mortality Database).

Rescaling and economic behavior: Retirement trends

As life expectancy rises and health at older ages improves, it seems natural that the age at retirement should rise as well. For example, in assessing the effects of longer life, Kotlikoff (1981) makes two alternative assumptions: that the age at retirement rises in proportion to life expectancy at birth, or alternatively that it rises more than in proportion to life expectancy at birth, to keep the years of retirement at the end of life constant. In fact, however, in industrial countries age at retirement and older men's labor force partici- pation rates have been dropping for more than a century, while life expect- ancy has risen by several decades (Costa 1998). In the United States in 1900, men retired in their early 70s, on average, compared with age 63 today (National Academy on an Aging Society 2000: 6). Labor supply at older ages has also declined sharply in developing countries (Durand 1977).

According to the US period life table for 1900, the ratio of the ex- pected years lived after age 70 to those lived during ages 20 through 69 is 0.10. For each year spent working, 0.1 years would have been spent re- tired.[10] If retirement in 1995 still occurred at 70, mortality decline since 1900 would have raised this ratio from 0.10 to 0.23. Given that the mean retirement age actually has fallen to 63 for men, the ratio in 1995 of ex- pected retirement years to expected work years has risen from 0.10 to 0.38, nearly quadrupling since 1900. In order to maintain the original ratio of 0.10 over the life cycle, the retirement age in 1995 would have to be moved to 78. If we were to allow for an earlier age at start of work in 1900, the results would be even more dramatic. While retirement age has stopped falling in the United States during the past decade, and has even modestly risen, the long-term trend has been strongly downward.

Ausubel and Grubler (1995) examined long-term trends, 1870 to 1987, in average hours worked over the life cycle for France, Germany, Great Britain, the United States, and Japan. They calculated disposable lifetime hours as $24 * 365e_{10}$, less $10 * 365e_{10}$ for physiological time (sleeping, eating, hygiene). For sexes combined in Great Britain from 1857 to 1981, they found that lifetime work hours declined from 124,000 to 69,000, while disposable nonwork hours increased from 118,000 to 287,000. Work as a share of to- tal disposable hours (that is, work as a share of total lifetime nonphysiological hours above the age of 10) declined from 50 percent to 20 percent.[11]

Of course, much has happened over this period besides the increase in longevity. Growth in financial institutions and in public and private pen- sions has made it easier to provide for consumption in retirement. Income has increased and educational attainment has risen. Given these changes, it is perhaps not surprising that retirement age has fallen. Leisure is presum- ably a luxury good. We would expect its share of the life cycle budget to

grow as income rises secularly. If life expectancy and health status had increased while hourly wages remained constant, then a decline in retirement age would seem unexpected. In this counterfactual case, we would
probably expect a proportional increase in both work time and leisure time,
to keep their marginal contributions to lifetime utility equal.

There is also abundant evidence that institutions and employers' practices have encouraged earlier departure from the labor force than individuals might have chosen otherwise. Discrimination against older workers used
to be common. In the United States it has been addressed by a series of
legislative acts of states going back to the 1930s and since the 1960s by
federal law (Neumark 2001). The incentives for early retirement in employer-provided defined-benefit pensions are also important (see e.g.,
Lumsdaine and Wise 1994; Wise 1997). Strong incentives are also built into
many defined-benefit public pensions, particularly when these are combined
with incentives arising from tax policies, long-term unemployment benefits, and disability benefits. In a striking cross-national study of 11 OECD
countries, Gruber and Wise (1997) found that in some countries the combined effect of such policies makes the net wage for continuing work after
age 60 years drop close to zero or even turn negative, a very heavy "implicit tax." They found that this implicit tax accounts for most of the variation across countries in labor force participation rates at older ages.[12]

In sum, a combination of economic and institutional change, distorted
incentives, and the behavioral response to these has caused the proportion
of the life cycle devoted to work to shrink dramatically.

Rescaling and the economy

Consider now the effect of proportional rescaling on the economy, taking a
doubling of life expectancy as a convenient, albeit implausible example.
Under stock-constrained proportional rescaling, all flow variables would be
only half as great, including wages, gross domestic product, and consumption and savings per unit time. Stock variables like capital would be unaf-
fected at corresponding ages over the life cycle (that is, at ages representing
the same proportion of life expectancy). The tax rate that would be required
for the working-age population to support the retirement of the elderly
through pay-as-you-go public pensions would remain the same at corresponding ages.

On reflection, however, this stock-constrained expansion of the life cycle
is neither realistic nor appealing. When we hypothetically changed the measure of time from years to six-month units, in fact nothing changed, although
all flow measures were halved. But if life expectancy were in fact to double,
our unit of measure would nonetheless remain the year. In this case, we
would not expect that output per year would be halved; nor wage rates, an-

nual consumption, or savings. If people were to work twice as many years consistent with the proportionality assumption, then their lifetime earnings would be roughly twice as great (more or less, depending on returns to experience versus obsolescence of skills and knowledge), as would their lifetime consumption and savings. The cumulation of savings over twice as many years would generate assets at retirement that would be twice as large—as they should be, in order to finance a retirement that would be twice as long.

Figure 5 illustrates the implications for capital accumulation and the capital/labor ratio. The age at beginning work is denoted A; it will vary when longevity varies, but is treated as fixed for purposes of this figure. In a stationary population with one person born into each generation, the size of the labor force will initially be the distance AB. Accumulated wealth, held as capital, will grow linearly over the life cycle to amount F, and then be spent-down to fund retirement until death at age C, when it is exhausted.[13] The total stock of capital in the economy will be the area of the triangle AFC. The capital/labor ratio is this area divided by the number of workers, AB. Now suppose life expectancy doubles from AC to AE, and the age at retirement doubles from B to D. The new labor force is AD (continuing to

FIGURE 5 The nonproportional consequences of proportional rescaling on the aggregate economy: Earnings, savings, and the capital/labor ratio

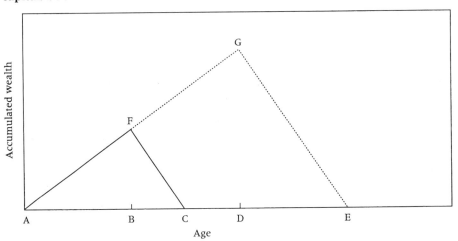

NOTE: In the short life cycle, people begin work at age A and retire at age B having accumulated amount F to support their retirement until death at C. The capital/labor ratio is the area of triangle AFC divided by the number of laborers AB in a stationary population. In the doubled life cycle, people retire at D having accumulated amount G to support them in their longer retirement until death at E. The capital/labor ratio is the area of the triangle AGE divided by the number of workers in the stationary population, AD. If the generation size remains constant, there are twice as many workers with the doubled life cycle, since length AD is twice AB. However, there is four times as much capital, since G is twice F, and E is twice C. Therefore, the capital/labor ratio is twice as great. This assumes life cycle savings behavior, with constant earnings by age and zero interest.

assume one person per generation), and the new capital stock is the area of the triangle AGE. Since both the base and the height of AGE are twice as great as the corresponding dimensions of AFC, its area is four times as great. However, the labor force is only twice as great, so the new capital/labor ratio is twice the old.

The productivity of labor will be higher because of the increased capital per worker, so wages and gross domestic product will also rise, but by less than a factor of two owing to diminishing returns. The higher wages will further raise the capital per worker. The total effect might be to raise per capita income and wage rates by about 40 percent.[15] More capital per worker will mean a lower marginal productivity of capital, and lower profit rates and interest rates. Lower interest rates will in principle alter decisions about how to allocate consumption over the life cycle and savings decisions.

It also is not plausible to expect the acquisition of knowledge by an enrolled student to occur at only half the rate after a rescaling of the life cycle as it did before, as strict proportional rescaling would require. If the time in school doubles, we would also expect the stock of knowledge acquired to roughly double. Longer education and greater knowledge would also raise the productivity of labor and wages throughout the life cycle, interacting with the effects of capital that were just discussed. What would then happen to the rate of return to education, which provides the incentive for people to invest in learning? Longer life would increase the payback period, tending to raise the rate of return. However, returns would be decreased by the greater quantity of education. It is also unclear how the life cycle trajectory of labor earnings would be affected, given that accumulation of experience contends with obsolescence of knowledge and skills.

In sum, the proportional expansion of the life cycle does not make sense once we take into account some basic economic ideas. Life cycle expansion, even with proportionality, would imply increased human and physical capital per worker, lower profit rates and interest rates, higher wages and per capita incomes, and altered returns to education and altered life cycle earning trajectories.

How might rising per capita incomes affect people's choices? Often economists assume that the preference functions governing choices among goods and activities are homothetic, meaning that as incomes rise, the utility tradeoffs (marginal rates of substitution) between items do not change if goods are consumed in the same proportion. Such an assumption here would preserve a proportional expansion of some aspects of the life cycle, despite rising incomes. When we think about saving behavior—trading off consumption today against consumption in old age, for example—this assumption may be defensible. However, it is more questionable in the context of the tradeoff between consumption and leisure, which will influence hours worked per day as well as the decision about when to retire.

Conclusions

We have explored the consequences of longer life for the timing of life cycle events between birth and death. We have used proportional rescaling as a baseline against which to compare the actual past and the potential future. Proportionality is not inevitable. In many cases, it is not even likely. In almost no case is the timing of life cycle events changing at exactly the same pace as longevity itself is increasing. Still, proportional rescaling provides a starting point, a simple framework, from which to view the largely unexplored consequences of increasing longevity for the timing of different segments of life. We do not suggest that all the changes we considered are directly or indirectly caused by increased longevity. In some cases there are plausible links, but in many cases the causes of change are not in any obvious way connected to mortality change.

In order to get the broadest picture, it is useful to set aside issues of the exact pace at which changes are occurring and to ask simply which life cycle stages are changing in a direction consistent with proportionality. As we live longer, we are indeed spending more time in school, staying single longer, delaying reproduction, delaying entry into the work force, and staying healthy longer. On the other hand, childhood at least as it is socially defined is not lengthening, and working life has shrunk. Biologically, human growth, maturity, and menopause have remained nearly fixed, even as life has been extended.

The barriers to proportional rescaling are behavioral, institutional, and biological. From a behavioral point of view, people have chosen to allocate increasing shares of life to leisure. As we have seen over the last century, longer and wealthier life has been accompanied by proportionately more years spent in leisure, not fewer. Although population aging may not allow the historic trend toward earlier retirement to continue, it is likely that leisure time, perhaps spent not only in retirement but also before entry into the labor force and perhaps during the working years, will increase at a faster than proportional rate with increased longevity.

Institutions also form barriers to proportional rescaling. For example, incentives for early retirement have grown with the generosity of pension programs in recent decades, amplifying the behavioral effects. Similarly, age-graded eligibility, whether it be for insurance, military service, or tax treatment, makes the timing of life cycle stages less elastic than it might otherwise be. Institutions may ultimately adapt to behavioral preferences, but they will do so slowly.

Finally, biological barriers to proportional rescaling appear. Because increased longevity is not a result of evolutionary forces that fundamentally change human biology, changes in mortality rates are not accompanied by changes in the timing of human growth, the length of time infants are physically dependent on their parents, sexual maturity, or menopause. Human aging may be slowing, but this is not yet influencing the timing of human development. However, biological constraints do not yet block proportional increases, at least in terms of weak proportionality, in the timing of reproduction. The mean age of childbearing could rise considerably before it reached current biological limits.

This chapter only touched on the wide range of issues that increasing longevity might imply for the reorganization of human life. Among the many worthwhile topics to pursue are the consequences of rescaling for a number of life cycle models. Mincer's (1974) formulation of human capital accumulation might be explored, looking specifically at the effect of proportional increases in education on entry into the labor force and on the age profiles of earnings. Under what conditions would the earnings peak itself move proportionally? What would happen to lifetime income? What would be the consequences for the opportunity costs of childbearing? A second area for research is the aggregate economic consequences of longer life. We have seen that even under proportionality, the consequences of longer life are not economically neutral, since human and physical capital per person would grow with the square of longevity. Further investigation of the equilibria implied by both proportional rescaling and different scenarios of nonproportional change would be revealing.

Another issue to consider is the plausibility of repetition as an alternative to elongation of life cycle states. To some extent, the increase in remarriage and in formation of "second" families suggests that many aspects of life may not be so constrained as we have suggested from our emphasis on stock constraints. Several distinct periods of schooling could be an alternative to simply adding years of schooling at the beginning of life. One could imagine that several careers, several families, and several hometowns could emerge with increased longevity. The viability of the repetitive life cycle as opposed to the elongated life cycle will depend on economic factors such as the depreciation rate of human capital with time, the earnings trajectory and value of experience in a particular career, and changing perceptions of social issues that make the unity of the life course a defining element of human identity. Perhaps the repeated life cycle may be psychologically unappealing, if the value of social networks of family, neighbors, and colleagues is so strong as to make full replacement impossible.

Finally, rescaling confronts measures of time that are external to the human life cycle. For many animals and plants, the length of the seasons is a fundamental unit of time, an underlying metronome that does not allow continuous rescaling. For humans, underlying pacemakers of life, external to human behavior, are some of the main obstacles to simple proportional change. Human skills degrade at some rate with nonuse, we learn at some rate, children grow up at some rate, social bonds are created and dissolved at some rate, technology changes at some rate, and so forth. To a large extent, the effect of longer life on the rescaling and reorganization of the life cycle will be determined by what happens to people's valuation of time in a general sense. The economic value of time in terms of productivity and earnings is one aspect of this. Another aspect is the various units of time: the workweek, the school year, and other rhythms of human life.

Notes

The first author's research for this chapter was funded by a grant from NIA, R37-AG11761.

1 As an example, consider the life table survival function, $l(x)$, and the density of deaths at age x, $d(x)$. Consider a proportional rescaling such that new age y equals x/c. If the new functions are l^* and d^*, then for $c=2$ we would have $l^*(100) = l(50)$. That is, under the new mortality regime, the proportion of people now survive to 100 that used to survive to 50. The density of deaths is given by $d(x) = -dl(x)/dx$ (where the d for derivative should not be confused with the d for deaths). It follows that $d^*(x) = (1/c)d(x/c)$. That is, not only is the $d(x)$ curve stretched out, but its level is also reduced by the factor $1/c$ at each x. $d(x)$ is stock constrained, because it must integrate across all ages to 1.0 or to the radix of the life table.

2 Another way to assess proportional stretching is to plot the old and new survival curves against the logarithm of age. The same horizontal displacement will then correspond to a proportionate increase in age from any starting point. Under strong proportionality, the horizontal distance between the two survival curves should be a constant, in our example equal to log(2). If the two curves being compared are rates, expressed per unit time, then the new curve should first be multiplied by c before plotting against the logarithm of age (see note 1 above).

3 Formally, for any s, $m(x,t) = m(x+\rho s/\theta, t+s)$.

4 The probability of surviving to age 50 in the 1995 period US life table for sexes combined was .925, and it is projected to be .965 in 2080.

5 Costa (2000) reports that from the early twentieth century to the 1990s, the average rate of functional disability for men aged 50 to 74 declined at 0.6 percent per year. Crimmins et al. (1997) have found similar rates of decline for recent decades, while Manton et al. (1997) and Freedman and Martin (1999) find considerably more rapid rates of decline, and Manton and Gu (2001) report accelerating rates of decline in chronic disability since 1982.

6 Analyses of data from the Social Security disability insurance program (DI) tell a different story, but those disability rates are dominated by behavioral responses to the incentives of the program and appear to be less relevant than direct measures of illness or functional status.

7 Couple formation, particularly in Sweden, is poorly measured by marriage alone, since cohabitation is so common. Levels of cohabitation are lower in the United States, and lower still in Japan.

8 For example, both Sweden and the United States lowered their voting ages from 21 to 18 during the 1960s and 1970s.

9 The baby boom years following World War II were accompanied by a dip in the mean age of childbearing.

10 Note that it is incorrect to base a calculation of this sort on the change in life expectancy at age 65. The probability of surviving from age 20 to age 65 increased from 0.52 in 1900 to 0.82 in 1995, for example. The correct calculation is based on T65/(T20−T65), which is the ratio of years lived after age 65 to those lived during ages 20–64 over the individual life cycle. These calculations do not take into account the distribution around the mean age of retirement.

11 Unfortunately, Ausubel and Grubler worked with life expectancy and not survival distributions, an approach that exaggerates the size of these proportional declines in work time.

12 The public pension programs in the United States and Japan stand out has having relatively little incentive for early retirement. In the US, at least, many employer-provided plans do have strong incentives, however.

13 Consumption in old age may also be funded in part by transfers from workers, as with pay-as-you-go public pension systems. The argument in this paragraph applies to the portion of consumption in retirement that is funded through private savings or employer-provided pensions. For simplicity the calculations ignore the return to investments in capital.

14 Under stock-constrained proportional stretching, the flow of births would actually be only half as great as before, so there would be only one-half person per generation. The total labor force size would be unchanged, but there would be twice as much capital, since the average worker holds twice as much capital as can be seen from Figure 5. All that really matters is the ratio of capital to labor, so the size of generations in the stationary population is irrelevant.

15 Suppose that per capita income is proportional to the capital/labor ratio raised to the 1/3 power, a standard assumption, and that the new capital/labor ratio equals the old times 2 times the ratio of new to old per capita income. Solving, we find that the ratio of per capita incomes equals the square root of 2, or about a 40 percent increase.

References

Ahlburg, Dennis and James Vaupel. 1990. "Alternative projections of the U.S. population," *Demography* 27(4): 639–652 (November).

Austad, Steven N. 1997. *Why We Age: What Science Is Discovering about the Body's Journey Through Life*. New York: J. Wiley & Sons.

Ausubel, Jesse and Anrulf Grubler. 1995. "Working less and living longer: Long-term trends in working time and time budgets," *Technological Forecasting and Social Change* 50(3): 195–213.

Bongaarts, John and Griffith Feeney. 1998. "On the quantum and tempo of fertility," *Population and Development Review* 24(2): 271–291.

Biddle, F. G., S. A. Eden, J. S. Rossler, and B. A. Eales. 1997. "Sex and death in the mouse: Genetically delayed reproduction and senescence," *Genome* 40: 229–235.

Carey, J. R. et al. 1998. "Dual modes of aging in Mediterranean fruit fly females," *Science* 281: 996–998.

Calhoun, Charles A. and Thomas J. Espenshade. 1988. "Childbearing and wives' foregone earnings," *Population Studies* 42(1): 5–37.

Charnov, Eric L. 1993. *Life History Invariants: Some Explorations of Symmetry in Evolutionary Ecology*. New York: Oxford University Press.

Costa, Dora. 1998. *The Evolution of Retirement: An American Economic History, 1880–1990*. Chicago: University of Chicago Press.

———. 2000. "Long-term declines in disability among older men: Medical care, public health, and occupational change," National Bureau of Economic Research, Working Paper Series W7605 (NBER, Cambridge, MA), pp. 1–40.

Crimmins, Eileen, Yasuhiko Saito, and Dominique Ingegneri. 1997. "Trends in disability-free life expectancy in the United States, 1970–90," *Population and Development Review* 23(3): 555–572.

Crimmins, Eileen, Sandra L. Reynolds, and Yasuhiko Saito. 1999. "Trends in the health and ability to work among the older working-age population," *Journal of Gerontology* 54B(1): S31–S40.

Durand, John. 1977. *The Labor Force in Economic Development*. Princeton, NJ: Princeton University Press.

Eveleth, Phyllis B. and J. M. Tanner. 1990. *Worldwide Variation in Human Growth*. Cambridge: Cambridge University Press.

Finch, C. E. 1990. *Longevity, Senescence, and the Genome.* Chicago: University of Chicago Press.

Freedman, Vicki and Linda Martin. 1999. "The role of education in explaining and forecasting trends in functional limitations among older Americans," *Demography* 36(4): 461–473 (November).

Fries, James. 1980. "Aging, natural death, and the compression of morbidity," *The New England Journal of Medicine* 303: 130–136.

Funatsuki, Kakuchi. 2000. "Big changes seen under new juvenile law," *Yomiuri Shimbun,* 2 Nov., p. 3.

Goldscheider, Frances and Calvin Goldscheider. 1999. *The Changing Transition to Adulthood: Leaving and Returning Home.* Thousand Oaks, CA: Sage Publications.

Goldstein, Joshua R. and Wilhelm Schlag. 1999. "Longer life and population growth," *Population and Development Review* 25(4): 741–747.

Gruber, Jonathan and David Wise. 1997. "Introduction and summary," *Social Security Programs and Retirement Around the World.* Cambridge, MA: National Bureau of Economic Research, Working Paper Series, W6134.

Japan National Institute of Population and Social Security Research (Kokuritsu Shakai Hosho Jinko Mondai Kenkyujo). 2000. *Latest Demographic Statistics* (Jinko tokei shiryoshu), Research Series No. 299, 20 September.

Japan Ministry of Health, Labor and Welfare (Koseisho Daijin Kambo Tokei Chosabu). 1999. *Vital Statististics of Japan 1999,* Volume 1 (Jinko dotai tokei).

Kaplan, Hillard. 1997. "The evolution of the human life course," in Kenneth W. Wachter and Caleb E. Finch (eds.), *Between Zeus and the Salmon: The Biodemography of Longevity* Washington, DC: National Academy Press, pp. 175– 211.

Kaplan, Hillard and D. Lam. 1999. "Life history strategies: The tradeoff between longevity and reproduction," paper presented at the Annual Meeting of the Population Association of America, New York, March.

Kotlikoff, Laurence. 1981. "Some economic implications of life span extension" in J. March and J. McGaugh (eds.), *Aging: Biology and Behavior.* New York: Academic Press, pp. 97–114. Reprinted as Chapter 14 in *What Determines Savings?* Cambridge, MA: MIT Press, pp. 358–375.

Lee, Ronald. 1994. "The formal demography of population aging, transfers, and the economic life cycle," in Linda Martin and Samuel Preston (eds.), *The Demography of Aging.* Washington, DC: National Academy Press, pp. 8–49.

Lee, Ronald and Lawrence Carter. 1992. "Modeling and forecasting U.S. mortality," *Journal of the American Statistical Association* 87(419): 659–671.

Lee, Ronald and Shripad Tuljapurkar. 1997. "Death and taxes: Longer life, consumption, and social security," *Demography* 34(1): 67–82.

Lin, K., J. B. Dorman, A. Rodan, and C. Kenyon. 1997. "Daf-16: An HNF-3/forkhead family member that can function to double the life span of *Caenorhabditis elegans*," *Science* 278: 1,319–1,332.

Lubitz, J. and R. Prihoba. 1984. "The use of Medicare services in the last two years of life," *Health Care Financing Review* 5: 117–131.

Lubitz, J., J. Beebe, and C. Baker. 1995. "Longevity and medicare expenses," *New England Journal of Medicine* 332: 999–1,003.

Lumsdaine, Robin L. and David A. Wise. 1994. "Aging and labor force participation: A review of trends and explanations," in Yukio Noguchi and David Wise (eds.), *Aging in the United States and Japan.* Chicago: University of Chicago Press, pp. 7–41.

Manton, Kenneth, Eric Stallard, and H. Dennis Tolley. 1991. "Limits to human life expectancy: Evidence, prospects, and implications," *Population and Development Review* 17(4): 603–638.

Manton, Kenneth, Larry Corder, and Eric Stallard 1997. "Chronic disability trends in elderly United States populations: 1982–1994," *Proceedings of the National Academy of Sciences* 94: 2,593–2,598.

Manton, Kenneth and XiLiang Gu. 2001. "Changes in the prevalence of chronic disability in the United States black and nonblack population above age 65 from 1982 to 1999," *Proceedings of the National Academy of Sciences* 98(11): 6,354–6,359.

Marini, M. M. 1987. "Measuring the process of role change during the transition to adulthood," *Social Science Research* 16: 1–38.

Millar, J. S. and R. M. Zammuto. 1983. "Life histories of mammals: An analysis of life tables," *Ecology* 64: 631–635.

Miller, Tim. 2001. "Increasing longevity and Medicare expenditures," *Demography* 38(2): 215–226.

Mincer, Jacob. 1974. *Schooling, Experience, and Earnings.* New York: Columbia University Press, for the National Bureau of Economic Research.

Modell, John, Frank F. Furstenberg, Jr., and Theodore Hershberg. 1976. "Social change and transitions to adulthood in historical perspective," *Journal of Family History* 1: 7–32.

National Center for Health Statistics. 2000. *Vital Statistics of the United States, 1997, Volume I, Natality, Third Release of Files* «http://www.cdc.gov/nchs/datawh/statab/unpubd/natality/natab97.htm» Table 1–5. Median age of mother by live-birth order, according to race and Hispanic origin: United States, selected years, 1940–97 (released 8/2000).

National Academy on an Aging Society. 2000. "Who are young retirees and older workers?" *Data Profiles: Young Retirees and Older Workers*, June, no. 1.

Neumark, David. 2001. "Age discrimination legislation in the United States," National Bureau of Economic Research, Working Paper Series 8152 (NBER, Cambridge, MA), pp. 1–44.

Rindfuss, Ronald R., C. Gray Swicegood, and Rachel A. Rosenfeld. 1987. "Disorder in the life course: How common and does it matter?" *American Sociological Review* 52(6): 785–801.

Schneider, Edward L. and Jack M. Guralnik. 1990. "The aging of America: Impact on health care costs," *Journal of the American Medical Association* 263(17): 2,335–2,340.

Shoven, John B., Michael D. Topper, and David A. Wise. 1994. "The impact of the demographic transition on government spending," in David Wise (ed.), *Studies in the Economics of Aging.* Chicago: University of Chicago Press, pp. 13–33.

Sweden. Central Statistical Bureau. 1995. *Population Statistics 1995* (Befolkningsstatistik 1995), Volume 4.

Tuljapurkar, Shripad, Nan Li, and Carl Boe. 2000. "A universal pattern of mortality decline in the G-7 countries," *Nature* 405: 789–792.

Vaupel, J. W. 1986. "How change in age-specific mortality affects life expectancy," *Population Studies* 40(1): 147–157 .

Willis, Robert J. 1973. "A new approach to the economic theory of fertility behavior," *Journal of Political Economy* 81(2): S14–S64.

Wilmoth, John R., Leo J. Deegan, Hans Lundström, and Shiro Horiuchi. 2000. "Increase of maximum life-span in Sweden, 1861–1999," *Science* 289: 2,366–2,368.

Wise, David. 1997. "Retirement against the demographic trend: More older people living longer, working less, and saving less," *Demography* 34(1): 83–96.

Survival Beyond Age 100:
The Case of Japan

JEAN-MARIE ROBINE

YASUHIKO SAITO

The highest reported age at death is now well above 110 years and appears to increase over time (Wilmoth and Lundström 1996; Wilmoth et al. 2000; Robine and Vaupel 2001; Wilmoth and Robine, in this volume). The number of centenarians has doubled every decade since 1960 in low-mortality countries (Vaupel and Jeune 1995), refuting the belief held two decades ago that this number could not increase (Fries 1980). Much controversy about the limits to life expectancy continues among demographers (Bonneux et al. 1998; Vaupel et al. 1998; Wilmoth 1998b; Horiuchi 2000; Tuljapurkar et al. 2000; Olshansky et al. 2001; Wilmoth 2002; Oeppen and Vaupel 2002). These past demographic changes and current uncertainties raise questions about the potential for human longevity and our collective future. The most important questions relate to the limits to life expectancy and to the future number of the oldest old, both nonagenarians and centenarians. In addition, there is particular interest in the health status of the oldest old. The levels of functional ability and robustness or frailty are major determinants of the quality of life of the oldest old. It is important to know whether life expectancy in good health can progress proportionally with total life expectancy (Robine, Romieu, and Cambois 1999). And it is important to better understand the demographic trends currently at work among the oldest old.

In this chapter we describe the emergence of centenarians in Japan. We tabulate the number of centenarians (and subsequently those aged 105 and older and 110 and older) and determine the effect of the size of birth cohorts on the trends in the number of centenarians. Second, we investigate trends in the number of deaths of centenarians (and subsequently of those aged 105 and older and 110 and older) and the maximum age at death. These data on deaths allow us to explore trends in the number of centenarians earlier in the twentieth century and to demonstrate the effect of past age misreporting on these trends. Finally, we combine the population and death data to examine death rates above the age of 100 in Japan since 1963. We also look at seasonal mortality as one way of evaluating the frailty of Japanese centenarians.

Background

Reliable data on the age at death above age 100 years depend upon a long-standing, reliable birth registration system. Only a few countries can provide useful data on age at death at advanced ages during recent decades. Swedish data, which provide the longest reliable records (Dupâquier and Dupâquier 1985), suggest that the secular increase in the highest age at death reported every year accelerated after 1970, resulting in an increase of three years for the last 30 years (Wilmoth et al. 2000). The International Database on Longevity (IDL), which includes persons having reached their 110th birthday, suggests an even steeper increase over the last 20 years, with the highest validated age at death increasing from 112 years in 1980 to 122 years at the end of the 1990s (Robine and Vaupel 2001 and 2002). French data, unfortunately censored at the age of 110 years for several years by the French Statistics office, suggest a similar acceleration of the increase in the highest age at death (Meslé et al. 2000). The emergence of centenarians has been extensively studied in England and Wales (Thatcher 1992, 1997, 1999a, 2001), Denmark (Kjaergaard 1995; Skytthe and Jeune 1995; Jeune and Skytthe 2001), Belgium (Poulain, Chambre, and Foulon 2001), and France (Vallin and Meslé 2001). These studies have been confined to a few countries in Western Europe, none of them very large. Japan can provide unique information on demographic trends for the extremely old population: currently its population is 14 times that of Sweden and its life expectancy is the highest in the world.

In 2000 life expectancy at birth in Japan was 77.6 years for males and 84.6 years for females (abridged life tables for Japan 2000), close to the limit of 85 years for life expectancy at birth hypothesized by James Fries two decades ago (Fries 1980). In Japan, increases in life expectancy have occurred without any observed change in its rate of increase over the past 15 years (National Institute of Population and Social Security Research 2002). There appears to be no slowing in the fall of mortality at the oldest ages as life expectancy has increased. For instance, life expectancy at birth increased by 1.3 years between 1995 and 2000 for males and by 1.8 years for females.[1] The probability of surviving from birth to age 80 increased from 48.4 percent in 1995 to 52.2 percent in 2000 for males and from 70.5 percent to 74.4 percent for females. Over the same time interval the probability of surviving from age 80 to age 100 increased from 1.0 percent to 1.7 percent for males and from 4.0 percent to 7.0 percent for females.

Data

A national registration law (*Koseki-Ho* in Japanese) was enacted in Japan in 1871 at the beginning of the Meiji Era, and a national registration system (*Jin-Shin Koseki*) was implemented in 1872. It was the first such system to cover the entire population of the country. Japan's earlier system did not include those who were classified as humble people and often did not register small children (Sekiyama 1948). The new system began with an enumeration of all living persons, and subsequently all births and deaths were registered. A chief of households was responsible for registering events for a specific number of households with the regional government office (Taeuber 1958). Individual records were kept in the local municipalities. The enactment in 1886 of a penalty for failing to report births or deaths suggests that it took some time for the registration system to become complete and accurate (Kitou 1997).

The national registration system had a mechanism to follow those who moved out of a municipality. As economic development in Japan produced an increase in the volume of migration, the national system was deemed insufficient for tracking migrants, and a resident registry system was established in 1914. The system carries information on place of original registration and date of birth. The date of death is reported to the municipality of the place of residence. Municipalities have compiled lists of centenarians (*Zenkoku koureisha meibo*) from the resident registration system since 1963.[2] Thus these data provide information on the number of centenarians for the last 40 years. Data on the number of births since 1872 were obtained from the Statistical Yearbook of the Empire of Japan (Statistics Bureau, various years), which has been published since 1882.

However, since the age of those born before 1872 was self-reported at the time of the implementation of the national registration system, the reported ages of those born before 1872 and in the early years following 1872 should be used with caution. Detailed statistics on Japanese mortality, published since 1899 by the Ministry of Health and Welfare (Vital Statistics of Japan), provide information on the number of deaths of centenarians and the highest reported ages at death.

The emergence of centenarians in Japan

The number of persons aged 100 years and older in Japan increased from 154 in September 1963 (20 men and 134 women) to 13,036 in September 2000 (2,158 men and 10,878 women). Thus, the number of centenarians grew by a factor of 100 in 38 years. The increase appears exponential (see Figure 1). In fact while there are large yearly fluctuations, the rate of increase itself tends to increase. The centenarian doubling time (CDT) decreased from nearly 6 years in the 1960s to around 4.8 years at the end of the 1990s.

FIGURE 1 Number of persons 100 years old and older,
Japan, 1963–2000, by sex

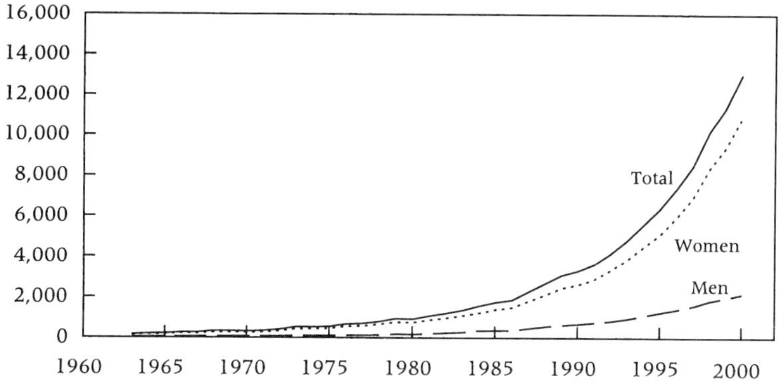

SOURCE: Ministry of Health and Welfare (1963–2000) *Zenkoku koureisha meibo* (A list of centenarians in Japan).

One possible reason for an increase in the number of centenarians would be an increase in the number of births 100 years earlier. The number of centenarians and births 100 years earlier are presented in Table 1. Figure 2 shows that the number of births increased markedly in Japan during the last years of the nineteenth century, from 569,034 births[3] in 1872 to 1,420,534 births in 1900 (that is, by a factor of 2.5), explaining some of the increase in the number of centenarians alive one century later. This increase in births contrasts with a relatively constant or even declining number of births in European countries such as France during the same period.

To assess the increase in the number of centenarians in Japan independent of the increase in births, we calculate the number of people aged 100 per 10,000 members of each birth cohort. We then calculate the num-

FIGURE 2 Number of births in Japan, 1872–1900

SOURCES: For 1872–99: Statistics Bureau (various years), *Statistical Yearbook of the Empire of Japan*; for 1900: National Institute of Population and Social Security Research (2002), *Latest Demographic Statistics 2001/2000*, NIPSSR, Tokyo.

TABLE 1 Number of centenarians in Japan on January 1st: Total number at age 100, 100+, 105+, and 110+, by sex and year, 1963–2000, and total number of births 100 years earlier, 1872–1900

Year	100 Total	100 Men	100 Women	100+ Total	100+ Men	100+ Women	105+ Total	105+ Men	105+ Women	110+ Total	110+ Men	110+ Women	Births 100 years earlier Year	Births 100 years earlier Number
1963	62	2	60	154	20	134	11	1	10	1	0	1		
1964	110	19	91	191	31	160	12	1	11	1	0	1		
1965	103	24	79	198	36	162	9	1	8	1	0	1		
1966	126	22	104	262	46	216	10	3	7	2	0	2		
1967	99	26	73	261	52	209	6	1	5	2	0	2		
1968	191	37	154	328	67	261	11	1	10	3	0	3		
1969	139	26	113	328	70	258	10	2	8	3	0	3		
1970	127	22	105	310	62	248	13	2	11	0	0	0		
1971	168	31	137	339	70	269	10	4	6	0	0	0		
1972	197	41	156	405	78	327	13	5	8	0	0	0	1872	569,034
1973	266	56	210	532	103	429	22	6	16	2	0	2	1873	809,487
1974	241	44	197	527	96	431	18	5	13	2	0	2	1874	836,113
1975	253	43	210	548	102	446	22	6	16	2	1	1	1875	869,126
1976	348	57	291	666	113	553	25	3	22	2	1	1	1876	902,946
1977	327	56	271	698	122	576	22	8	14	1	1	0	1877	890,518
1978	386	68	318	792	132	660	25	8	17	1	1	0	1878	874,883
1979	451	101	350	937	180	757	27	7	20	1	1	0	1879	876,719
1980	402	70	332	913	165	748	23	4	19	1	1	0	1880	883,584
1981	499	98	401	1,072	202	870	34	5	29	2	1	1	1881	941,343
1982	529	108	421	1,200	233	967	37	6	31	1	1	0	1882	922,715
1983	624	132	492	1,354	269	1,085	44	4	40	1	1	0	1883	1,004,989
1984	732	178	554	1,563	347	1,216	60	10	50	1	1	0	1884	975,252
1985	774	157	617	1,740	359	1,381	67	12	55	1	0	1	1885	1,024,574
1986	796	168	628	1,851	361	1,490	65	8	57	1	0	1	1886	1,050,617
1987	1,088	225	863	2,271	462	1,809	72	12	60	1	0	1	1887	1,058,137
1988	1,259	298	961	2,668	562	2,106	104	18	86	2	0	2	1888	1,172,729
1989	1,390	302	1,088	3,078	630	2,448	123	24	99	6	0	6	1889	1,209,910
1990	1,443	314	1,129	3,298	680	2,618	119	14	105	5	0	5	1890	1,145,374
1991	1,620	373	1,247	3,625	749	2,876	133	19	114	5	0	5	1891	1,086,775
1992	1,871	388	1,483	4,152	822	3,330	140	24	116	4	0	4	1892	1,207,034
1993	2,130	430	1,700	4,802	943	3,859	185	30	155	4	0	4	1893	1,178,428
1994	2,494	509	1,985	5,593	1,093	4,500	249	40	209	8	1	7	1894	1,208,983
1995	2,982	640	2,342	6,378	1,255	5,123	237	30	207	8	1	7	1895	1,246,427
1996	3,168	619	2,549	7,373	1,400	5,973	290	45	245	10	1	9	1896	1,282,178
1997	3,642	722	2,920	8,491	1,570	6,921	331	50	281	10	1	9	1897	1,334,125
1998	4,498	837	3,661	10,158	1,812	8,346	400	54	346	11	1	10	1898	1,369,638
1999	4,776	885	3,891	11,346	1,973	9,373	458	58	400	6	1	5	1899	1,386,981
2000	5,493	945	4,548	13,036	2,158	10,878	582	80	502	9	0	9	1900	1,420,534

SOURCES: Number of centenarians: Ministry of Health and Welfare (1963–2000), *Zenkoku koureisha meibo* (a list of centenarians in Japan). Number of births: For 1872–99: Statistics Bureau (various years), *Statistical Yearbook of the Empire of Japan*; and for 1900: National Institute of Population and Social Security Research (2002). *Latest Demographic Statistics 2001/2000*, NIPSSR, Tokyo, Japan.

ber of people aged 100 in each year assuming the number of births remained constant at the level of 1873.[4] According to this scenario the number of persons 100 years old would increase from 266 in 1973 to 3,130 in the year 2000 (almost a 12-fold increase) compared to the 5,493 persons actually counted that year (an increase by a factor of almost 21). This means that the greater part of the increase since 1973 in the number of persons 100 years old currently living in Japan (55 percent) is due to an increased probability of surviving, while the rest of the increase (45 percent) is due to the increase in the size of the birth cohorts.[5]

Figure 3 shows that the number of persons 100 years old increased from 3.5 per 10,000 births one century earlier for the 1873 cohort (1.4 for men and 5.6 for women) to 38.7 for the 1900 cohort (13.0 for men and 65.6 for women).[6] This increase, independent of the size of the successive cohorts, appears to be linear on a log scale. While there are large yearly fluctuations in the percentage change in the number of centenarians standardized to the 1873 birth cohort (see Figure 4), the rate of increase is higher at the end of the period. Estimates from the linear trend fitted to the whole period indicate a rate of increase of 7.1 percent in 1975, 9.6 percent in 1985, and 13.4 percent in 2000. When the trend from 1985 onward is calculated, the slope is found to be quite similar to that for the whole period, confirming an accelerating rate of increase.

Using the fitted rates of increase from Figure 4, we can determine the trend in centenarian doubling time. Results indicate that the doubling time decreased from around 10 years in 1973 to about 5 years in 2000, assuming the number of births from 1873 to 1900 to be constant at the 1873 level (Figure 5). The doubling times observed for other countries are comparable to those for Japan in the 1970s (Vaupel and Jeune 1995; Vaupel et al. 1998), but to date no other country has shown such a rapid decrease in centenarian doubling time. This evolution suggests a significant accelera-

FIGURE 3   Number of persons aged 100 years per 10,000 births 100 years earlier, Japan, 1973–2000

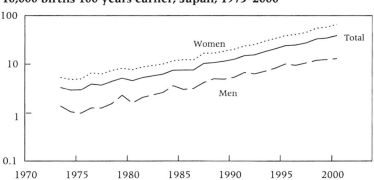

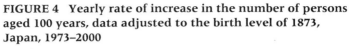

FIGURE 4   Yearly rate of increase in the number of persons aged 100 years, data adjusted to the birth level of 1873, Japan, 1973–2000

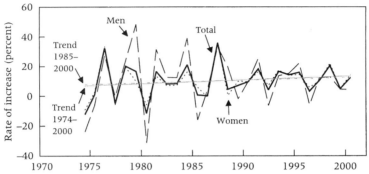

FIGURE 5   Centenarian doubling time in years for the whole population aged 100 years and older (raw data) and for the number of persons aged 100 years (data adjusted to the birth level of 1873), Japan, 1973–2000

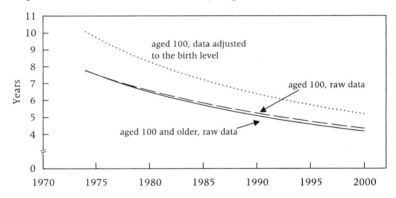

tion in the emergence of the centenarian population in Japan attributable to increased survival.

Population 105 years old and older

The rise in the number of centenarians in Japan was accompanied by an even more remarkable increase in the number of persons aged 105 years and older (Figure 6). This number increased from 11 in September 1963 (1 man and 10 women) to 582 in September 2001 (80 men and 502 women), an increase by a factor of 53 in 37 years.

Assuming the number of births constant at the 1873 level would result in an increase in the number of persons aged 105 years from 11 in 1978 to

FIGURE 6 Number of persons aged 105 years old and older, Japan, 1963–2001, by sex

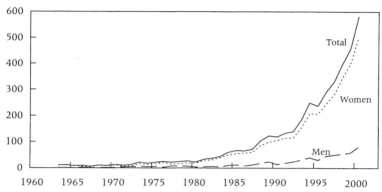

SOURCE: Ministry of Health and Welfare (1963–2001), *Zenkoku koureisha meibo* (A list of centenarians in Japan).

209 in 2000 (a 19-fold increase) compared to 321 persons actually counted that year (about a 29-fold increase). This means that 64 percent of the increase since 1973 in the number of persons 105 years old currently living in Japan is attributable to an increased probability of surviving and 36 percent is attributable to an increase in the size of the birth cohorts. The number of persons 105 years old per 10,000 births one century earlier increased from 0.14 for the 1873 cohort (0.05 for men and 0.23 for women) to 2.58 for the 1895 cohort (0.77 for men and 4.47 for women).[7]

Again there are large observed annual fluctuations in the rate of increase in the number of persons aged 105 years, but a linear regression indicates an almost constant rate of increase of nearly 20 percent over the last 22 years. This would correspond to a doubling time of 3.5 years for those aged 105. With data adjusted for birth cohort size (see Figure 7),[8] the annual increase in the rate is about 17 percent, corresponding to a doubling time of 4 years. This means that the number of persons 105 years old doubled every 3.5 years in Japan for the last two decades and would have doubled every 4 years had the size of the birth cohorts remained constant since 1873.

Supercentenarians

Supercentenarians are persons who have reached their 110th birthday. The list of centenarians alive, published by the Ministry of Health and Welfare of Japan since 1963, has generally included one or two supercentenarians each year from 1963 to 1986, although 1968 and 1969 have 3 cases and the period 1970–72 had none. Since 1983, persons celebrating their 110th

FIGURE 7   Yearly rate of increase in the number of persons aged 105 years, data adjusted to the birth level of 1873, Japan, 1978–2000

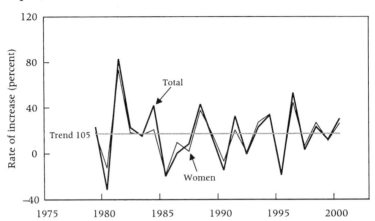

birthday should have been registered at birth. The supercentenarians listed before 1983 were born before the implementation of the national registration system, and reports of their age must be considered with caution. Figure 8 shows a sudden increase in the number of supercentenarians in Japan after 1989, to a number currently around 10.

The number of supercentenarians per 10,000 births one century earlier increased from none for the 1873 cohort (0.012 for the 1875 cohort) to 0.055 for the 1891 cohort (see Figure 9).[9] This series is too short to allow us to assess the trend statistically, but it does suggest an accelerating rate of increase.

FIGURE 8   Number of persons aged 110 years old and older, Japan, 1963–2000, by sex

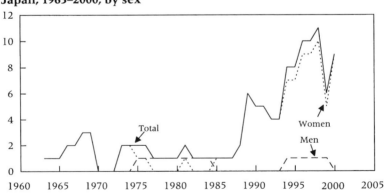

SOURCE: Ministry of Health and Welfare (1963–2000), *Zenkoku koureisha meibo* (A list of centenarians in Japan).

FIGURE 9   Number of persons aged 110 years per
10,000 births 110 years earlier, Japan, 1983–2001

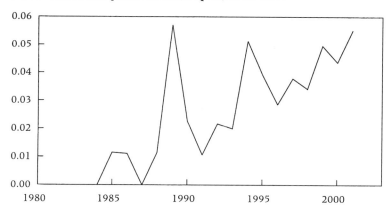

Deaths of centenarians in Japan

The emergence of centenarians in Japan in the early part of the twentieth
century can be measured by the change in the number of deaths of persons
100 years old and older from 1899 onward (Table 2). Although indirect,
the number of deaths gives a robust estimate of the number of centenarians for the period before 1963 when the published list of centenarians begins. Figure 10 indicates not only the expected increase in the number of
centenarians after World War II but also an unexpected rise in the number
of deaths of centenarians roughly between 1905 and 1925. At the peak, the
deaths of 193 female centenarians were recorded in 1913, corresponding to
persons aged 59 and older in 1872 when the national registration system
was implemented.

   Examination of the deaths occurring at age 110 years and older suggests substantial misreporting among the deaths of recorded supercentenarians before the 1950s (Figure 11). Maximum age misreporting occurs
roughly between 1915 and 1935, corresponding to the unexpected increase
in the number of deaths of centenarians described above. Figure 11 also
shows a rise in deaths at age 110 and older in the 1990s. This increase is
based on ages recorded from the established registration system and is thus
more reliable.

   Figure 12 shows the highest reported age at death since 1899. It clearly
demonstrates that the effect of age misreporting at the beginning of the
national registration system continued to affect the data during a large part
of the twentieth century. The highest reported age at death rises steeply
from 1900 to the 1930s and then fluctuates around age 120 years until the
1950s. After this, there is a slow decline with outliers still present in the

TABLE 2   Number of deaths in Japan at age 100, 100–104, and 105+ by sex and year, 1963–98

Year	100			100–104			105+		
	Total	Men	Women	Total	Men	Women	Total	Men	Women
1963	79	16	63	73	16	57	6	0	6
1964	106	12	94	102	12	90	4	0	4
1965	106	25	81	103	25	78	3	0	3
1966	121	26	95	116	26	90	5	0	5
1967	127	18	109	123	16	107	4	2	2
1968	137	28	109	135	28	107	2	0	2
1969	168	41	127	168	41	127	0	0	0
1970	202	38	164	199	37	162	3	1	2
1971	171	42	129	163	41	122	8	1	7
1972	190	54	136	183	49	134	7	5	2
1973	307	80	227	293	77	216	14	3	11
1974	320	64	256	308	61	247	12	3	9
1975	363	82	281	346	76	270	17	6	11
1976	315	61	254	305	60	245	10	1	9
1977	370	69	301	353	66	287	17	3	14
1978	404	78	326	386	73	313	18	5	13
1979	418	84	334	402	80	322	16	4	12
1980	497	100	397	481	99	382	16	1	15
1981	515	95	420	491	95	396	24	0	24
1982	537	108	429	517	104	413	20	4	16
1983	647	137	510	613	128	485	34	9	25
1984	703	151	552	672	146	526	31	5	26
1985	825	192	633	789	182	607	36	10	26
1986	806	184	622	756	175	581	50	9	41
1987	898	182	716	866	179	687	32	3	29
1988	1,183	282	901	1,148	274	874	35	8	27
1989	1,282	299	983	1,215	287	928	67	12	55
1990	1,569	367	1,202	1,493	350	1,143	76	17	59
1991	1,585	377	1,208	1,510	367	1,143	75	10	65
1992	1,704	380	1,324	1,622	370	1,252	82	10	72
1993	2,006	428	1,578	1,922	413	1,509	84	15	69
1994	2,309	511	1,798	2,194	493	1,701	115	18	97
1995	2,780	639	2,141	2,623	601	2,022	157	38	119
1996	2,757	653	2,104	2,630	635	1,995	127	18	109
1997	3,262	734	2,528	3,067	692	2,375	195	42	153
1998	3,750	748	3,002	3,540	725	2,815	210	23	187

SOURCE: Ministry of Health and Welfare (1899–1998), Vital Statistics Annual issue since 1899.

FIGURE 10 Number of deaths of persons aged 100 years old and older, Japan, 1899–1998, by sex

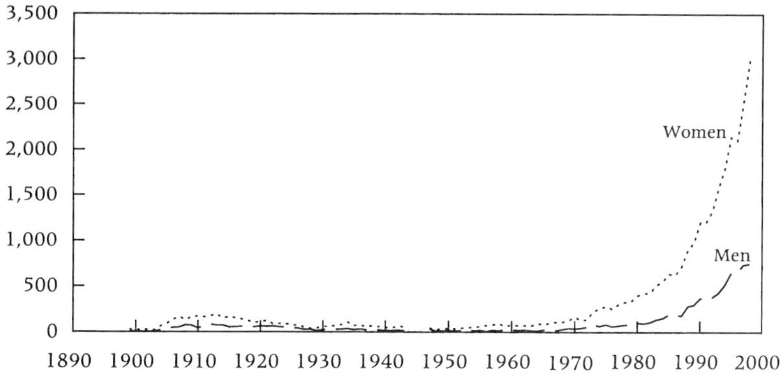

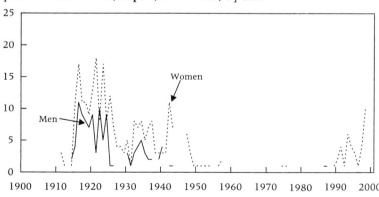

SOURCE: Ministry of Health and Welfare (1899–1998), Vital Statistics Annual issue since 1899.

1960s and in the 1970s for women and the famous case of a Japanese male, Shigechiyo Izumi, who was reported to have been born in 1865 and died in 1986. An age at death of 120 years between 1930 and 1950 corresponds to persons aged between 42 and 62 years in 1872.

Taken together, Figures 10–12 suggest the influence of age misreporting for those aged roughly 40 to 70 years in 1872 when the national registration system was implemented, leading to a maximum recorded number of 265 centenarians in 1913 and of 28 supercentenarians in 1916 and to a highest reported age at death of 125 years in 1943. On the other hand, the three figures also suggest a regular increase not only in the number of centenarians since the 1960s but also in the highest reported age at death.

FIGURE 11 Number of deaths of persons aged 110 years old and older, Japan, 1899–2000, by sex

SOURCE: Ministry of Health and Welfare (1899–1998), Vital Statistics Annual issue since 1899.

FIGURE 12 Highest reported age at death, Japan,
1899–1998, by sex

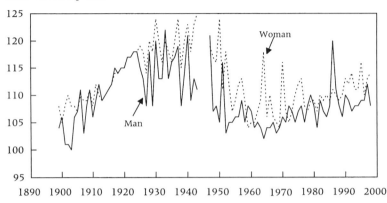

SOURCE: Ministry of Health and Welfare (1899–1998), Vital Statistics Annual issue since 1899.

Highest age at death

When the highest ages at death are plotted together with the tenth highest
ages at death since 1899 to eliminate the possible outliers, they seem to
show the real trend in the extreme age at death during the second half of
the twentieth century (Figure 13). A steady increase in the tenth highest
age at death for men is shown from around 100 years in 1950 to around
103 years in 1980 and to around 105 years at the end of the century with a
peak of 107 years in 1997. Going back to the 1940s, this trend in the highest age at death and tenth highest age at death suggests some acceleration
in the increase in the tenth highest age at death and also suggests that the
highest age at death is about three years higher than the tenth highest age
at death.

    For women, an acceleration of the increase in the tenth highest age at
death is clearly shown from 102 years in 1950 to 105 years in 1980 and 110
years in 1998. The highest age at death is about four years higher than the
tenth highest. These results confirm previous results in Sweden (Wilmoth
et al. 2000) and elsewhere (Wilmoth and Robine, in this volume).

Mortality above age 100 years in Japan

Figure 14 shows the change in annual death rates at ages 100–104 years
and at age 105 and older since 1963 in Japan by sex. The fluctuations mostly
disappear after 1973 if we eliminate the death rate at age 105 and older for
men. For women, the death rate at ages 100–104 decreases from around 50
percent in 1975 to about 35 percent in 1998 (see Figure 15). Despite larger
fluctuations, Figure 15 also suggests a decrease in the death rate for women

FIGURE 13 Highest and 10th highest reported age at death, Japan, 1899–1998, by sex

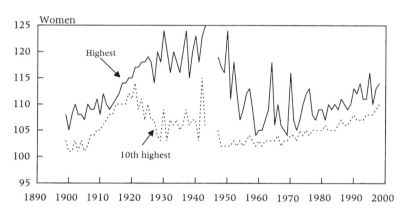

SOURCE: Ministry of Health and Welfare (1899–1998), Vital Statistics Annual issue since 1899.

at age 105 and older from about 70 percent in 1975 to around 50 percent at the end of the twentieth century. During the same period, the death rate at ages 100–104 for men decreased from around 58 percent in 1975 to about 42 percent in 1998.

The seasons of death in Japan

The final two figures indicate seasonality of deaths using information by month and then by season for the deaths of all centenarians in Japan since 1951. Figure 16 shows a clear contrast in the number of deaths between June, which appears to be the most favorable month, and the three winter months of December, January, and February. Twenty percent of the deaths of the male centenarians occurred in summer versus nearly 32 per-

FIGURE 14   Annual death rate at age 100–104 years and at age 105 and older, Japan, 1963–98, by sex

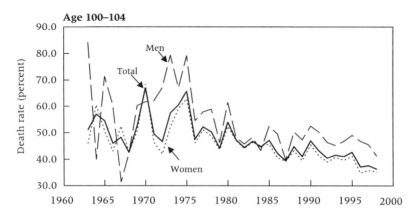

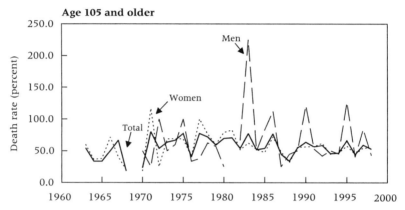

NOTE: In theory the annual death rate can exceed 100 percent for the highest ages. However, the values reached by men aged 105 and older in 1983 are due to age inaccuracy combined with small numbers.

FIGURE 15   Annual death rate at age 100–104 years and at 105 and older, Japan, 1975–1998, females

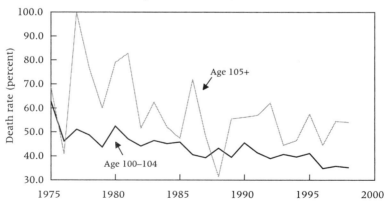

cent in winter when 25 percent are expected each season; these figures are 21 percent and 31 percent respectively for female centenarians (see Figure 17 and Table 3).

The figures suggest that centenarians are frail. Their level of mortality varies considerably with environmental change associated with the passing of seasons, which is of little threat to young adults. On the other hand, the figures suggest that the mortality level of centenarians could still decrease significantly with better control of the detrimental environmental changes related to the excess seasonal mortality. If mortality could be kept at its summer level year round, the death rate of Japanese centenarians would decrease by 20 percent for males and by 16 percent for females.

FIGURE 16   Number of deaths of persons aged 100 years old and older according to month, Japan, 1951–99, by sex

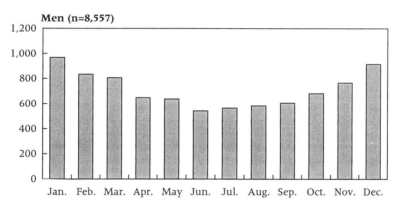

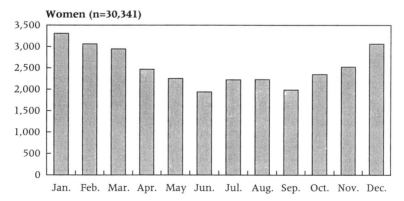

SOURCE: Ministry of Health and Welfare (1951–99), Vital Statistics Annual issue.

FIGURE 17   Number of deaths of persons 100 years old and older according to season: Observed percentage deviation from expected proportion of deaths, Japan, 1951–1999, by sex

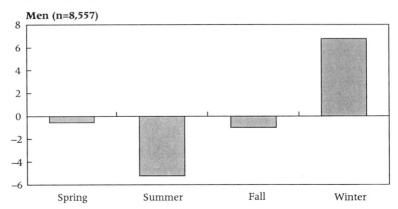

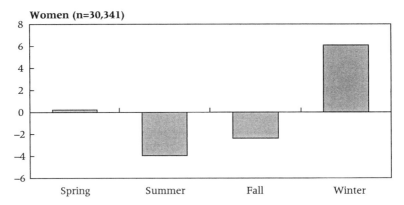

SOURCE: Ministry of Health and Welfare (1951–99), Vital Statistics Annual issue.

TABLE 3   Percent distribution of deaths according to season, Japan, 1970s–1990s, by sex

	Spring (March–May)	Summer (June–August)	Fall (September–November)	Winter (December–February)
Men				
1990s	24.7	19.8	24.2	31.3
1980s	24.2	20.4	24.0	31.4
1970s	23.8	21.0	23.3	31.9
Women				
1990s	25.0	21.3	22.9	30.8
1980s	26.2	21.3	21.8	30.7
1970s	24.2	20.3	22.8	32.8

SOURCE: Ministry of Health and Welfare (1970–99), Vital Statistics Annual issue.

Discussion

Our analysis uses unique data from the country where life expectancy at birth is the highest in the world, with females approaching the limit of 85 years suggested by Fries (1980). The data allow careful monitoring of demographic changes for at least the last quarter of the twentieth century, a time when almost all living Japanese have been registered at birth through the national system that was implemented in 1872.

The future trend in mortality at the highest ages in low-mortality countries, such as Japan, still provokes controversy. Some researchers conclude that future gains in life expectancy will be measured in days or months rather than years, unless the aging process itself can be brought under control (Olshansky, Carnes, and Désesquelles 2001); others expect longer gains (Vaupel and Lundström 1996; Vaupel et al. 1998; Oeppen and Vaupel 2002). Still other researchers have hypothesized that in a country like Japan, which is moving rapidly through the health transition, a period life expectancy higher than 85 years is possible at the end of the transition but only for a while. Mortality is expected to increase after the effect of high mortality selection early in life disappears as birth cohorts exposed to high pretransitional death rates die out (Bonneux, Barendregt, and Van der Maas 1998). Wilmoth (1998a) questions how long Japan will keep its avant-garde position and whether the rate of increase in life expectancy in Japan will converge toward that of other low-mortality countries.

Our results suggest a strong and continuing acceleration in the emergence of the centenarian population in Japan, with the adjusted centenarian doubling time being halved in just 25 years. Inaccurate data and age misreporting in the past are responsible for large fluctuations in several low-mortality countries, although more recent trends are clearer (Jeune and Skytthe 2001). In Japan, the introduction of a national registration system in 1872 makes it easier to distinguish periods where age misreporting and inaccurate data may be a problem. In recent years, with reliable data, there appears to be a steady increase in the highest reported age at death, with an acceleration of this phenomenon for women. Death rates at ages 100–104 significantly decreased for men and women during the last 25 years; this was also true at age 105 and older for women. There is no sign that the avant-garde country is losing its position.[10] Our results also suggest that the mortality level of centenarians could still decrease significantly if Japan were able to control the detrimental environmental changes responsible for the seasonal excess of mortality, suggesting an important plasticity in longevity with regard to environmental conditions (Robine 2001a).

Recently, several approaches have been proposed to elucidate further the determinants and limits of human longevity. These include studying the causes of the increase in the number of centenarians in terms of the relative contributions of increase in the number of births, increase in the probability of survival to age 80, and increase in the probability of survival from age 80 to 100 (Thatcher 1999a). Additional information can be gathered from clarification of the mortality trajectory of the oldest old (Vaupel et al. 1998; Robine and Vaupel 2001, 2002), from the rate of aging implied by this trajectory (Thatcher 1999b), from the shape of the survival curve (Robine 2001b), and from the concentration of life durations using different indicators such as the interquartile range (Wilmoth and Horiuchi 1999), the concentration of deaths around the mode (Kannisto 1999), or the life expectancy at the mode of the life durations (Kannisto 2001).[11] Japanese data offer the opportunity to follow each of these avenues in the country whose life expectancy is the highest in the world.

Notes

We thank Shiro Horiuchi, Carol Jagger, and Roger Thatcher for their remarks and comments on an earlier version of this chapter, and Hiroshi Kito at Sophia University, Japan, for his expertise on the historical demography of Japan. This research was initiated while the first author was a visiting scholar at Nihon University Population Research Institute, Tokyo, Japan.

1   For this comparison we use a special life table for 1995 that excludes the influence of the Great Hanshin-Awaji Earthquake in January 1995 (The 18th life tables of Japan, Additional tables) in order to minimize the gaps between 1995 and 2000.

2   From 1963 to 1967 and 1969 to 1973 the information is listed as of 30 September each year. For 1968, the information is listed as of 31 October. From 1974 to 2001, the list was compiled on 1 September but information is listed as of 30 September.

3   The number of births for 1872 was from the end of January to 2 December in the lunar calendar. December 3, 1872 became January 1, 1873 in the solar calendar owing to the change in the calendar system in Japan.

4   We dropped the year 1872 because this first year in which the national registration system was formally implemented is obviously incomplete.

5   It is quite possible that the registration of births was incomplete during the first years of the registration system and that the rate of increase in the number of births from 1872, shown in Figure 2, is exaggerated. In this case

the calculated ratios of centenarians per 10,000 of their birth cohort, plotted from 1973 to 2000 in Figure 3, are too high at the beginning of the period. This possible bias minimizes the slope of the increase in the number of persons aged 100 for every 10,000 births that occurred 100 years earlier. Thus the share (45 percent) of the number of centenarians due to the increase in the size of the birth cohorts is a maximum. We compute this share with a "what if" scenario, keeping constant the number of births at the level of 1873. Formal decomposition procedures would have given different results, but the main problem is the choice of the decomposition method (see Horiuchi 1991).

6   The number of persons aged 100 years old for every 10,000 births is computed by dividing the number of persons aged 100 years alive at each successive 30 September by the number of births that have occurred between 1 January and 31 December one century earlier. For instance, the number of persons aged 100 years on 30 September 2000 is divided by the number of births occurring between 1 January and 31 December 1900. Thus the ratio computed is not exactly a ratio by cohort of persons reaching their 100th birthday for every 10,000 births but a reasonable estimation of this ratio. The number of births by sex is estimated using a ratio of 105 males per 100 females at birth, a ratio that is verified for the years when the information by sex is available.

7   The number of persons aged 105 years old per 10,000 births is computed by dividing

the number of persons aged 105 years alive at each successive 30 September by the number of births occurring between 1 January and 31 December 105 years earlier. See note 6.

8   The fluctuations for males are too large to be displayed in Figure 7.

9   See note 6.

10 More precisely, John Wilmoth's argument is that the pace of mortality change in Japan should slow down once it became the avant-garde country, not that Japan would nec-

essarily lose its position in the future (Wilmoth 1998a).

11 According to Väinö Kannisto, modal age at death is a good indicator of the normal life duration. To measure whether the survival curve becomes more rectangular when the normal life duration increases, Kannisto proposed to compute life expectancy at the mode. A decrease of the life expectancy at the mode of the life durations would indicate a rectangularization of the survival curve.

References

Bonneux, Luc, Jan J. Barendregt, and Paul J. Van der Maas. 1998. "The expiry date of man: A synthesis of evolutionary biology and public health," *Journal of Epidemiology and Community Health* 52: 619–623.

Dupâquier, Jacques and Michel Dupâquier. 1985. *Histoire de la démographie*. Paris: Pour l'Histoire, Perrin.

Fries, James F. 1980. "Aging, natural death, and the compression of morbidity," *New England Journal of Medicine* 303: 130–135.

Horiuchi, Shiro. 1991. "Assessing the effects of mortality reduction on population ageing," *Population Bulletin of the United Nations* 31/32: 38–51.

———. 2000. "Greater lifetime expectations," *Nature* 405(15 June): 744–745.

Jeune, Bernard and Axel Skytthe. 2001. "Centenarians in Denmark in the past and the present," *Population* 13(1): 75–94.

Kannisto, Väinö. 1999. "Measuring the compression of mortality," paper distributed at European Population Conference, EAPS, IUSSP, NIDI, SN, NVD, 1999 (Abstract: 137).

———. 2001. "Mode and dispersion of the length of life," *Population* 13(1): 159–172.

Kitou, Hiroshi. 1997. "Meiji-Taishoki jinko tokei ni okeru shussho" (Number of birth in population statistics during Meiji and Taisho era), *Jyochi Keizai Ronshu* 43(1): 41–65.

Kjaergaard, Thorkild. 1995. "Alleged Danish centenarians before 1800," in Bernard Jeune and James W. Vaupel (eds.) *Exceptional Longevity: From Prehistory to the Present.*" Odense: University Press of Southern Denmark, pp. 47–54.

Meslé, F., J. Vallin, and J-M. Robine. 2000. "Vivre plus de 110 ans en France," *Gérontologie et Société* 94: 101–120.

Ministry of Health and Welfare. 2001. *Abridged Life Tables for Japan 2000*. Statistics and Information Department, Minister's Secretariat, Ministry of Health and Welfare. Tokyo.

———. 1998. *The 18th Life Tables*. Statistics and Information Department, Minister's Secretariat, Ministry of Health and Welfare. Tokyo.

———. 1899–1998. Vital Statistics Annual issue since 1899. Tokyo.

———. 1963–2000. *Zenkoku koureisha meibo* (A list of centenarians in Japan). Tokyo.

National Institute of Population and Social Security Research. 2002. *Latest Demographic Statistics 2001/2002*. Tokyo, Japan.

Olshansky, S. Jay, Bruce Carnes, and Aline Désesquelles. 2001. "Prospects for human longevity," *Science* 291: 1491–1492.

Oeppen, Jim and James W. Vaupel. 2002. "Broken limits to life expectancy," *Science* 296: 1029–1031.

Poulain, Michel, Dany Chambre, and Michel Foulon. 2001. "Survival among Belgian centenarians," *Population* 13(1): 117–138.

Robine, Jean-Marie. 2001a. "A new biodemographic model to explain the trajectory of mortality," *Experimental Gerontology* 36(4-6): 899–914.

————. 2001b. "Redefining the stages of the epidemiological transition by a study of dispersion of life spans: The case of France," *Population* 13(1): 173–194.

Robine, Jean-Marie, Isabelle Romieu, and Emmanuelle Cambois. 1999. "Health expectancy indicators," *Bulletin of the World Health Organization* 77: 181–185.

Robine, Jean-Marie and James W. Vaupel. 2001. "Supercentenarians, slower ageing individuals or senile elderly?" *Experimental Gerontology* 36(4–6): 915–930.

————. 2002. "Emergence of supercentenarians in low mortality countries," *North American Actuarial Journal* 6(3): 54–63.

Sekiyama, Naotaro. 1948. *Kinsei Nippon jinko no kenkyu* (A study of Japanese population in modern times). Tokyo: Ryngin-sha.

Skytthe, Axel and Bernard Jeune. 1995. "Danish centenarians after 1800," in Bernard Jeune and James W.Vaupel (eds.) *Exceptional Longevity: From Prehistory to the Present*. Odense: University Press of Southern Denmark, pp. 55–66.

Statistics Bureau. Various years. *Statistical Yearbook of the Empire of Japan*. Tokyo.

Taeuber, Irene B. 1958. *The Population of Japan*. Princeton, New Jersey: Princeton University Press.

Thatcher, A. Roger. 1992. "Trends in numbers and mortality at high ages in England and Wales," *Population Studies* 46: 411–426.

————. 1997. "Trends and prospects at very high ages," in John Charlton and Mike Murphy (eds.), *The Health of Adult Britain: 1841–1994*. London: The Stationery Office, Chapter 27.

————. 1999a. "The demography of centenarians in England and Wales," *Population Trends* 162 (Part 1): 5–43.

————. 1999b. "The long-term pattern of adult mortality and the highest attained age," *Journal of the Royal Statistical Society* Series A 162: 5–43.

————. 2001. "The demography of centenarians in England and Wales," *Population* 13(1): 139–156.

Tuljapurkar, Shripad, Li Nan, and Carl Boe. 2000. "A universal pattern of mortality decline in the G7 countries," *Nature* 405(15 June): 789–792.

Vallin, Jacques and France Meslé. 2001. "Vivre au-delà de 100 ans," *Population et Société* 365: 4.

Vaupel, James W. and Bernard Jeune. 1995. "The emergence and proliferation of centenarians," in Bernard Jeune and James Vaupel (eds.), *Exceptional Longevity: From Prehistory to the Present*. Odense: University Press of Southern Denmark, pp. 109–116.

Vaupel, James W. and Hans Lundström. 1996. "The future of mortality at older ages in developed countries," in Wolfgang Lutz (ed.), *The Future Population of the World: What Can We Assume Today?* Laxenburg, Austria: International Institute for Applied Systems Analysis.

Vaupel, James W. et al. 1998. "Biodemographic trajectories of longevity," *Science* 280: 855–860.

Wilmoth, John R. 1998a. "Is the pace of Japanese mortality decline converging toward international trends?" *Population and Development Review* 24(3): 593–600.

————. 1998b. "The future of human longevity: A demographer's perspective," *Science* 280: 395–397.

————. 2002. "How long can we live? A review essay," *Population and Development Review* 27(4): 791–800.

Wilmoth John R., Leo J. Deegan, Hans Lundström, and Horiuchi Shiro. 2000. "Increase of maximum life-span in Sweden, 1861–1999," *Science* 289: 2366–2368.

Wilmoth, John R. and Shiro Horiuchi. 1999. "Rectangularization revisited: Variability of age at death within human populations," *Demography* 36(4): 475–495.

Wilmoth, John R. and Hans Lundström. 1996. "Extreme longevity in five countries," *European Journal of Population* 12: 63–93.

Life Course, Environmental Change, and Life Span

JEAN-MARIE ROBINE

Biodemographic models, incorporating biological, social, and environmental heterogeneity, are essential to understanding the determinants of human longevity (Carnes and Olshansky 2001). Yet these conventional factors are not sufficient, even in an interactive way, to explain the life history and the longevity of species, including the human species. Finch and Kirkwood (2000) have proposed the addition to the conventional models of intrinsic chance, known to physicists as chaos. I suggest the addition of the life course.

In recent publications (Robine and Vaupel 2001; Robine 2001a), I have proposed a new biodemographic model to explain the trajectory of mortality currently observed for humans. It is generally believed that life span is genetically determined and that the environment is responsible for individual differences. I propose to consider the opposite hypothesis: that the environment plays an essential part in defining the limits of the life span while genetic heterogeneity explains a large part of the individual differences in the duration of life. The aim of this chapter is to elaborate this hypothesis in which human longevity is seen as a "plastic" outcome, especially in its ability to explain the evolution of the mortality trajectory observed through the twentieth century. In this approach, environment is understood to include the built environment, living and working conditions, and changes in medical knowledge such as the discoveries of vaccines and antibiotics.

For human populations the notion of transition is important (Meslé and Vallin 2000). The demographic transition allowed the human species to move from high mortality and high fertility to the current demographic situation characterized by low mortality and low fertility. The epidemiological transition, paralleling the demographic changes (Omran 1971; McKeown 1976), allowed the human species to move from a situation dominated by mortality resulting from infectious diseases to the current epidemiological situation where most deaths are associated with the aging process. Myers and Lamb (1993) extended the notion of transition to changes in functional health status that accompanied the demographic and the epidemiological changes. In the same way, the notion of transition must be extended to gerontological changes. Before entering the epidemiological transition, humans age though rarely become old; apart from a few exceptional cases, they die of accidents and infectious diseases before reaching old age. During the transition, mortality attributable to infectious disease recedes, and humans have increasing access to old age—becoming sick and then old or, alternatively, old and then sick before dying. After the transition, if the pathological causes of death are controlled, virtually all humans will reach old age and will eventually die of frailty. The mortality trajectory observed through the twentieth century corresponds to the transitional phase.

Before the transition, that is, before the eighteenth century in Europe, the environment was relatively homogeneous and changed very little for an individual through his or her life course. During the transition the environment became more complex, with an increase in the number of specific environments corresponding to the main life periods. The environmental change can be seen, at an individual level, as a series of breaks in the life course (school conditions, higher education conditions, working conditions, retirement conditions, nursing home conditions). Further into the transition, the number of environments could be much higher (village or residence for seniors; residence associated with a nursing home; or specialized nursing home), and the changes between environments could be seen as a more nearly continuous process. During the life course individuals enter different environments (i.e., life conditions) with associated theoretical life spans and with a final change (institutionalization) before dying. I will discuss these changes in relation to the mortality trajectory and provide some specifications to develop the mathematics of such a model.

The "ages" of life are not a new notion. Already in the sixteenth century Shakespeare distinguished the seven ages of man: infancy, school days, the young man, the soldier, the grave citizen, the retired old man, and finally second childhood. But in practice, before the demographic transition, this complete life course was applicable to only a small portion of the population. Moreover, the associated environments were more constant across the life span in the past than today.

The biodemographic model of mortality trajectory

For biologists, mortality primarily serves to measure the aging process. For demographers, mortality measures the quality of the current ecological and social environment, the current conditions. In view of the fact that human beings spend the greater part of their time improving the quality of their physical and social environment, making it more and more favorable for the realization of their potential longevity, I proposed a biodemographic model to explain the trajectory of mortality for the human species that takes into account the combined effects of the quality of the environment and of aging (Robine 2001a).

Starting with youth, the age when individuals are the most robust and the most resistant to environmental hazards, the increase in mortality measures the aging process translating young, vigorous individuals into frail, senile elders. When individuals become frail and are no longer able to resist environmental hazards or resist them extremely weakly, the mortality rate becomes constant. The lowest mortality rate recorded at the starting point and the level of the final plateau of mortality measure the quality of the ecological and social environment. Between these two measures, the mortality trajectory measures the aging process.

Excluding infant and child mortality, we can summarize the mortality trajectory corresponding to this model by (1) a lowest point, similar to the initial mortality rate proposed by Finch and his collaborators (Finch et al. 1990), followed by (2) an increase in the mortality rate and (3) a plateau of mortality, corresponding to the three stages of the developmental transition: young and robust/aging/old and frail. Nothing is fixed in this model: neither the age with the lowest mortality rate, the level of the lowest mortality rate, the rate of mortality increase during the aging phase, the age at which the mortality increase reaches a plateau, nor the level of the final plateau. The model does not imply a constant rate of increase for the mortality rate during the aging phase, but that phase may be a "Gompertzian segment" (Finch and Pike 1996). One can also imagine a slower rate of increase at both ends of the aging phase: at the beginning because initially robust young people can resist the transition and at the end because aged people can fight to retain resources to continue resisting, even very weakly, the environmental hazards. This model is complementary to the frailty model (Vaupel et al. 1979) since the aging trajectory, the central part of the mortality trajectory, may incorporate differential aging trajectories characterizing different homogeneous groups within the population.

Life course and environmental changes

The specific contribution of the new model is to incorporate the life course and the environmental changes as part of the biological aging process to explain the mortality trajectory and its transition over the twentieth century. One explanation could be the substantial improvement, since 1945, in the general environment in countries with low mortality, leading to a fall of mortality at the oldest ages. At each year of age, individuals who should be higher on their theoretical trajectory of mortality are living in an environment that is increasingly favorable to the realization of their potential for longevity. Another explanation could be that when individuals become too fragile to face the difficulties of the common environment, they are placed in a protected environment, such as that provided by a nursing home, where their aging is slowed considerably. Such a model with two contrasted environments supports representations of a mortality trajectory with two successive segments (whether Gompertzian or not). Variations between individuals in the ages at which the environment changes could support the existence of a progressive transition between the two segments.

The gerontological transition

The next question is how the three elements summarizing the mortality trajectory (lowest point, increase, plateau) could have changed during the demographic and epidemiological transition. First, the mortality level at the lowest mortality point, correctly observed since the eighteenth century, has diminished considerably since the beginning of the demographic transition—unambiguous testimony to a substantial improvement in the physical and social environment, specifically in conditions during childhood. With the transition, children have benefited from more-protected environments—being excluded from the labor force, for example. It is noteworthy that no change is detectable in the age with the lowest mortality rate—suggesting no environmental impact on the time needed to reach the maximum survival from birth.[1] Second, we have no evidence that the rate of mortality increase with age during the aging phase has slowed during the demographic transition. Whatever the model used to summarize the mortality trajectory during the aging phase, the rate of mortality increase with age appears to be constant, or even to increase, through the demographic transition (Thatcher 1999). This suggests that the demographic and epidemiological transitions have had little if any impact on the aging process itself, although they may have contributed toward its acceleration. The latter hypothesis could be explained by a decrease in selection in infancy before the starting point of the aging process and also by the fact that the climb to the plateau now starts at a lower level of minimum mortality. However, during the aging phase, mortality rates double approximately every eight years for the human species. Third, the existence of a plateau of mortality when individuals become frail and cease resisting environmental hazards is still a hypothesis, although it is substantiated by the models that best fit current mortality data (Vaupel et al. 1998; Thatcher et al. 1999; Lynch and Brown 2001) and by empirical observations among centenarians and supercentenarians (Robine and Vaupel 2001, 2002). In the future the mortality plateau, if it exists, will be better observed when the size of the population reaching the age where it appears increases. Alternative models have been proposed, however, such as quadratic trajectories that involve a decrease in the mortality rate after a maximum is reached (Vaupel et al. 1998). Only time will tell.

The evolution of the mortality trajectory through the twentieth century

Figures 1 and 2 summarize the mortality trajectories of low-mortality countries (i.e., the countries most advanced in the transition) through the twentieth century. The first trajectory was extrapolated by Vincent (1951) using data from France, the Netherlands, Sweden, and Switzerland circa 1900–45. He observed regularly increasing death rates (i.e., the quotient of mortality) to ages beyond 100 years that exceeded the value 0.6. By simple linear extrapolation of the death rates plotted on a semi-logarithmic graph, he found that the mortality trajectory reached a death rate of 1.0 at age 110 for both sexes. He therefore concluded that human life is limited to 110 years (Vincent 1951). The second trajectory was extrapolated by Dépoid (1973), using the same method as Vincent and with data from the same countries but for a later period, circa 1945–70. Dépoid found limit values of 117 years for men and 119 years for women. Surprisingly, there was no further discussion about the eight years separating the limits found using the same method but with a 50-year difference in measurement period. The third trajectory was extrapolated by Thatcher et al. (1998) with a logistic model fitted to the mortality data of the 13 countries with the best human mortality data for the period 1980–90 (Austria, Denmark, England and Wales, Finland, France, West Germany, Iceland, Italy, Japan, the Nether-

FIGURE 1 Mortality trajectories beyond the age of 95 years, arithmetic scale: France, Netherlands, Sweden, and Switzerland (circa 1900–45 and 1945–70); 13 countries (1980–90); French centenarians (1980–90); supercentenarians (1960–2000)

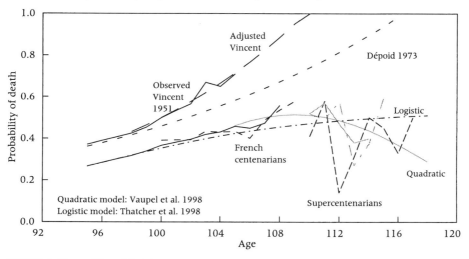

SOURCE: Robine and Vaupel (2001).

FIGURE 2 Mortality trajectories beyond the age of 95 years, logarithmic scale: France, Netherlands, Sweden, and Switzerland (circa 1900–45 and 1945–70); 13 countries (1980–90); French centenarians (1980–90); supercentenarians (1960–2000)

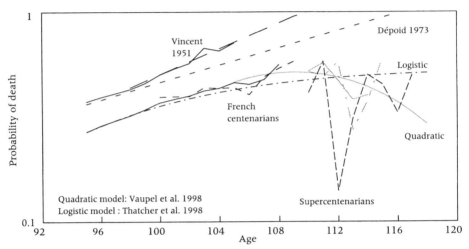

SOURCE: Robine and Vaupel (2001).

lands, Norway, Sweden, and Switzerland). Thatcher et al. found that a logistic model is the best to fit the observations since 1960 and that this model suggests that the mortality quotient will never increase above about 0.6 (Thatcher et al. 1998; Kannisto 1999). This does not mean that man will become immortal but that, even at the highest ages, the few survivors have a small chance to survive to the next year. Similarly, if man does not age, he would not become immortal. The fourth and last trajectory was extrapolated by Vaupel et al. (1998), using the same data base for the period 1980–90. Vaupel et al. found that a quadratic model fits the most recent data very well. After having reached a maximum at age 109 years with a mortality level slightly above 0.5, the mortality trajectory goes down (Vaupel et al. 1998). In addition to these four trajectories, Figures 1 and 2 display the observed rates used by Vincent in 1951 and by Thatcher et al. in 1998. The figures also display the mortality rates of French centenarians for the period 1980–90 (Kannisto 1996) and the mortality rates of French supercentenarians (Robine and Vaupel 2001). A more recent study confirms the mortality rates of supercentenarians but with greater precision (Robine and Vaupel 2002).

The plasticity of longevity

Altogether the extrapolated trajectories and the observed data support the hypothesis of substantial changes in the mortality trajectory of the human species during the twentieth century. These observations provide strong support for the new concept of plasticity of aging (Finch 1998; Finch and Tanzi 1997). But what has changed, the shape of the trajectory or its level? My first hypothesis is that only the level of the trajectory has changed with the demographic and epidemiological transition because of the substantial improvement in the physical and social environment. My second hypothesis is that the shape of the mortality trajectory has always appeared constant since it is for the most part determined by the biological life history of humans. A classic assumption in demography is that the age pattern of mortality is independent of the level of mortality, this being the rationale for model life tables (Coale and Demeny 1983). Before the transition the mortality plateau (i.e., the level of mortality reached by frail persons who can no longer resist environmental hazards or have only a weak resistance) was much higher than today, although the level was below unity and was reached much earlier in the life course. So few people reached this stage that its observation was impossible. With the demographic transition and the fall of mortality among the aging population, more people are getting old and frail, thus increasing the chance of observing a mortality plateau, even if it occurs at an older age. At the same time, the level of mortality at the plateau has decreased because of improvements in the physical and social environment during the transition. Good historical data, such as the French-Canadian data for the seventeenth century, show an exponential increase in mortality with age without any evidence of a leveling off at extreme ages (Bourbeau and Desjardins 2002). However, the extreme ages involved, 85 to 95 years, as well as the level of mortality, make the results of these studies compatible with our model. Before the demographic transition, the level of mortality was so high that only the aging segment could be observed.

At an individual level, aging today is often a long retirement process. When individuals reach age 60 or 65 years, the majority retire from work. With time they reduce or stop various elements of their activity and begin to reduce mobility to the confines of their community and then their neighborhood. These successive environments could be associated with specific conditions more favorable to their survival, but the changes are generally gradual. When a person becomes too frail to survive in the common environment, he/she moves to a radically different environment—a nursing home or home confinement—that provides more secure conditions in relation to the individual's health status. This series of changes may not only change the rate of increase in mortality rates but may also eventually lead to a plateau of mortality for those living in a nursing home or confined to their own home. Although some time may be required to acknowledge the necessity of a move to a new environment to improve the chance of survival, the move can immediately produce lower mortality conditions. In addition if the price to pay for the benefit of better mortality conditions is high in terms of quality of life (confinement, social isolation), the move can be delayed as long as possible, increasing the gap between the mortality conditions before and after the move and leading eventually to a decrease in the mortality level.

Three concomitant factors can contribute to lowering the level of the plateau of mortality reached by the frail oldest old in the future. First, the recognition of the syndrome of frailty itself and its association with a higher risk of mortality (Fried et al. 2001) will help in the development of specific interventions. Second, general improvement and the development of specialized nursing homes, better adapted to the specific health status of the residents, will increase the level of safety of such homes. Third, the growing recognition that the preservation of the quality of life of the oldest old is an important social goal will help to increase the quality and variety of the services and possibilities offered by the nursing homes, making them more attractive. Acting in synergy these three factors—specific interventions targeted to frail persons and general improvement of the safety and quality of the environment provided by nursing homes—have an enormous potential to lower the final plateau of mortality.

A note on mathematical specification

According to the proposed biodemographic model, the mortality trajectory can be summarized by three elements, (1) a lowest mortality point, (2) an aging segment, and (3) a mortality plateau, corresponding to the three stages of the developmental transition: young and robust/aging/old and frail. We have seen that nothing is fixed in this model. The associated mathematical model must allow these elements to occur at different ages and thereby to shape the aging segment. Recent work on mortality deceleration suggests several solutions for such an approach, including an arctangent model (Lynch and Brown 2001). However, the logistic model suggested by Thatcher (1999) seems to be the most promising mathematical approach as it includes three parameters: one governing the lowest mortality rate, one governing the rate of mortality increase during the aging phase, and the third governing the age at which mortality starts to decelerate. The logistic approach does not strictly fit a mortality plateau but determines through its parameters a ceiling, which gives the highest possible value for the probability of death (i.e., the quotient of mortality) as well as ceilings for the force of mortality and central death rate. Changes over time in the basic parameters of such a model would allow us both to reconstruct past trajectories when the plateau was invisible and to forecast future trajectories.

Notes

I thank Carol Jagger and Roger Thatcher for comments on an earlier version of this chapter.

1 This point deserves to be carefully reassessed, as at first glance a retrospective French yearbook suggests that the age with the lowest mortality rate moved from ages 12 years and 11 years for males and females in 1898– 1903 to age 10 years for both sexes in 1952– 56 (INSEE 1961). However, a special tabulation made for us by INSEE from 1890–94 to 1990–94 (Robine 2001b) shows no trend in France in the lowest mortality rate through the twentieth century. It is exceptionally stable for females between ages 10 and 11 years.

References

Bourbeau, Robert and Bertrand Desjardins. 2002. "Dealing with problems in data quality for the measurement of mortality at advanced ages in Canada." *Society of Actuaries, Living to 100 and Beyond: Survival at Advanced Ages* «www.soa.org/research/living.html»

Carnes, Bruce A. and S. Jay Olshansky. 2001. "Heterogeneity and its biodemographic implications for longevity and mortality," *Experimental Gerontology* 36: 419–430.

Dépoid, Françoise. 1973. "La mortalité des grands vieillards," *Population* 755–792.

Coale, Ansley and Paul Demeny. 1983. *Regional Model Life Tables and Stable Populations.* New York and London: Academic Press.

Finch, Caleb E. 1998. "Variations in senescence and longevity include the possibility of negligible senescence," *The Journals of Gerontology: Biological Sciences and Medical Sciences* 53: B235–B239.

Finch, Caleb E. and Thomas B. L. Kirkwood. 2000. *Chance, Development and Aging.* New York: Oxford University Press.

Finch, Caleb E. and Malcolm C. Pike. 1996. "Maximum life span predictions from the Gompertz mortality model," *The Journals of Gerontology: Biological Sciences and Medical Sciences* 51A: B183–B194.

Finch, Caleb E., Malcolm C. Pike, and Matthew Witten. 1990. "Slow mortality rate accelerations during aging in animals approximate that of humans," *Science* 249: 902–905.

Finch, Caleb E. and Rudolph E. Tanzi. 1997. "Genetics of aging," *Science* 278: 407–411.

Fried, P. Linda et al. 2001. "Frailty in older adults: Evidence for a phenotype," *The Journals of Gerontology: Biological Sciences and Medical Sciences* 56A(3): M146–M156.

INSEE. 1961. *Annuaire statistique de la France. Retrospectif.* Paris: Institut National de la Statistique et des Etudes Economiques.

Kannisto, Väinö. 1996. *The Advancing Frontier of Survival.* Odense: Odense University Press.

———. 1999. "Discussion on the paper by Thatcher," *Journal of the Royal Statistical Society: Series A* 162: 33.

Lynch, Scott M. and J. Scott Brown. 2001. "Reconsidering mortality compression and deceleration: An alternative model of mortality rates," *Demography* 38(1): 79–95.

McKeown, Thomas. 1976. *The Modern Rise of Population.* London: Edward Arnold.

Meslé, France and Jacques Vallin. 2000. "Transition sanitaire: Tendances et perspectives," *Médecine/Sciences* 16: 1,161–1,171.

Myers, George C. and Vicki L. Lamb. 1993. "Theoretical perspectives on healthy life expectancy," in Jean-Marie Robine, Colin D. Mather, Margaret R. Bone, and Isabelle Romieu (eds.), *Calculation of Health Expectancies: Harmonisation, Consensus Achieved and Future Perspectives.* Montrouge: John Libbey Eurotext, pp. 109–119 (Colloque INSERM no. 226).

Omran, Abdel R. 1971. "The epidemiologic transition: A theory of the epidemiology of population change," *Milbank Memorial Fund Quarterly* 49: 509–538.

Robine, Jean-Marie. 2001a. "A new biodemographic model to explain the trajectory of mortality," *Experimental Gerontology* 36: 899–914.

Robine, Jean-Marie. 2001b. "Redefining the stages of the epidemiological transition by a study of the dispersion of life spans: The case of France," *Population* 13(1): 173–194.

Robine, Jean-Marie and James W. Vaupel. 2001. "Supercentenarians, slower aging individuals or senile elderly?" *Experimental Gerontology* 36: 915–930.

———. 2002. "Emergence of supercentenarians in low mortality countries," *North American Actuarial Journal* 6(3): 54–63.

Thatcher, A. Roger. 1999. "The long-term pattern of adult mortality and the highest attained age," *Journal of the Royal Statistical Society: Series A* 162: 5–43.

Thatcher, A. Roger, Väinö Kannisto, and James W. Vaupel. 1998. *The Force of Mortality at Ages 80 to 120.* Odense: Odense University Press.

Vaupel, James W., James R. Carey, Kaare Christensen, Thomas E. Johnson, Anatoli I. Yashin, Niels V. Holm, Ivan A. Iachine, Väinö Kannisto, Aziz A. Khazaeli, Pablo Liedo, Valter D. Longo, Yi Zeng, Kenneth G. Manton, and James W. Curtsinger. 1998. "Biodemographic trajectories of longevity," *Science* 28: 855–860.

Vaupel, James W., Kenneth G. Manton, and Eric Stallard. 1979. "The impact of heterogeneity in individual frailty on the dynamics of mortality," *Demography* 16: 439–454.

Vincent, Paul. 1951. "La mortalité des vieillards," *Population* 181–204.

The World Trend in Maximum Life Span

JOHN R. WILMOTH
JEAN-MARIE ROBINE

The maximum observed life span is known to have been rising slowly for at least 140 years in Sweden and has presumably been rising in other industrialized countries as well (Wilmoth et al. 2000; Wilmoth and Lundström 1996). If we consider men and women together, the Swedish maximum age at death in a calendar year rose from around 102 years in the 1860s to around 109 in the 1990s. This result is significant because no other country offers such a long series of high-quality data that can be used to study maximum life span. An obvious question is whether the Swedish trend is typical of the world, or at least some part of the world.

Only four Swedes are known to have achieved the age of 110 years. In the world as a whole, however, the number of "validated super-centenarians," or individuals aged 110 or older for whom adequate documentation is available, is increasing (Robine and Vaupel 2001). Several of these cases lie well above the Swedish trend, and the difference is large enough to require a rigorous explanation. Perhaps in some important way Sweden is not representative of the rest of the world, or even of the industrialized world. Or perhaps the world records are higher than the Swedish ones merely because they are drawn from a larger population. If the latter explanation (or some variant of it) is true, then it should be possible to speculate about the world trend in maximum life span by extrapolation from the Swedish trend.

Empirical evidence regarding trends in the human life span

It is well known that age is often misreported in human populations, especially at older ages. Furthermore, it is often impossible to confirm or disprove the accuracy of a given claim absent documentary evidence. Today's centenarians and super-centenarians were born, obviously, more than 100 or 110 years ago, when record keeping was less complete than it is today. In only a handful of populations are complete birth records available from the late nineteenth century that would permit an exhaustive check of all cases of extreme longevity reported today. A complete system of Swedish national statistics dating from 1749 helps to explain the accuracy of death records at extreme ages beginning in the 1860s. For countries with less complete records, it may nevertheless be possible to certify the accuracy of an individual case of extreme longevity by assembling the available evidence, carefully checking its consistency, and, whenever possible, interviewing the person in question as a final check on the coherence of the case (Jeune and Vaupel 1999).

Keeping these remarks about data quality in mind, we begin with a review of the empirical evidence. The unmistakable conclusion is that the maximum observed life span has been increasing for decades and possibly for one or more centuries, not only in Sweden but also for the human population as a whole. We must first resolve, however, whether the slope of the increase is similar in different populations. For this purpose, we review the Swedish evidence since 1861, analyze individual records of confirmed super-centenarians, and study trends in the maximum age at death for seven countries that have complete and validated records since at least 1950.

Sweden

The increase in the maximum age at death by sex in Sweden is shown in Figure 1. This graph, published in Wilmoth et al. (2000), serves as a reminder of current knowledge about trends in maximum human life span. The slope of the parallel regression lines (expressed in years per decade) is 0.4 before 1969 and 1.1 afterward. The major components of this increase can be studied using a statistical model, as reviewed below.

Individual cases

The increasing number of super-centenarians in the world has only recently been documented (Robine and Vaupel 2001). Figure 2A shows the maximum age at death for calendar years since 1959 based on Robine and Vaupel's published list of validated super-centenarians (thus, by definition the distribution is truncated below age 110). For every year since 1977, the oldest validated death in the world was at age 110 or above. Given this information, it appears that maximum age at death in the world increased at a pace of 3.1 years per decade during 1977–2000, thus much more quickly than the Swedish trend. We must, however, consider two sources of bias.

First, it is possible that some of these "validated" cases are not genuine. Robine and Vaupel (2001) distinguish different levels of validation and propose that further analyses should be based on "3-star validated" cases,

FIGURE 1 Maximum age at death as annually observed, by sex, and trend lines, Sweden 1861–1999

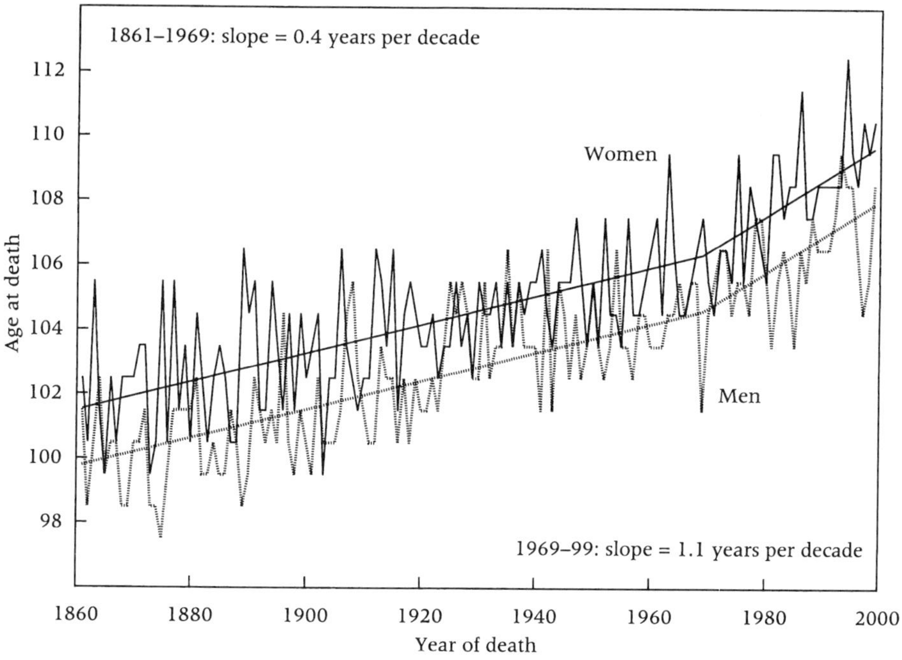

SOURCE: Wilmoth et al. (2000).

indicating that birth and death records (or photocopies thereof) have been brought together and compared side by side to verify the reported age. Figure 2B shows that when we limit the analysis to such cases, the slope of increase is slightly less, or about 2.7 years per decade during 1977–2000.

Other countries with complete and valid data

Second, it is possible that earlier cases of super-centenarians have escaped notice, introducing another source of upward bias to the slope shown in Figures 2A and 2B. Robine and Vaupel (2001) suggest therefore that it may be appropriate to limit our attention to countries with an exhaustive system for validating apparent super-centenarians. There are at present only seven such countries: Belgium, Denmark, England and Wales,[1] Finland, the Netherlands, Norway, and Sweden. As shown in Figure 2C, the slope is reduced further if we limit ourselves to such cases. In fact, at 1.4 years per decade during 1977–2000, the rate of increase in the maximum age at death for these seven countries is only slightly higher than the value of 1.1 noted earlier for Sweden during 1969–99.

Obviously, the truncation of age at 110 years limits the time period of the analysis as well, since super-centenarians were still quite rare before the

FIGURE 2 Maximum age at death among super-centenarians

A. Validated cases (15 countries)

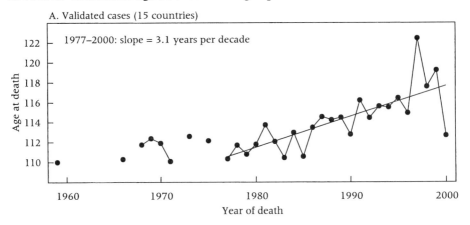

B. 3-star validated cases (10 countries)

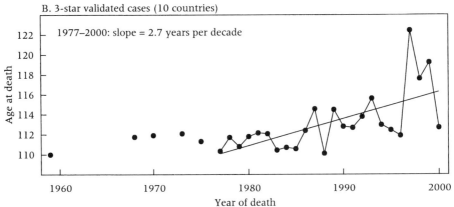

C. Only countries with complete and 3-star validated data (7 countries)

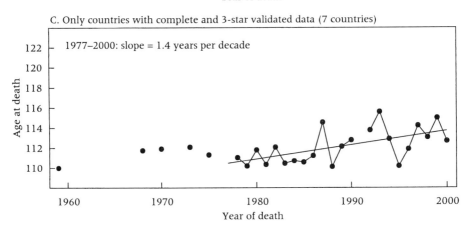

NOTES: (1) Seven countries provide the data shown in 2C: Belgium, Denmark, England and Wales, Finland, the Netherlands, Norway, and Sweden. Although data are complete only for these seven, three additional countries contributed at least some cases of "3-star validated" super-centenarians used to construct 2B: Canada, France, and the United States. Using a less rigorous selection criterion, six more countries contributed cases of "validated" super-centenarians shown in 2A: Australia, Italy, Japan, Scotland, South Africa, and Spain. Thus, since England, Wales, and Scotland are all part of the United Kingdom, a total of 15 countries have contributed data to this analysis. (2) Trend lines for Figure 2 were fit by ordinary least squares. (3) In 2C, missing values for 1977 and 1991 were omitted from the regression.
SOURCE: Robine and Vaupel (2001).

1970s. For six of these seven countries (with the exception of Belgium), however, we are able to fill in the missing parts of the trend using national data. Thus, Figure 3 shows the maximum age at death, for men and women combined, in Sweden and in the group of "complete and validated" countries (omitting Belgium). For the latter group, the trend in Figure 3 is based on data from the super-centenarian database whenever possible (i.e., a death that occurred above age 110 in a given calendar year) and on national death statistics otherwise. The two trend lines, based on data for 1950–99, are separated by about 3.1 years but are very nearly parallel, each with a slope around 1.1 years per decade.

In light of this analysis, we conclude that the slope of the trend in maximum life span has been the same in Sweden and in the other "complete and validated" countries since at least 1950. Furthermore, the increase accelerated over this time period and was faster in the last quarter of the twentieth century. It remains to be seen, however, whether the observed difference in level (of about 3.1 years) can be explained simply as a function of population size, or whether other factors also affect the difference.

FIGURE 3 Maximum age at death with trend lines, Sweden 1861–1999 and "complete and validated" cases (6 countries), 1950–2000

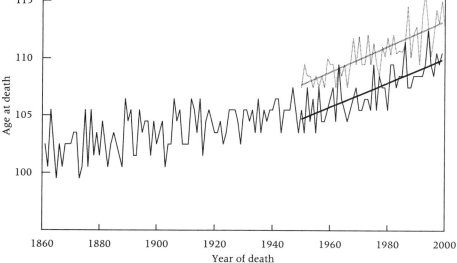

NOTES: (1) Six countries underlie the trend shown here for 1950–2000: Denmark, England and Wales, Finland, the Netherlands, Norway, and Sweden. Although Belgium contributes to the universe of "complete and validated" super-centenarians (Figure 2C), it lacks mortality data above age 100 in the Kannisto–Thatcher database, which is used here to fill in the trend below age 110. (2) Trend lines for 1950–99 were fit by ordinary least squares.
SOURCES: Wilmoth et al. (2000); Robine and Vaupel (2001); Kannisto–Thatcher Database on Old-Age Mortality.

Statistical models of extreme longevity

The maximum age at death can be modeled statistically by treating it as an extreme value of a random sample from some probability distribution (Gumbel 1937, 1958; Aarssen and de Haan 1994; Wilmoth et al. 2000). With such models, it is possible to "predict" the maximum age at death for a cohort of individuals born in the same year. Furthermore, it is possible to compute the complete distribution of the maximum age at death for a given cohort. In other words, the observed value is only one observation from a range of possible values, and the probability of observing any value in that range is represented by the said distribution. Figure 4 shows the trend in the Swedish maximum age at death by cohort, for men and women combined, along with quantiles of the estimated distribution of the extreme value. For example, there is a 50/50 chance that the maximum age at death observed in a given calendar year lies above or below the line corresponding to the median (i.e., the 0.5 quantile). This figure also shows all of the 3-star validated super-centenarians in the world. Clearly, these cases lie well above the Swedish trend.

We attempt to explain this difference using statistical models of extreme values. We discuss such models in general, describe their application

FIGURE 4 Maximum age at death in Sweden with predicted quantiles, compared to all 3-star validated super-centenarians in the world, birth cohorts of 1756–1884

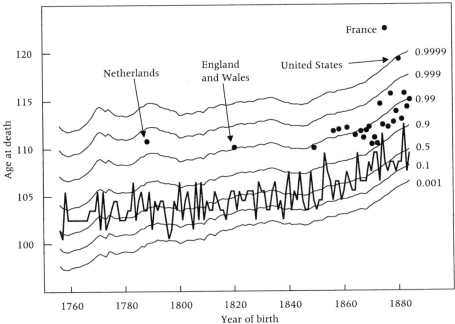

to the Swedish trend in the maximum age at death, and review the principal results from an earlier study (Wilmoth et al. 2000). All of the new calculations presented here treat the Swedish example as a baseline, so it is
important to understand how the earlier results were derived.

Extreme value distributions

Suppose that $S(x)$ is the probability of survival from birth to age x for an
individual chosen at random from a cohort of N births.[2] Then, the probability that the maximum age at death for this cohort lies above age x is given
by $S^N(x) = 1 - [1 - S(x)]^N$. Accordingly, the maximum age at death is itself a
random variable with a probability distribution, and this distribution is determined by N and the $S(x)$ function or, alternatively, by N and the probability distribution of ages at death, given by the function

$$f(x) = \frac{-dS(x)}{dx}.$$

In the earlier analysis, the mortality experience of the Swedish cohort born
in year t was described by a vector of age-specific death rates, $\mathbf{M_t} = (M_{0t},
M_{1t}, ..., M_{79t}, \hat{M}_{80t}, ..., \hat{M}_{119t})$, which was transformed into cohort survival prob-
abilities, $S(x,t)$.[3]

Below age 80 years, death rates were observed values for cohort t, whereas
at higher ages they were derived from a logistic model.[4] Wilmoth et al. (2000)
provides further information on how this model was applied in the Swedish
case. However, one detail requires special attention. The logistic model implies that death rates approach a fixed upper limit, denoted here by r. However, the level of r is unknown and difficult to determine. In the analysis of
Swedish data, the upper asymptote was fixed at 1.25 because that value maximized a global measure of goodness-of-fit. Although such an assumption has
no strong justification, it was convenient since cohort-specific estimates of r
are highly unstable owing to random fluctuations in death rates at older ages.
Moreover, sensitivity analysis confirmed that the main results of interest in
the Swedish study did not vary significantly with alternative choices of r.

In effect, the value of r in the previous study was used as a kind of
"tuning constant." A lower value of r would have moved the entire distribution of the maximum age at death upward, whereas a higher value would
have pushed it downward. The value of 1.25 was chosen because it placed
the distribution at the correct level for the observed data. Since a formal
criterion was used for choosing the optimal value, the methodology used in
the Swedish study offers a new means of estimating r based on extreme
values rather than on the shape of the age curve of mortality at lower ages.
The latter method was used, for example, by Thatcher et al. (1998), who

typically found much lower values of r, often below 1.[5] It is also worth noting that the simple model of old-age mortality proposed by Kannisto is in fact the logistic model with an asymptote of 1.[6]

Thus, although it describes the Swedish collection of extreme values quite well, an asymptote of 1.25 in the logistic model seems impossibly high. Before attempting to extrapolate from the Swedish trend to the rest of the world, we first examine whether the risk of mortality at extremely high ages in Sweden is typical of other industrialized countries. We begin by reviewing one other key result of the earlier Swedish study, which will help to guide our further analysis.

Explaining the rise in maximum life span

What factors determine the level of the maximum age at death and its changes over time? The Swedish study addressed this question by examining the components of the increase in the median of the estimated distribution of the maximum age for each cohort. The conclusion was that 72.5 percent of the rise in the Swedish maximum age at death from 1861 to 1999 is attributable to reductions in death rates above age 70, with the rest due to increased numbers of survivors to old age (both larger birth cohorts and increased survivorship from infancy to age 70).

Even though mortality below age 70 fell precipitously over this time period, that change accounts for only 16 percent of the rise in the maximum age at death. Death rates at older ages fell less markedly, but the impact of this change on extreme values was much greater. The annual number of births in Sweden increased only slightly over the relevant range of cohorts and, for that reason, had a relatively minor impact on the trend in the maximum age at death. However, a larger change in the number of births would have had a much more significant effect, as demonstrated below. Thus, to explain the differences seen earlier in the level of the maximum age at death between populations, we must consider primarily any differences in population size and in patterns of old-age mortality.

Comparing trends in the maximum age at death

We have demonstrated that the level of the trend in the maximum age at death differs depending on the population being considered. On the other hand, in all cases where we have examined complete and validated records for a population, the slope of this trend appears to be similar. Thus, our task is to explain differences in the level of the trend between populations. As noted already, two major factors must be considered: the size of the "eligible" population, and the level and pattern of old-age mortality.

The "eligible" population

It is difficult to define the universe from which selected cases of extreme longevity have been drawn. We must consider the fact that any person who is eligible to become a validated super-centenarian must live in a country or region in which the necessary documentation is available. One strategy, therefore, is to add up the populations of all such areas and to apply the same calculations made earlier for Sweden to this larger population. Of course, it is difficult to determine such a collection of countries, since all included areas should have reliable vital statistics not only today but also more than 100 years ago. Furthermore, we would ideally need to compare not the size of the total populations, but rather the size of birth cohorts from over 100 years ago. To simplify the task (with minimal loss of accuracy), we derive ratios of relative population size from estimates of national populations in the year 2000.

For the present purpose, we consider three populations: (1) the population that yielded the "complete and validated" list of super-centenarians, (2) the population that produced the 3-star validated super-centenarians, and (3) the total population of the world with a mortality experience that could plausibly yield cases of world record longevity. The first two populations are used for a comparison between the Swedish trend in maximum life span and the data on super-centenarians presented earlier (supplemented by national data in the case of the "complete and validated" countries). The last population is used to speculate about the world trend in maximum life span.

For each of these cases, we specify a preferred scenario. In the second and third cases, we also investigate alternative scenarios. The population sizes for each scenario are based on Table 1, which gives estimates of the year 2000 population for all countries included in either the super-centenarian database or the original Kannisto–Thatcher Database on Old-Age Mortality (Kannisto 1994). The reasoning used for each scenario is described in a later section.

Mortality variation at older ages

Mortality rates vary across national populations, even among low-mortality countries. Table 2 summarizes mortality estimates derived from the Kannisto–Thatcher (KT) database for countries with reliable data covering the age range above 80 years.[7] From this table, it is clear that Sweden has an atypical mortality pattern at older ages. As summarized in Table 3, Swedish mortality is relatively low at ages 80–89 but relatively high at age 100+. This is true whether the point of comparison is an aggregate of 13 countries with complete, high-quality data (see Table 2) or our smaller collection of "complete and validated" countries (excluding Belgium in this instance because it lacks data above age 100 in the KT database). These differences

TABLE 1 Countries included in the super-centenarian and in the Kannisto–Thatcher database

Country	Super-centenarian database			Kannisto-Thatcher database	Population in 2000 (millions)
	Full list	3-star validated	Complete and validated		
Australia	X			X	19.2
Austria				X	8.1
Belgium	X	X	X	X	10.2
Canada	X	X			30.8
Czech Republic				X	10.3
Denmark	X	X	X	X	5.3
England and Wales	X	X	X	X	52.7
Estonia				X	1.4
Finland	X	X	X	X	5.2
France	X	X		X	59.4
Germany				X	82.1
Hungary				X	10.0
Iceland				X	0.3
Ireland				X	3.8
Italy	X			X	57.8
Japan	X			X	126.9
Latvia				X	2.4
Luxembourg				X	0.4
Netherlands	X	X	X	X	15.9
New Zealand				X	3.8
Norway	X	X	X	X	4.5
Poland				X	38.6
Portugal				X	10.0
Scotland	X			X	5.1
Singapore				X	4.0
Slovakia				X	5.4
South Africa	X				43.4
Spain	X			X	39.5
Sweden	X	X	X	X	8.9
Switzerland				X	7.1
United States	X	X			275.6
Total population in 2000 (millions)	760.4	468.5	102.7	598.5	946.3

NOTES: (1) Following Robine and Vaupel (2001), the "full list" used in this analysis includes cases of "validated" super-centenarians, whose age was confirmed using some form of *prima facie* documentary evidence. Cases designated as "3-star validated" have passed a more rigorous test, consisting of a side-by-side comparison of official birth and death records (or photocopies thereof). In the seven countries designated as "complete and validated," it has been possible to apply the 3-star criterion to all cases of alleged super-centenarians, of which only confirmed cases are analyzed here. (2) The original Kannisto–Thatcher database included 28 countries (see Note 1 at the end of this chapter). The list shown here has the same number: although the former Czechoslovakia has been split in two, East and West Germany have been combined. (3) The Kannisto–Thatcher database includes only the non-Maori population of New Zealand and the Chinese population of Singapore. However, population figures shown here refer to the full populations of both areas.
SOURCES: (1) All population estimates (except for areas within the United Kingdom) refer to mid-2000 and are from the Population Reference Bureau, *World Population Data Sheet 2001*, http://www.prb.org/pubs/wpds2000/. (2) Population estimates for parts of the United Kingdom (England and Wales, and Scotland) refer to mid-1999 and are from http://www.statistics.gov.uk/ (3) The super-centenarian database used here is described by Robine and Vaupel (2001). The Kannisto–Thatcher database is described by Kannisto (1994).

TABLE 2   Death rates by sex above age 80 for 13 countries with complete, high-quality data, circa 1960–98

Country	Women				
	80–89	90–99	100+	105+	110+
Austria	0.123	0.282	0.550	0.716	0.462
Denmark	0.103	0.247	0.512	0.704	0.429
England and Wales	0.107	0.238	0.460	0.643	0.684
Finland	0.117	0.265	0.500	0.687	0.429
France	0.100	0.240	0.480	0.611	0.485
Germany, West	0.116	0.269	0.514	0.576	0.921
Iceland	0.092	0.227	0.488	0.643	na
Italy	0.114	0.268	0.488	0.573	1.091
Japan	0.093	0.223	0.425	0.513	0.446
Netherlands	0.100	0.245	0.501	0.648	2.571
Norway	0.103	0.250	0.489	0.717	0.857
Sweden	0.100	0.244	0.500	0.623	0.909
Switzerland	0.099	0.245	0.504	0.650	6.000
13 countries	0.105	0.246	0.474	0.601	0.613
(95% C.I.)	(0.105, 0.105)	(0.246, 0.246)	(0.471, 0.476)	(0.589, 0.614)	(0.521, 0.705)

Country	Men				
	80–89	90–99	100+	105+	110+
Austria	0.158	0.326	0.601	0.857	na
Denmark	0.141	0.298	0.596	0.821	na
England and Wales	0.151	0.297	0.533	0.635	0.429
Finland	0.153	0.307	0.620	0.909	na
France	0.139	0.292	0.576	0.585	0.794
Germany, West	0.156	0.322	0.602	0.571	1.103
Iceland	0.119	0.253	0.659	6.000	na
Italy	0.146	0.311	0.560	0.645	0.857
Japan	0.131	0.273	0.499	0.663	0.783
Netherlands	0.140	0.292	0.568	0.855	na
Norway	0.137	0.295	0.532	0.516	0.375
Sweden	0.137	0.301	0.581	0.712	na
Switzerland	0.136	0.295	0.567	0.721	na
13 countries	0.143	0.298	0.554	0.638	0.729
(95% C.I.)	(0.143, 0.143)	(0.297, 0.298)	(0.548, 0.559)	(0.606, 0.671)	(0.417, 1.040)

NOTES: (1) Data for all countries are available during 1960–98 (inclusive) with the following exceptions: Data for Iceland begin in 1961, data for Italy end in 1993, data for Denmark end in 1999, and data for West Germany are missing above age 100 during 1960–63 and above age 105 in 1975. (2) Confidence intervals for death rates in the 13-country aggregate population were computed by assuming Poisson variability of death counts (Brillinger 1986) and using a normal approximation. (3) Death rates above age 110 are marked as "na" in the table if there were zero person-years of exposure at these ages over the given time period.
SOURCE: Original calculations using the Kannisto–Thatcher Database on Old-Age Mortality.

TABLE 3   Ratio of death rates for two groups of countries compared to Sweden, by age and sex, circa 1960–98

Age group	13-country aggregate			Complete and validated		
	Women	Men	Total	Women	Men	Total
80–89	1.05	1.04	1.03	1.05	1.06	1.04
90–99	1.01	0.99	0.99	0.99	0.99	0.97
100+	0.95	0.95	0.94	0.95	0.95	0.94

NOTES: (1) The group of 13 countries is listed in Table 2. (2) The "complete and validated" countries are listed in Table 1. However, this comparison excludes Belgium, which lacks mortality data above age 100 in the KT database.
SOURCE: Original calculations using the Kannisto–Thatcher Database on Old-Age Mortality.

should be taken into account when modeling the trend in maximum life span for other groups of countries based on the Swedish experience.

Scenarios consistent with "validated" cases

We now define a variety of scenarios in an attempt to explain the validated cases of extreme longevity in comparison to the Swedish trend. We build on the earlier model for Sweden (see Figure 4). Our strategy is to specify scenarios in which population size and the level of old-age mortality are expressed relative to comparable values for the Swedish population. However, we define explicitly the level of r, the asymptotic upper bound on death rates at older ages. The various scenarios are summarized in Table 4.[8]

The first row of Table 4 shows the baseline values for each of the parameters in the current model. All parameters except r equal one, because they are expressed relative to the baseline (Swedish) model. The next three rows investigate the effect of changing population size alone. The effect is expressed in terms of the median increase, for the most recent 40 cohorts, in the 50th percentile of the estimated distribution of the maximum age at death. For a population 10 times as large, the median maximum age at death should be 2.86 years higher. For a population 100 or 1,000 times as large, the median should be 5.47 and 7.90 years higher, respectively. It is essential to note the diminishing importance of increasing population size. Also, we should remember that a population 1,000 times as large as that of Sweden (with 8.9 million people in 2000) would be larger than the current population of the world.

The next row of Table 4 defines a scenario for countries with "complete and validated" super-centenarian data. The population in this situation is well defined. We note that there were 102.7 million persons in these countries in 2000, almost 12 times the population size of Sweden.[9] Because the relative population size is unambiguous and because we have direct information on the relative level of mortality, there is no alternative scenario in this case. In addition to a 12-fold increase in population size, we assume a 4 percent increase in mortality at age 84 (based on the information for ages 80–89 in the

TABLE 4   Scenarios to predict the trend and distribution of the maximum age at death for four populations

Scenario	Compared to Swedish baseline for cohort t, relative level of:				Upper limit of force of mortality	Median difference in the median maximum age at death (in years)	
	Annual number of births	Probability of survival to age 80	Force of mortality at age 84	Mortality increase with age			
	$\dfrac{N(t)}{N_0(t)}$	$\dfrac{S(80,t)}{S_0(80,t)}$	$\dfrac{\mu(84,t)}{\mu_0(84,t)}$	$\dfrac{\theta(t)}{\theta_0(t)}$	r	Predicted	Observed
Baseline (Sweden)	1	1	1	1	1.25	0	na
Population size (1)	10	1	1	1	1.25	2.86	na
Population size (2)	100	1	1	1	1.25	5.47	na
Population size (3)	1,000	1	1	1	1.25	7.90	na
Complete and validated	12	1	1.04	1	1.18	3.01	3.00
3-star validated A	18	1	1.03	1	1.02	4.51	4.50
3-star validated B	42	1	1.06	1	1.15	4.47	4.50
World I	85	1	1.03	1	1.02	6.45	na
Size only	85	1	1	1	1.25	5.29	na
Mortality only	1	1	1.03	1	1.02	0.62	na
World II	106	1	1.06	1	1.02	6.44	na
World III	106	1	1.06	1	1.15	5.53	na

NOTES: (1) The baseline scenario, denoted by the subscript 0, depicts the evolution of the maximum age at death in Sweden for cohorts born from 1756 to 1884 (Wilmoth et al. 2000). See text for an explanation of the other scenarios. (2) The second-to-last column records the median difference (between a given scenario and the baseline) in the 0.5 quantile of the estimated probability distributions of the maximum age at death for the most recent 40 cohorts (i.e., those born 1845–84). (3) For the "complete and validated" countries (excluding Belgium), the last column gives the observed median difference (compared to Swedish data) in the actual maximum age at death for calendar years 1950–99. For countries with 3-star validated super-centenarians, the last column gives the median difference (compared to Swedish data) in the maximum age at death for all cohorts born after 1849, with missing values (for cohorts with no validated super-centenarians) set to age 109. (4) See Notes 4 and 8 at the end of this article for definitions of key mathematical symbols. (5) As noted in the text, population size is taken as a proxy for size of birth cohort.

total population of such countries, as seen in Table 3). However, the value of the asymptote, r, is chosen to obtain a good match between data and model. Thus, the value $r = 1.18$ yields a close fit. This assumption is not implausible, given that observed mortality rates above age 100 in these countries are about 6 percent lower than in Sweden (see Table 3).

Next, we attempt to reconstruct the population that yielded the 3-star validated centenarians. We include all 102.7 million persons residing in the "complete and validated" countries, plus a fraction of the populations for other countries where 3-star validated centenarians have been found. This fraction equals the number of 3-star validated super-centenarians as a proportion of the total number expected in each country (based on the prevalence of super-centenarians in the "complete and validated" countries). The expected numbers are taken from Robine and Vaupel (2001). For example, for the United States only two super-centenarians are included on this list, whereas 184 would have been expected. Therefore, only about 1 percent of the US population, or 3.0 million, is included in the population from which these super-centenarians have been drawn. Making similar calculations for Canada and France, and adding those numbers to the 102.7 million for the other seven countries, yields a population of 157.6 million, or about 18 times the size of Sweden's population.

To complete the scenario, we must also specify the relative mortality level at age 84. The three countries that have been added to the list (France, Canada, and the United States) all have relatively low death rates at older ages, similar to or slightly below Sweden. However, these three countries contribute only about a third of the total population under this scenario (an 18-fold increase as opposed to a 12-fold increase in the previous case). Therefore, we set the relative mortality level at 1.03 in this scenario, only slightly lower than for the "complete and validated" countries alone. As before, the value of the asymptote, $r = 1.02$, was chosen to obtain a close match to the data. As noted earlier, however, a value of r around 1 is closer to previous estimates of this parameter by other authors (Thatcher et al. 1998).

Alternatively, one might argue that all of the remaining populations in the Kannisto–Thatcher database might also be eligible to contribute to the population of 3-star validated super-centenarians. Some of these populations, like Japan, have super-centenarians on the full list but not on the 3-star validated list (this may change as the validation process moves forward). Since such cases are not included in the 3-star list of super-centenarians, it does not seem correct to include the population of these countries in this part of the analysis. At most, however, we might include a fraction of such populations. Thus, as an alternative scenario for the 3-star super-centenarians, we have added one-half of the population of the remaining countries in the Kannisto–Thatcher database, yielding a total of 376.7 million persons from whom the population of 3-star validated super-centenarians might be drawn, or about 42 times as large as the population of Sweden.

However, if the population were enlarged in this way, it would probably have a higher level of old-age mortality. Thus, we set the relative level of mortality at age 84 to 1.06, rather than 1.03. To match the data, we must increase r to around 1.15. In short, if the population that yielded the pool of super-centenarians were larger, then the mortality levels of this population would have to be higher in order to have yielded the observed data. However, we find the previous scenario more plausible and believe that the 3-star validated super-centenarians are drawn from a population that is only about 18 times as large as the Swedish population, with slightly higher mortality at ages 80–89 but substantially lower mortality above age 100 and especially above age 110.

Percentiles of the trend from the first (preferred) scenario for 3-star validated super-centenarians are shown in Figure 5. Remembering that the

FIGURE 5   3-star validated super-centenarians with predicted quantiles, birth cohorts of 1756–1884

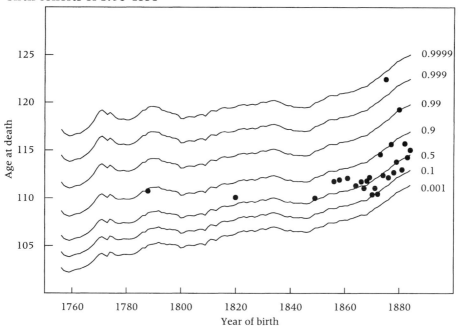

NOTE: Predicted quantiles are based on the "3-star validated A" scenario of Table 4, which is our preferred model of the maximum age at death in the population from which such cases might arise.

data are truncated at age 110, the fit appears quite plausible. It is notable, however, that the case of Jeanne Calment (who died at age 122.44 in 1997) is truly exceptional. According to these calculations, the chance that the oldest member of her cohort (born in 1875) should live to this age or higher is a minuscule 2.19×10^{-4}. Thus, Jeanne Calment's exceptional longevity, relative to members of her own cohort, is the sort of event that should occur only about once every 4,500 years.[10]

Speculations about the world trend in maximum life span

Finally, we have tried to specify a scenario to depict the world trend in maximum observed life span. Our preferred scenario includes the population of all countries with at least one validated super-centenarian (whether in the 3-star or in a lesser category of validation). Presumably, this list includes most areas with mortality conditions and record keeping that are capable of producing validated world record longevity. As seen in Table 1, this group of countries has a population of 760.4 million people, or about 85 times the size of Sweden. We chose the same mortality parameters used in our preferred scenario for the 3-star validated super-centenarians. As shown in Table 4, this scenario leads to our speculation that the world trend in maximum life span may lie about 6–7 years higher than the Swedish trend.

Alternatively, we might include a larger population in this calculation, as shown in the second and third scenarios for the world in Table 4. In this case, we include all countries in either the super-centenarian or Kannisto–Thatcher databases, for a total population of 946.3 million, or about 106 times the Swedish population. However, it seems likely that such a larger population would have relatively higher levels of mortality at older ages. As seen in Table 4, if the mortality level is increased along with the population size, the median level of the maximum age at death does not increase and may even decrease. In short, for populations that are already so large, fluctuations in size play a relatively small role, whereas changes in mortality patterns are quite important.

As illustrated in Table 4, most of the difference in maximum life span between Sweden and the world is attributable to the difference in population size. Beneath the World I scenario, two additional rows show the median increase in maximum predicted life span if we vary population size or mortality levels alone. The 85-fold increase in population size, with no change in mortality, still yields more than a five-year increase in the maximum age at death for the most recent 40 cohorts. On the other hand, the difference in old-age mortality implied by the World I scenario (slightly higher among people in their 80s, but slightly lower among the very old) has little effect on its own.

The dominance of the population size effect is important for our final conclusion. We know that the world trend should lie about 5.3 years above the Swedish trend simply because of the larger size of the "eligible" population. However, we also know that a population 1,000 times as large as the Swedish population—thus larger than the current world population—would yield only about an 8-year increase in maximum life span. Absent a large difference in mortality between Sweden and the "eligible" population of the world (as defined earlier), the real difference should lie somewhere between these two extremes. Although mortality in Sweden is quite low by international standards across most of the age range, the evidence reviewed here suggests that Swedish mortality at very high ages (above age 100) is somewhat higher than in other industrialized countries. This mortality differential, when added to the population size effect, quite plausibly drives up the difference in maximum life span by about 6–7 years.

Conclusion

We have shown that it is possible to reconcile the Swedish trend in maximum observed life span with validated cases of super-centenarians in other populations, although the case of Jeanne Calment remains exceptional. The world trend in maximum life span is higher than in Sweden because of two factors: a larger population from which the world records are drawn, and atypically high mortality rates among Swedish centenarians. It is reasonable to speculate that the world trend in maximum life span may lie above the Swedish trend by about 6–7 years. If so, the maximum age at death in a calendar year would have been around 108 years in the 1860s (with most cases between 106 and 111) and about 115 or 116 years in the 1990s (with most cases between 114 and 118).

Notes

Several people gave helpful comments and suggestions about how to improve this chapter, including Kenneth Wachter, David Steinsaltz, and other participants in the Bay Area Colloquium on Population of September 2001. The authors also thank the many individuals who have contributed to this project by researching and validating the cases of extreme longevity discussed here, in particular Robert Bourbeau, Graziella Caselli, Dany Chambre, Bertrand Desjardins, Viviana Egidi, Louis Epstein, Bernard Jeune, France Meslé, Michel Poulain, Yasuhiko Saito, Axel Skytthe, Roger Thatcher, Jacques Vallin, and Robert Young.

1 Although England and Wales are part of the United Kingdom, they have historically maintained a separate statistical system. For this reason, it is convenient to treat them here as a separate country. Likewise, Scotland is presented scparately in Table 1.

2 The model used here is correct whether the population is homogeneous or heterogeneous with respect to mortality risks. In a homogeneous cohort derived from N births, the probability that the maximum age at death is less than some age x is simply $F(x)^N$, where $F(x) = 1 − S(x)$ is the cumulative density function for age at death. This calculation, like all subsequent ones, assumes that the life spans of individuals are independent of one another (given an overall level of mortality).

On the other hand, assume that the above cohort is composed of k homogeneous groups of individuals, each having its own cumulative density and survival functions: $F_i(x) = 1 − S_i(x)$, for $i = 1,\dots, k$. Let p_i be the probability that an individual chosen at random from the full population (of N births) belongs to group i, where

$$\sum_{i=1}^{k} p_i = 1.$$

Let X be a random variable denoting the survival time of a randomly chosen individual. It follows immediately that

$$F(x) = \Pr(X \le x) = \sum_{i=1}^{k} p_i F_i(x)$$

and

$$S(x) = \Pr(X > x) = \sum_{i=1}^{k} p_i S_i(x)$$

are the cumulative density and survival functions for the full, heterogeneous population. Observed death rates for the population as a whole can be used to derive estimates of $F(x)$ and $S(x)$.

In this model of a heterogeneous population, the sizes of the subgroups are not fixed in advance, but they must still sum to N. If Y_i is a random variable representing the size of group i, then the collection of group sizes, $(Y_1,\dots, Y_k)$, has a multinomial distribution with parameters N and p_i, for $i = 1,\dots, k$. For the full population, the probability that the maximum age at death is less than some age x is

$$\sum_{y_1, y_2, \dots, y_k} [F_1(x)]^{y_1} [F_2(x)]^{y_2} \dots [F_k(x)]^{y_k} \cdot$$
$$\Pr(Y_1 = y_1, Y_2 = y_2, \dots, Y_k = y_k),$$

where the summation extends over all possible combinations of group sizes. Substituting the formula for a multinomial probability in this equation, it is not difficult to show that the above expression equals $F(x)^N$.

Therefore, for this particular model of heterogeneity (which is itself quite flexible), it makes no difference whether we (i) compute probabilities associated with the maximum of the k maxima observed for homogeneous subgroups of the population (using death rates and survival curves computed separately for each group) or we (ii) compute such probabilities for the population as a whole, paying no attention to individual differences in survival. Using either method, the answer is identical.

3 Death rates were smoothed across cohorts to minimize the effects of random fluctuations.

4 The logistic model can be written

$$\mu(x,t) = \frac{r \cdot B(t)e^{\theta(t)x}}{1 + B(t)e^{\theta(t)x}},$$

where $\mu(x,t)$ is the force of mortality at exact age x for cohort t. The parameter, $B(t)$, defines the overall level of mortality relative to r, which is the upper limit to mortality at the highest ages. The increase of mortality with age is controlled by the parameter, $\theta(t)$. Some authors define the logistic model to be the above formula plus an additive constant, usually denoted as c (Thatcher et al. 1998). The additional term (also known as Makeham's constant) is useful when modeling mortality across the entire adult age range, but it is not necessary for studies of old-age mortality alone.

The logistic model has been recommended, in one form or another, for use in modeling old-age mortality because it is consistent with patterns observed in populations known to have high-quality data. Although there is no complete theoretical explanation for why the logistic curve fits old-age mortality patterns so well, it is possible that the process of physiological aging slows down considerably at very advanced ages (see Robine, this volume). An alternative explanation is that the logistic mortality curve is a result of population heterogeneity with respect to mortality risks. Specifically, if mortality for individuals follows a Gompertz pattern—but with proportional differences between individual mortality curves, where the constant of proportionality is distributed across the population according to a gamma distribution—then mortality for the population as a whole will follow a logistic pattern (see Thatcher et al. 1998).

Aside from these considerations, our choice to use only the logistic curve for mod-

eling mortality at advanced ages in this study is based on a fairly simple observation: among the models of old-age mortality most commonly discussed in the literature, the logistic is the only one that both fits observed data well and does not yield impossible predictions about the level of mortality at very high ages. For example, the Gompertz model implies an overly rapid increase in old-age mortality and is thus inconsistent with known data. On the other hand, the quadratic model fits observed experience from age 80 to 110 quite well, but it yields decreasing and eventually negative death rates at higher ages.

5 Referring to Table 4 of Appendix D in Thatcher et al. (1998), it is possible to compute the upper asymptote of the mortality curve according to the formula

$$r = c + \frac{b}{\sigma^2},$$

where c, b, and σ^2 are parameters in the authors' version of the logistic model (which includes Makeham's constant, c, as mentioned in Note 4). After correcting an obvious typographical error in the value of one instance of σ, the range of r is consistently below 1 for all of the female data sets. Although it is higher than 1 for males in three out of four cases, it is the female mortality pattern that matters more when discussing extreme longevity for a population as a whole.

6 Kannisto proposed that observed death rates at older ages, M_x, adhere closely to a linear model after performing a logit transformation:

$$\ln\left(\frac{M_x}{1 - M_x}\right) = a + bx + \varepsilon_x,$$

where ε_x is an error term.

7 The raw material of the Kannisto–Thatcher Database on Old-Age Mortality consists of death counts at ages 80 and older classified by sex, age, year of death, and year of birth. The database also provides age-specific annual population estimates (on 1 January) over the full range of the data series for each country. These population estimates were derived from raw death counts using the methods of extinct cohorts (Vincent 1951) and survivor ratios (Thatcher et al. 2001). Such information was used in the current study to compute death rates over age and time for individual countries or groups of countries.

8 For purposes of discussion and comparison, it is convenient to express the level of mortality in the logistic model in terms of mortality around ages 80–89. Therefore, in Table 4 and elsewhere, we use $\mu(84,t)$, the force of mortality at exact age 84 for cohort t, instead of the parameter, $B(t)$, that appears in the formula for the logistic curve in Note 4. It is easy to verify that

$$B(t) = e^{-84\theta(t)} \frac{\mu(84,t)}{r - \mu(84,t)}.$$

9 For the scenarios involving the "complete and validated" countries, or larger groups that include this set of countries, we have counted Belgium in calculations of population size, even though we lack comparable mortality data for this country above age 100 and have omitted Belgium from Tables 2 and 3. In the super-centenarian database, Belgian cases have never attained world-record status even for a single calendar year. Given this fact plus its small size, the choice to include or exclude this country in these rough calculations of relative population size has no relevance to our conclusions.

10 If the trend in maximum life span continues to increase, it is possible and even likely that Jeanne Calment's record longevity will be surpassed within the next few decades. It is important to emphasize that the probability given here estimates the chance of observing such an extreme value for an individual born in 1875, not in any other year.

References

Aarssen, Karin and Laurens de Haan. 1994. "On the maximal life span of humans," *Mathematical Population Studies* 4(4): 259–281.

Brillinger, David R. 1986. "The natural variability of vital rates and associated statistics," *Biometrics* 42(4): 693–734.

Gumbel, Emil J. 1937. *La durée extrême de la vie humaine*. Paris: Hermann.

———. 1958. *Statistics of Extremes*. New York: Columbia University Press.

Jeune, Bernard and James W. Vaupel (eds.). 1999. *Validation of Exceptional Longevity*. Odense, Denmark: Odense University Press.

Kannisto, Väinö. 1994. *Development of Oldest-Old Mortality, 1950–1990: Evidence from 28 Developed Countries*. Odense, Denmark: Odense University Press.

Robine, Jean-Marie and James W. Vaupel. 2001. "Supercentenarians: Slower ageing individuals or senile elderly?" *Experimental Gerontology* 36(4-6): 915–930.

Thatcher, A. Roger, Väinö Kannisto, and Kirill Andreev. 2001. "The survivor ratio method for estimating numbers at high ages," *Demographic Research* [Online], Vol. 6, available at http://www.demographic-research.org/volumes/vol6/1

Thatcher, A. Roger, Väinö Kannisto, and James W. Vaupel. 1998. *The Force of Mortality at Ages 80 to 120*, Odense, Denmark: Odense University Press.

Vincent, Paul. 1951. "La mortalité des vieillards," *Population* 6(2): 181–204.

Wilmoth, John R., Leo J. Deegan, Hans Lundström, and Shiro Horiuchi. 2000. "Increase of maximum life-span in Sweden, 1861–1999," *Science* 289: 2366–2368.

Wilmoth, John R. and Hans Lundström. 1996. "Extreme longevity in five countries: Presentation of trends with special attention to issues of data quality," *European Journal of Population* 12(1): 63–93.

Post-Darwinian
Longevity

James W. Vaupel

The force of evolution peters out with age. The age-specific intensity of natural selection depends on the proportion of individuals who survive and on these individuals' remaining contribution to reproduction. Few individuals reach advanced ages. For species in which individuals grow to a fixed size, fertility falls with age. In many social species the elderly contribute to reproduction by nurturing their younger relatives (Carey and Gruenfelder 1997). The net contribution of the elderly, however, diminishes with decrepitude, and the dilution of their genes in successive generations weakens the action of natural selection.

Fisher (1930), Haldane (1941), Medawar (1946, 1952), Williams (1957), Hamilton (1966), and Charlesworth (1994, 2001) developed the notion that senescence results from the declining intensity of natural selection with age. Mutations that are harmful at older ages accumulate because only weak selection acts against them. At postreproductive ages there is no Darwinian culling to impede the spread of mutations that are lethal at those ages but have no effect (or a positive effect) at younger ages (Haldane 1941; Charlesworth and Partridge 1997; Partridge and Mangel 1999). Hence, as Shripad Tuljapurkar critically remarked, there should be a black hole of mortality at the age when reproduction ceases.

Such a wall of death has not been observed in any species comprised of individuals that reproduce at more than one age. Indeed, the increase in death rates tends to decelerate with age in many species, and for some species death rates decline after a given age (Vaupel 1997; Vaupel et al. 1998). It is possible that late-acting deleterious mutations are rare or nonexistent. If such mutations do exist, then, as Charlesworth (2001) concludes, they must have negative effects on fitness at younger ages and hence be quelled by natural selection.

Evolutionary geneticists have an impressive armamentarium of concepts and methods for thinking about mortality trajectories at reproductive ages. The research by Charlesworth and Partridge (1997), Partridge and Mangel

(1999), and Charlesworth (2001), together with studies by Tuljapurkar (1997), Pletcher and Curtsinger (1998, 2000), Pletcher, Houle, and Curtsinger (1998, 1999), Promislow and Tatar (1998), and others, suggests the vitality of recent efforts to explain why survival does not plummet to zero when reproduction ceases. This line of research, however, does not and probably cannot answer detailed questions about mortality in the postreproductive span of life.

In this chapter I venture into the reaches of age that lie beyond the force of natural selection, to shed some light on what determines the duration of postreproductive survival. The general approach I suggest may also prove useful in addressing such related questions as the following. What determines the trajectory of mortality at postreproductive ages? Under what circumstances does mortality reach a plateau? Under what circumstances does mortality decline after some age? What determines the level of postreproductive mortality?

The postreproductive ages are, to use James Carey's phrase, post-Darwinian, in the sense that there is no longer any age-specific pressure from natural selection. Nevertheless, the health and vitality of individuals of some species when they enter the postreproductive period of life are determined by evolutionary forces operating at younger ages. An analogy helps explain the steps involved.

The speed and trajectory of a ball are governed by the pitcher's strength and skill up to the moment when the ball leaves the pitcher's hand. Thereafter, the ball's course is determined by the force of gravity acting on the momentum of the ball. Similarly, the course of life up until the end of reproduction is determined by evolutionary forces. After reproduction ceases, the remaining trajectory of life is determined by forces of wear, tear, and repair acting on the momentum produced by Darwinian forces operating earlier in life. Reliability engineers who study wear, tear, and repair refer to the failure of equipment and use failure-time as a synonym for life span. Hence, I employ the phrase "force of failure" as a shorthand for the action of the various forces that determine the failure of complicated systems, including living organisms. The force of failure is analogous to the force of gravity in the metaphor about a ball's path. The force of gravity acting on the momentum of the ball determines how far it will travel. The force of failure acting on the post-Darwinian momentum of an organism determines how long the organism will survive.

The dynamics of the force of failure hinge on the design of a complicated system. At postreproductive ages the design is set: evolution can no longer modify it. At younger ages, however, the powerful creativity of evolution produces miracles of innovative design. Thus at younger ages it may be reasonable to focus attention on evolutionary forces rather than on considerations of reliability engineering. Nonetheless, evolution has to operate under constraints dictated by physical and chemical laws and engineering principles.

Furthermore, evolution maximizes fitness rather than design reliability. As in design of equipment by humans, designs favored by natural selection usually economize on materials and allow imperfections and variation. A full explanation of the trajectory of mortality at ages before the end of reproduction is likely to build largely on evolutionary theory, but it will be based partially on considerations of reliability engineering. Shiro Horiuchi's pathbreaking chapter in this volume demonstrates this. Earlier research on reliability analyses of the aging of living organisms includes contributions by Abernathy (1979) and Gavrilov and Gavrilova (2001); the Gavrilovs' references are helpful but their purported facts and findings are unreliable.

My thesis is that the force of natural selection governs the length of life of most individuals of a species in a given environment, but that the force of failure governs the length of the outer tail of longevity. The start and end of the tail can be quantified in many ways. Here I use a simple expedient. I assume that the tail of longevity starts at the age when only 10 percent of a birth cohort is still alive. For the species for which data on sizable populations are available, reproduction at this age is negligible. Although this is a very rough indicator, I consider this age to be the start of the post-Darwinian span of life. I assume that the tail of longevity ends at the maximum life span attained by the cohort being studied. The length of the tail of longevity is given by record life span minus the top tenth percentile of life span. The relative length of the tail is the length divided by this percentile.

The elusive concept of life span

The life span of an individual is the duration of that individual's life. When, however, does life start? For humans, life spans are usually measured from birth. If they were measured from conception and if nine of ten conceptions end in miscarriage, then human life expectancy would be cut by a factor of ten. The life span of other species is often measured from the start of a particular stage of life, for example emergence from the pupa for some insects.

For some species, age at death can be difficult to determine, especially in the wild. Furthermore, individuals of some species spend long periods in hibernation or some other stage in which the flame of life barely flickers.

Environmental conditions can drastically influence the duration of life for a species. Human life expectancy in some populations today is well under 50 years, and in some historical populations it was under 25, whereas for Japanese women today it is 85 (Oeppen and Vaupel 2002; Japanese Ministry of Health, Labor and Welfare 2002). Depending on diet and other conditions, the life span of medflies can be substantially extended (Carey et al. 1998). Queen bees can live an order of magnitude longer than workers, even though queens and workers are genetically identical (Finch 1990). Point mutations of one or two genes can double or triple life spans (Johnson 1997).

These well-known complexities of the concept of life span imply that life span data must be treated with caution, especially when comparing individuals from different species or from different environments. Here I highlight an additional difficulty that is less well understood. Record life span, which is usually the only life span measure available for a species (Carey and Judge 2000), can vary considerably depending on population size and the trajectory of mortality at advanced ages. This trajectory, as I argued above, is determined by the force of failure operating on post-Darwinian momentum. It is often assumed that the average duration of life for individuals in a species is some more-or-less fixed fraction of the maximum life span observed for individuals in the species. If this were true, then record life span could be used as a surrogate for average life span. The facts, however, are very different.

It used to be believed that for most species death rates tend to increase exponentially with age at adulthood. It is now known that there is a deceleration of the increase in death rates at older ages and sometimes a decline after some age. For humans, a simple exponential curve (the Gompertz curve) fits mortality data for most populations serviceably well from age 35 or so up to age 95 or so. After age 95, a marked deceleration of the rate of mortality increase is observed in populations with reliable data (Thatcher et al. 1998). After age 110, human mortality may reach a plateau or even start to fall (Vaupel et al. 1998; Robine and Vaupel 2002). For other species for which large populations have been followed from birth to death, deceleration of the rise of mortality is also observed, usually with a plateau that is reached when a few individuals are still alive and sometimes with a strong decline in mortality after some age (Vaupel 1997; Vaupel et al. 1998).

It is useful, then, to consider three cases: (1) death rates increase exponentially with age; (2) death rates reach a plateau; and (3) death rates decline with age after some age. For simplicity, assume that only 10 percent of the population is alive at the age when death rates reach 10 percent per unit of time (year, month, week, or day depending on the species). In Table 1, three specific formulas are used to model mortality after this post-Darwinian age. In the first model, death rates rise exponentially. In the second model, death rates remain at 10 percent. And in the third model, death rates gradually fall from 10 percent to 1 percent. In the table, four population sizes are considered—populations with ten observations of old-age life span, populations with a thousand observations, populations with a million observations, and populations with a billion observations.

Note that if mortality increases exponentially, then the post-Darwinian span of life is only moderately influenced by population size. If, however, mortality declines, then the post-Darwinian tail lengthens dramatically with increases in population size. Also note that at any given population size, the duration of life from the onset of old age to the longevity record is

TABLE 1 Median record life span and 98 percent range of record life spans for small and large cohorts and for increasing, constant, and declining age-specific mortality

Population size	Increasing mortality	Constant mortality	Declining mortality
10	13 7 to 21	27 10 to 69	180 21 to 600
1,000	21 19 to 25	73 54 to 115	637 394 to 1,061
1,000,000	27 26 to 30	142 123 to 184	1,276 1,139 to 1,752
1,000,000,000	31 29 to 33	211 192 to 253	2,019 1,728 to 2,443

SOURCE: Author's calculations. The record life spans are computed from the age when the force of mortality reached 10 percent. From this starting age, the trajectory of increasing mortality is given by $\mu(x)=.1\exp(.1x)$. In the constant mortality case, the force of mortality was held at 10 percent per unit of time. The trajectory of decreasing mortality is $\mu(x)=.09\exp(-.1x)+.01$. In a cohort of n individuals, the probability that all are dead by age x is $(1-s(x))^n$, where

$$s(x) = \exp(-\int_0^x \mu(a)da)$$

is the probability of surviving from the starting age to age x. Hence the pth fractile of the distribution of maximum life span can be calculated as the age at which $s(x)=1-p^{1/n}$. The medians in the table were computed by setting $p=0.5$; for the range, values of $p=0.01$ and $p=0.99$ were used.

very long for species for which mortality declines with age and quite long for species for which mortality has leveled off.

Table 2 provides some empirical data about tails of longevity for several species. In modern human populations with low levels of mortality, the tail of longevity is relatively short. As indicated in Table 2, it used to be longer. For genetically identical lines of rats, *Drosophila*, and nematode worms, even longer tails are observed. For one-celled yeast and for large, genetically heterogeneous populations of invertebrates, the tails are very long.

Caution must be used in interpreting the table because population sizes vary greatly. As shown in Table 1, population size can have a substantial impact on maximum life span, especially when mortality levels off or declines. Furthermore, regardless of population size, the trajectory of mortality at advanced ages has a major impact on record life span. As mentioned earlier, for modern human populations for which reliable data are available, mortality increases more or less exponentially until age 95 or so and does not level off until age 110. For the genetically identical populations of flies and worms in Table 2, a mortality plateau is reached relatively earlier in life (see Horiuchi's chapter in this volume). Yeast mortality fluctuates, rising and falling and rising again (Vaupel et al. 1998). For the genetically heterogeneous populations of insects in Table 2, mortality falls sharply at older ages (Vaupel et al. 1998). The length of the tails in Table 2 is thus consistent with the general thrust of Table 1.

TABLE 2   The relative length of the tail of longevity for several species

Population	Size of cohort	Age when 10 percent survive, x_{10}	Maximum age attained, x_{max}	Relative length of the tail of longevity, $\tau = (x_{max} - x_{.10})/x_{.10}$
Humans				
Japanese females, 1999 period life table	$\approx 10^6$	97 years	113 years	0.16
Swedish females, cohort of 1870	60,000	86 years	110 years	0.28
Swedish females, cohort of 1760	30,000	77 years	102 years	0.32
Roman gravestones in Hungary, both sexes, AD 0–399	184	62 years	100 years	0.61
Genetically identical				
Rats, both sexes	770	698 days	1,427 days	1.04
Drosophila	16,000	52 days	95 days	0.83
Nematode worms, wild-type	20,000	19 days	32 days	0.68
Nematode worms, age-1 mutant	10,000	27 days	54 days	1.00
Yeast (S88C)	$\approx 10^6$	27 days	119 days	3.41
Yeast (EG103)	$\approx 10^6$	10 days	67 days	5.90
Genetically heterogeneous				
Medflies, both sexes	1,200,000	33 days	171 days	4.18
Anastrepha ludens, both sexes	1,600,000	50 days	163 days	2.26
Anastrepha serpentina, both sexes	350,000	33 days	90 days	1.73
Anastrepha obliqua, both sexes	300,000	32 days	75 days	1.34
Parasitoid wasps, both sexes	30,000	17 days	70 days	3.12

NOTES: Japanese and Swedish data are from the Human Mortality Database maintained by the University of California, Berkeley, and the Max Planck Institute for Demographic Research (http://www.demogr.mpg.de/databases). The data for Hungary are from a life table, based on gravestone epitaphs, estimated by Acsadi and Nemeskéri (1970); these data may not be reliable. In particular, it is unlikely that the maximum age actually was 100 (Jeune and Vaupel 1995, 1999). More likely it was 85 or 90, which would give a value of τ of 0.37 or 0.45. Numerous life tables based on age estimation of skeletal remains have been published, but the methods are highly questionable (Hoppa and Vaupel 2002). The nonhuman data are all from the Nonhuman Mortality Database maintained at the Max Planck Institute for Demographic Research. The rat data were supplied by Vladimir Anisimov; they pertain to a Wistar-derived line called LIO. The *Drosophila melanogaster* data are from James W. Curtsinger, the nematode worm (*C. elegans*) data are from Thomas E. Johnson, and the yeast (*S. cerevisiae*) data are from Nadège Minois at the Max Planck Institute for Demographic Research. The medfly (*Ceratitis capitata*), *Anastrepha*, and parasitoid wasp (*Diachasmimorpha longicaudtis*) statistics are from experiments conducted in Tapachula, Mexico, under the direction of James R. Carey. Note that x_{max} is the empirical maximum age attained and hence is different from the theoretical x^*. Thus τ in this table is subtly different from the τ that is a function of x^*.

In any case, the tails of longevity in Table 2 are astounding. For Japanese women today, the maximum span of life is a mere 16 percent higher than the top tenth percentile. For yeast, medflies, and parasitoid wasps, however, maximum life span is more than triple the advanced old age that only a tenth of the population attains. So long for some invertebrates and so short for modern humans: this is a tail on which to hang a tale—of reliability.

Determinants of the length of the tail of longevity

Using some simple models, in my research I have started to explore the impact of various design features and environmental characteristics on the remarkably dissimilar gaps between record life span and the top tenth percentile of life span. I have looked at three aspects of a species' physical design: repair, redundancy, and individual variability. In addition, I have begun to consider the impact of environmental variability.

Consider first an organism that suffers some constant hazard of death μ at all ages. The chance of survival to age x is then given by $s(x)=\exp(-\mu x)$. Hence the age at which only 10 percent of the population is still alive, $x_{.10}$, is given by $.1=\exp(-\mu x_{.10})$, which implies $x_{.10}=-\ln(.1)/\mu$. In a cohort of n individuals, the chance that all are dead by age x is $(1-\exp(-\mu x))^n$. Therefore, x^*, the median of the maximum life span, is given by $.5=(1-\exp(-\mu x^*))^n$, which implies $x^*=-\ln(1-.5^{1/n})/\mu$. The relative length of the tail of longevity, τ, can thus be calculated by

$$\tau = \frac{\ln(1-.5^{1/n})}{\ln(.1)} - 1.$$

Note that μ, the force of mortality, drops out of this formula. If the hazard of death is 5 percent, then 10 percent of the population can be expected to survive to age 46. If the cohort has 1,000 members, then the median age at which the last individual dies is 145.5. The tail of longevity is 2.16 times longer than the age to which only a tenth of the population survives. (The life expectancy of this species, by the way, is the inverse of μ, or 20 when μ is 5 percent. If μ were 1 percent, then life expectancy would be 100. The relative length of the longevity tail, however, would be 2.16 regardless of the constant value of μ.)

Consider another species that has two "systems" such that death results only when both systems fail. As a simple analogy, consider human eyesight. Blindness occurs if both eyes fail. How does such redundancy affect the longevity tail? The age to which 10 percent of the population survives is given by $.1=1-(1-\exp(-\mu x_{.10}))^2$, so $x_{.10}=-\ln(1-.9^{1/2})/\mu$. The median of the maximum life span achieved is given by $.5=((1-\exp(-\mu x^*))^2)^n$, so $x^*=-\ln(1-.5^{1/2n})/\mu$. Thus the relative length of the tail is given by

$$\tau = \frac{\ln(1-.5^{1/2n})}{\ln(1-.9^{1/2})} - 1.$$

Note again that μ drops out of the formula for the tail. Hence the force of mortality for the redundant species could be set so that the life expectancy (or any other index of survival) for this species was the same as for the simpler species: τ will not be affected. In a population of size 1,000 with μ equal to 5 percent, the relative length of the tail of longevity is 1.68 for the redundant species compared with 2.16 for the simpler species.

This finding—redundancy reduces the relative length of the tail of longevity—holds for all the species designs I have explored. It is easily shown that putting more and more systems in parallel, such that all have to fail before the organism dies, reduces τ. I have also looked at more complicated designs, with both parallel and serial elements, using computer simulation to estimate τ. It may be possible to prove the result under general conditions, perhaps using the theory of phase-type distributions, but so far I have only considered some simple systems with independent elements that suffer constant mortality.

I have also explored the impact of allowing repair. When an individual is about to die, the individual can be granted a second chance. Or individuals can be given a new lease on life with some probability. Individuals could be given nine chances or more, with some probabilities. "Repaired" or "resuscitated" individuals could have the same or worse survival chances than they faced before. Anatoli Yashin and I explored such models in the context of human life expectancy (Vaupel and Yashin 1986, 1987), and some of our results could be extended to analyses of longevity tails. My preliminary mathematical and computer-simulation results suggest that the more repair allowed, the shorter the relative tail of longevity. This may be a result that holds under general or fairly general conditions.

Instead of assuming that all individuals face the same mortality schedule, individual variability could be allowed. One way to model such heterogeneity is the frailty model introduced by Vaupel, Manton, and Stallard (1979). If frailty is assumed to be gamma distributed with mean of 1 and variance σ^2, then the chance of surviving to age x is given by $s(x)=(1-\sigma^2 H(x))^{-1/\sigma^2}$, where $H(x)$ is the cumulative hazard for an individual of frailty 1. If the force of mortality follows a Gompertz curve, $\mu(x)=a\exp(bx)$, then $H(x)=(a\sigma^2/b)(\exp(bx)-1)$. The age to which a tenth of the population survives is given by $s(x_{.10})=.1$ and the median of the maximum age attained is given by $((1-s(x^*))^n=.5$. It is possible to solve for $x_{.10}$ and x^* in terms of the parameters a, b, and σ^2 and then to compute the derivative of τ with respect to σ^2. This derivative is always positive, which implies that the greater the heterogeneity in individual frailty, the longer the relative length of the tail of longevity. This result may also hold for trajectories of mortality other than the Gompertz curve, for distributions of frailty other than the gamma distribution, and for more complicated models of individual variability than the frailty model. Computer-simulation experiments I have done suggest that the result may hold under a wide range of plausible conditions.

Horiuchi, in this volume, estimates values for σ in models with gamma-distributed frailty. He finds that individual variability is lowest for humans, higher in genetically identical lines of *Drosophila* and nematode worms, and highest in genetically heterogeneous populations of medflies, parasitoid wasps, and bean beetles. This ordering is consistent with the values of τ, the relative length of the tail of longevity, presented above in Table 2.

Finally, I have explored the impact of environmental variability on the longevity tail. I did this by allowing the level of mortality to vary stochastically from one time period to the next, in simple computer-simulation models. The greater the environmental variability, the longer the relative length of the tail of longevity. Often in my experiments, however, the effect was small, especially when mortality was allowed to jump randomly to some level above or below its mean level from one short time interval to the next. The mortality fluctuations tended to cancel each other out by the time older ages were reached, so that the age at which a tenth of the population survives and the highest age attained were not much affected.

To summarize, the less redundancy, the less repair, the more individual variability, and the more environmental variability, the longer the relative length of the tail of longevity. These results are preliminary, but suggestive. They seem consistent—or at least not inconsistent—with the empirical results in Table 2.

The combination of redundancy, repair, and low variability among individuals might be referred to as the "reliability" of a species. Humans are a reliable species in a steady environment; medflies are an unreliable species in an uncertain environment. Horiuchi, in this volume, rightly emphasizes the "quality control" of individual variability. Redundancy and repair are also important. Humans have more design redundancy than worms and worms have more than single-celled yeast. The insects and worms in Table 2 are postmitotic: they cannot replace cells and hence have limited repair capabilities.

The basic proposition set forth in this chapter is that reliable species have short tails of longevity and unreliable species have long tails of longevity. Because this may seem counterintuitive or even paradoxical, it is worth further consideration. If the tail of longevity is short relative to a species' average life span, then mortality at advanced ages is high compared with mortality at younger ages. That is, death rates before old age are relatively low. Hence, the fundamental thesis of this chapter also can be expressed as follows: reliable species enjoy low death rates at younger ages and experience relatively high death rates at older ages. In unreliable species, the gap between mortality at advanced ages and at younger ages is smaller and sometimes even negative.

In principle an unreliable species could have low mortality and long life expectancy. Consider, for instance, a system consisting of a single element with no repair. If the force of mortality for this element were low, life expectancy would be long—and the relative tail of longevity would be very long, as indicated in the first example above. So, an unreliable species is not necessarily the same as a low-quality species. In most cases, however, unreliable species suffer high death rates. Furthermore, an uncertain environment could, on average, be a favorable one.

Conclusion

In addition to mobilizing the concepts and methods of evolutionary thinking to address the black-hole problem of the theories of mutation accumulation and antagonistic pleiotropy, researchers can address a stimulating new question. What degree of reliability maximizes Darwinian fitness in different environments and ecological niches? A species' reliability is determined by natural selection operating during the reproductive period of life. A species' reliability, interacting with the uncertainty of the environment, determines the length of its postreproductive life span. The biology of longevity has to be considered in the light of evolution, but it also has to be considered in the light of reliability engineering. And both evolution and reliability engineering have to be considered in the light of population thinking, that is, demography.

   Hence, demographic perspectives on the comparative biology of longevity can produce illuminating insights that augment the research of evolutionary biologists. This chapter and others in this volume provide examples. Evolutionary biologists have devoted some attention to postreproductive life (Wachter and Finch 1997), but much more research is warranted. The trajectory of mortality at advanced ages is of fundamental scientific interest to researchers interested in aging. The dramatic rise of human life expectancy and the rapid aging of human populations make understanding the outer reaches of survival highly relevant. Following Lotka's lead, demographers can continue to make substantial contributions to knowledge about the forces that govern life.

Note

The author thanks Annette Baudisch, James R. Carey, Maxim Finkelstein, Jutta Gampe, and Shiro Horiuchi for helpful comments.

References

Abernathy, J. D. 1979. "The exponential increase in mortality rate with age attributed to wearing-out of biological components," *Journal of Theoretical Biology* 80: 333–354.

Acsádi, G. and J. Nemeskéri. 1970. *History of Human Life Span and Mortality.* Budapest: Akadémiai Kiadó.

Carey, J. R. and C. Gruenfelder. 1997. "Population biology of the elderly," in K. W. Wachter and C. E. Finch (eds.), pp. 127–160.

Carey, J. R. and D. S. Judge. 2000. *Longevity Records: Life Spans of Mammals, Birds, Amphibians, Reptiles, and Fish.* Odense, Denmark: Odense University Press.

Carey, J. R., P. Liedo, H.-G. Müller, J.-L. Wang, and J. W. Vaupel. 1998. "Dual modes of aging in Mediterranean fruit fly females," *Science* 281: 996–998.

Charlesworth, Brian. 1994. *Evolution in Age-Structured Populations.* Cambridge: Cambridge University Press.

———. 2001. "Patterns of age-specific means and genetic variances of mortality rates predicted by the mutation-accumulating theory of ageing," *Journal of Theoretical Biology* 210: 47–65.

Charlesworth, B. and L. Partridge. 1997. "Levelling of the grim reaper," *Current Biology* 7: 440–442.

Finch, C. E. 1990. *Longevity, Senescence, and the Genome.* Chicago: University of Chicago Press.

Fisher, R. A. 1930. *The Genetical Theory of Natural Selection.* Oxford: Oxford University Press.

Gavrilov, L. A. and N. S. Gavrilova. 2001. "The reliability theory of aging and longevity," *Journal of Theoretical Biology* 213: 527–545.

Haldane, J. B. S. 1941. *New Paths in Genetics.* London: Allen and Unwin.

Hamilton, W. D. 1966. "The moulding of senescence by natural selection," *Journal of Theoretical Biology* 12: 12–45.

Hoppa, R. D. and J. W. Vaupel. 2002. *Paleodemography: Age Distributions from Skeletal Samples.* Cambridge: Cambridge University Press.

Japanese Ministry of Health, Labor and Welfare. 2002. http://www.mhlw.go.jp/english/database/db-hw/lifetb00/part1.html

Jeune, B. and J. W. Vaupel (eds.). 1995. *Exceptional Longevity: From Prehistory to the Present.* Odense, Denmark: Odense University Press.

———. 1999. *Validation of Exceptional Longevity.* Odense, Denmark: Odense University Press.

Johnson, T. E. 1997. "Genetic influences on aging," *Experimental Gerontology* 12: 11–21.

Medawar, P. B. 1946. "Old age and natural death," *Modern Quarterly* 1: 30–56.

———. 1952. *An Unsolved Problem in Biology.* London: H. K. Lewis.

Oeppen, J. and J. W. Vaupel. 2002. "Broken limits to life expectancy," *Science* 296(5570): 1029–1031. Available online at http://www.demogr.mpg.de/publications/files/brokenlimits.htm

Partridge, L. and M. Mangel. 1999. "Messages from mortality: The evolution of death rates in the old," *Trends in Ecology and Evolution* 14: 438–442.

Pletcher, S. D. and J. W. Curtsinger. 1998. "Mortality plateaus and the evolution of senescence: Why are old-age mortality rates so low?" *Evolution* 52: 454–464.

———. 2000. "The influence of environmentally induced heterogeneity on age-specific variance for mortality rates," *Genetic Research* 75: 321–329.

Pletcher, S. D., D. Houle, and J. W. Curtsinger. 1998. "Age-specific properties of spontaneous mutations affecting mortality in *Drosophila melanogaster*," *Genetics* 148: 287–303.

———. 1999. "The evolution of age-specific mortality rates in *Drosophila melanogaster* among unselected lines," *Genetics* 153: 813–823.

Promislow, D. E. L. and M. Tatar. 1998. "Mutation and senescence: Where genetics and demography meet," *Genetica* 102/103: 299–313.

Robine, J.-M. and J. W. Vaupel. 2002. "Emergence of supercentenarians in low mortality countries," *North American Actuarial Journal* 6(3).

Thatcher, A. R., V. Kannisto, and J. W. Vaupel. 1998. *The Force of Mortality at Ages 80 to 120.* Odense, Denmark: Odense University Press. Available online at www.demogr.mpg.de/Papers/Books/Monograph5/ForMort.htm

Tuljapurkar, S. 1997. "The evolution of senescence," in K. W. Wachter and C. E. Finch (eds.), pp. 65–77.

Vaupel, J. W. 1997. "Trajectories of mortality at advanced ages," in K. W. Wachter and C. E. Finch (eds.), pp. 17–37.

Vaupel, J. W., J. R. Carey, K. Christensen, T. E. Johnson, A. I. Yashin, N. V. Holm, I. A. Iachine, V. Kannisto, A. A. Khazaeli, P. Liedo, V. D. Longo, Y. Zeng, K. G. Manton, and J. W. Curtsinger. 1998. "Biodemographic trajectories of longevity," *Science* 280: 855–860.

Vaupel, J. W., K. G. Manton, and E. Stallard. 1979. "The impact of heterogeneity in individual frailty on the dynamics of mortality," *Demography* 16: 439–454.

Vaupel, J. W. and A. I. Yashin. 1986. "Targeting lifesaving: Demographic linkages between population structure and life expectancy," *European Journal of Population* 2: 335–360.

———. 1987. "Repeated resuscitation: How lifesaving alters life tables," *Demography* 24: 123–135.

Wachter, K. W. and C. E. Finch (eds.). 1997. *Between Zeus and the Salmon.* Washington, DC: National Academy Press.

Williams, G. C. 1957. "Pleiotropy, natural selection and the evolution of senescence," *Evolution* 11: 398–411.

Hazard Curves and
Life Span Prospects

Kenneth W. Wachter

When we see centenarians among us, singing, swimming, driving, and hear of super-centenarians, ten years or more beyond a hundred, it seems to auger well for continued steady progress against mortality. Can the fact that most people in developed societies die in their 70s and 80s be altogether resistant to change, if some people are living so much longer?

One may ask about specific causes, changing environments, social support, medical advances, but one must also ask about underlying biological potential. Humans are a special case, but also an example of the general case, molded like other species by principles of evolution, whose application to longevity is imperfectly understood. Comparing human survival at extreme ages with survival in other species, biodemography has been fostering a new view of biological potential.

Early achievements of biodemography are recounted in the volume *Between Zeus and the Salmon*, edited by Wachter and Finch (1997). The chapters in the present volume show how much has been learned since then, and how much more we stand to learn today. This chapter looks at the biodemography of longevity through the prism of connections between the forms of hazard curves and the prospects for future progress against old-age mortality. Hazard curves are mathematical summaries of the dependence of mortality on age. Understanding how hazard curves assume their shapes helps tell us how much resistance to extended survival may be programmed into our biology. The mathematics of hazard modeling intertwines with evolutionary theory and empirical research.

It is not the intention here to provide a synopsis of the contributions to this volume. Rather, the aim is to take up one theme central to biodemography and draw selectively from the chapters, reacting to suggestions that they offer and bringing out cross-cutting ideas and ties among them.

The opening contribution, by James Carey, sets the stage for our discussion as it does for the rest of the collection. Carey outlines the context within which life span comes to be viewed as an evolutionary adaptation.

He reviews the ecological correlates of life span and the roles of the elderly in social species, and he gives a roster of behavioral and scientific developments likely to propel the biological future of human life spans, under the four headings of healthful living, disease prevention and cure, organ replacement and repair, and aging arrest and rejuvenation. Such specifics provide an essential backdrop for the general consideration of constraints and opportunities in the present chapter. It is important to keep in mind that the "life span prospects" under discussion here are prospects for gradual, incremental extensions of life spans across large populations, progress of the kind seen around the world over the last century. Drastic bioengineering and visionary advances like those surveyed in Michio Kaku's (1997) *Visions*, however enticing, are beyond our chosen scope.

The mathematical representation of mortality to which we mainly refer, the "hazard function" or "hazard curve," is a graph of the force of age-specific mortality as a continuous function of age. Equal to the downward slope of the logarithm of the survivorship curve, it is discussed in detail in Shiro Horiuchi's contribution to this volume. Hazard curves that rise very steeply at advanced ages seem to suggest formidable obstacles to broad-based extensions of life. Hazard curves whose steepening tapers off at advanced ages seem to suggest more manageable challenges. The evolutionary theory of senescence, developed by biologists over the last half-century, encourages the idea of a strong connection between the shapes of hazard curves by age and their resistance to reduction over time. The theory and the connection will be described shortly. What is important to say first is that under this theory the same general principles operate to shape the hazard curves of other species as to shape our own.

In this chapter, the word "tapering" is used as shorthand for reductions in the slope of the logarithm of the hazard curve at advanced ages. Using logarithms puts measurements of slopes on a proportional basis and turns exponential curves into straight lines. Tapering hazard curves are often called mortality "plateaus," but hazards may taper so far as to bend downward, stretching the analogy with terrestrial plateaus.

As long as it was still widely believed that hazards in other species typically rose exponentially with age, free of tapering, such beliefs fostered the view that diminishing returns in the struggle against human mortality at older ages could not be far away. Then came the discoveries initiated by Carey et al. (1992) and Curtsinger et al. (1992) of tapering hazards at extreme ages in a number of other species, first Mediterranean fruit flies and *Drosophila*, later nematode worms and yeast. Data may be found in Vaupel et al. (1998) and in Figures 1, 2, and 3 of Horiuchi's contribution to this volume. Robine and Vaupel (2001) and the contribution of Robine and Saito to this volume show that the record human life spans confirmed over the last few years demand human hazard functions that also taper beyond age 95 and level out or drop beyond 108.

Guiding forecasts

It is fitting that a recent meeting of biodemographers took place on the Greek island of Santorini, a home of ancient augury. Biodemography speaks to demographers in their role as augurers. The regularity of past trends in itself can scarcely indicate whether extrapolation is justified or whether a change point, abrupt or gradual, lies around the corner. Structural understanding has to set the context. This truism applies palpably to predicting the course of mortality over the next half-century. Official forecasts almost all impose assumptions of diminishing returns to progress against mortality at older ages. In contrast, researchers impressed by the long-sustained pace of progress extrapolate unabating gains. An airing of the issues may be found in *Beyond Six Billion*, edited by Bongaarts and Bulatao (2000). The policy ramifications are obvious, as the different forecasts put very different constraints on the political decisions required now to preserve the solvency of social insurance systems and the economic health of nation states.

Choice between such forms of forecasts is not a technical issue. It is a matter of judgment, context, and plausibility, as are the ultimately subjective uncertainty bounds to be attached to all the forecasts. Here, in making possibilities plausible, biodemography can have its say.

The relevance of underlying biology to future trends in mortality, however, is not incontestable. It may turn out that features of the human environment, configurations of life-extending technology, and behaviors unique to humans will govern trends over the next century in ways that are hardly clarified by understanding our genetic heritage and our commonalities with other species. It may also be that concerns about pushing up against biological limits to longevity will come to seem like *fin de siècle* luxuries, if recurring acts of terrorism in the new century elevate mortality and unhinge the security and liberty on which our long run of progress has depended.

If civilization is successfully defended in the future, as it has been, often but not always, in the past, then several kinds of evidence do argue in favor of attention to fundamental biology. Notwithstanding all the human paraphernalia of dwellings, supermarkets, and hospitals, over a broad range the functional form of adult human mortality as a function of age across the globe in 2003 is the same as the functional form for worms, flies, deer, and most of the complex species that we know. From middle adulthood up to moderately old ages, exponential functions enshrined in the model of Benjamin Gompertz (1825) give serviceable approximations to hazard curves in species after species, as Finch (1990: 13–23) describes. Exponential curves have straight lines for their logarithms, that is, lines with unchanging slopes. Progress over time has shifted the curves but preserved their shape, strongly suggesting that the general biological principles behind the common shape are still operative for humans.

The salience of underlying biology is also supported by the steadiness of the gains in developed countries over several generations. Oeppen and Vaupel (2002) have shown that "best-practice" life expectancy— e_0 in the country leading the world in e_0 in each given year—has been increasing quite nearly linearly at a rate of about three months per year over 160 years. Limits on life expectancy announced by demographers have been broken shortly after or even before they have been stated. Steady progress in the face of enormous differences in the character of advances during different stages of this long period argues for the existence of some deeply rooted structure of opportunity that humans can exploit.

One might have expected heroic medical breakthroughs like the introduction of antibiotics to show up sharply in the historical record. One might also have expected fitful starts and stops in the gains at different ages, as different diseases and conditions fatal at different ages successively yield to medical treatment and health-enhancing behaviors. In fact, however, age-specific mortality rates for different age groups have been declining in tandem, each with its own exponential rate of decline, well represented by Lee and Carter's (1992) model. For the leading industrialized countries since at least 1945 (Tuljapurkar, Li, and Boe 2000), these rates of decline have stayed quite nearly constant. The exponential functional form for the changes over time recapitulates Gompertz's exponential for changes over age. Age groups hold on to their shares in the broad march forward.

The regularities just described are features of a broad-brush picture. With close scrutiny, deviations can be seen. Horiuchi directs attention in this volume to upward bending that often appears in the logarithms of human hazard curves before tapering sets in, consistent with demographers' longstanding use of Makeham models in place of Gompertz models for refined work. Change points appear when Lee–Carter models are fit to longer stretches of historical data. Even after 1945 the shares of age groups in overall mortality reduction are not strictly constant. The trend in maximum recorded life span and the burgeoning number of centenarians described in the contributions to this volume by Wilmoth and Robine and by Robine and Saito signal an expanding share in overall progress for the oldest old. Reflecting developments over the last five years, many of the chapters in this volume press beyond broad regularities to examine nuanced differences in hazard curves across species and across time and age. But the existence of the regularities remains as a strong argument that long-term mortality forecasting is not solely a matter of anticipating particular medical and health advances, but also a matter of understanding the dynamics of resource deployment and biological potential.

How many battles?

Classical evolutionary theories of senescence are recounted in this volume by Steven Orzack (pp. 20–23), who introduces us to "mutation accumulation," "antagonistic pleiotropy," and the "disposable soma." Overviews accessible to demographers may also be found in Wachter and Finch (1997), Finch (1990: 36–42), Charlesworth (1994), and Rose (1991). Here we focus on aspects bearing on connections between hazard curves and potential progress. All the variant theories have the same starting point: the less individuals of a certain age contribute to reproducing and sustaining the next generation, the less the force of Darwinian natural selection will work to suppress mortality at that age. "A corollary," as Finch (1990: 38) puts it, "to the proposition that the force of selection diminishes with age…is that genetically based manifestations of senescence should vary widely among species and within species."

This idea has pessimistic implications when one asks how many separate battles will have to be fought by medicine or lifestyle modification to repeat at ever older ages the victories over mortality won at medium-old age. The implications are particularly striking for simpler forms of mutation accumulation theory. These posit the existence of large numbers of mutations, each with small deleterious effects on survival at specific ages or over bounded age ranges. Mutations that affect young ages are weeded out of the population quickly by natural selection. Those affecting ages that reproduce less or contribute less or nothing to the next generation are weeded out less rapidly, or not at all.

All the while the weeding proceeds, more small mutations are being introduced into the population. Mutations accumulate to the point where there is a balance between the force of mutation and the force of selection. Mortality rates rise with age in proportion as the force of selection weakens with age. At any time, mutations with specific age effects are to be found scattered independently across the individuals in a population, combining in large numbers to dictate each individual's internal susceptibilities to causes of death.

In this picture, many things go wrong more and more often at older ages. Countless battles loom against modes of failure dispersed across the population in different forms more and more numerous in later life. Hard-won progress at younger ages only presages harder challenges at older ages. Once mortality is low, further increases in human life expectancy depend on progress at ages well beyond the ages that could plausibly have been contributing to reproductive fitness over evolutionary time. Without further provisos, one would expect mutation accumulation to drive mortality rates toward infinity at postreproductive, postnurturant ages. Such considerations abetted pessimistic mortality projections several decades ago. The discovery of hazard curves tapering at extreme ages brings the simple theory into doubt. As discussed below, current thinking focuses on whether revising the theory to accommodate tapering does or does not mean abandoning its essentially pessimistic implications.

Antagonistic pleiotropy is a process that could in principle drive up hazard curves at older ages even more dramatically than mutation accumulation and spell even harder challenges to extension of human life span. The theory posits mutations that promote reproduction or survival at some ages but incur a penalty at other ages, depressing survival or reproduction. Under natural selection, survival at old ages is readily sacrificed for benefits at younger ages. Genes with such pleiotropic actions could reflect the tradeoffs between investment in reproduction and investment in repair and maintenance described in Kirkwood's theory of disposable soma, reviewed by Orzack (p. 21) in this volume.

Theories of antagonistic pleiotropy, however, need not carry all the pessimistic implications of mutation accumulation theory. Antagonistic pleiotropy might act on hazard functions through the dynamics of muta-tion-selection balance as just described, or it might act through the gradual fixation of mutations in populations in which the net effect on fitness of the mix of positives and negatives was positive. In the latter setting, hetero-geneity could arise mainly from heterogeneous interactions between genes and the environment and life histories of individuals rather than from dif-fering individual portfolios of mutations. A limited number of mutations with substantial effects rather than a large number of mutations with small effects might be determinative, softening the prospect of an endless array of future battles against mortality at ever older ages.

The large issue with which biodemographers are struggling today con-cerns the conclusions to be drawn from the observations of tapering hazard curves and the experimental results that have come in their wake. Are the new discoveries consistent with versions of the classical theories, or do they demand revisions that remove the main pessimistic implications?

Some biodemographers think that the tapering primarily reflects com-positional change of the group of survivors in the presence of persistent heterogeneous frailty, without the need to appeal to detailed age-specific programs or far-reaching revisions of classical theories. Other biodemo-graphers think that the tapering of hazard functions at the population level reflects programs of age-based change at the individual level that do need to be grounded in critically revised theories. The leading contender for a compositional explanation will be called "simple culling" in this chapter. We discuss it first, and afterward take up versions of revisionism.

Simple culling

The compositional explanations for tapering involve demographic selection in the presence of heterogeneity in probabilities of survival. In considering these explanations, it is important to distinguish age-free heterogeneity from age-bound heterogeneity. The frailty models of Vaupel, Manton, and Stallard (1979) and Vaupel and Yashin (1985), which have guided hazard analysis over the last 20 years, posit "age-free" heterogeneity: differences from individual to individual are represented by a personal multiplier that operates independently of age, remaining fixed over the whole relevant segment of the life course. The variation of hazards with age, that is, the age profile of mortality, is the same for all individuals. The power of this framework lies in its simplicity.

One can also envision models for age-bound or age-structured heterogeneity in which individuals differ in the age profile of expected mortality. For example, individuals may differ in their inherent vulnerability to certain causes of death, and causes of death may each have a characteristic age pattern. Another example would be models in which the Gompertz slope parameter or Horiuchi's life table aging rate (pp. 129–133 in this volume) are significantly heterogeneous. In models with age-free and age-bound heterogeneity alike, more vulnerable individuals tend to die earlier, survivors are progressively selected from the less and less vulnerable, and this culling causes hazard rates to taper. However, unlike the age-free models, in the age-bound models culling is not the whole story: the age profiles in the mixture modulate whatever tapering occurs, and prospects for reductions in mortality are interwoven with the determinants of the causes and profiles.

By simple culling we mean survival of the less vulnerable within the age-free, fixed-frailty framework. If nothing but simple culling is behind the observed tapering of hazard rates in species studied so far, then the classical accounts of mutation accumulation and antagonistic pleiotropy, with their pessimistic overtones, are easier to maintain. No one doubts that something like simple culling plays some role. The question is whether simple culling suffices to explain all the tapering.

Data underlying this debate are found in Figures 1, 2, 3, and 6 of Horiuchi's contribution to this volume and in Vaupel et al. (1998). Tapering shows up in the graphs of the logarithms of hazard curves versus age as they bend away from straight lines. The log-hazard curves for medflies and yeast actually bend over and drop downward beyond the domains shown in Horiuchi's figures. For humans, the record life span of Jeanne Calment, at 122, may be a hint that the human curve also ultimately drops.

The models in mind when simple culling is under discussion posit a Gompertz baseline hazard function with the same slope parameter for each individual. The frailty multiplier scales the level of the hazard function up or down by the same factor at all ages. The distribution of the frailty multiplier is most often taken from the gamma family of distributions, partly because this choice produces convenient closed-form formulas, but also because simple culling keeps the frailty distribution within the gamma family. However, gamma frailty with Gompertz hazards leads to a logistic model for aggregate survival that forbids any drop in extreme-age hazards. To account for drops observed with medflies and yeast and possibly with humans, a simple culling model has to have a small concentration of individuals more robust than any distribution from the gamma family would allow.

The early work in biodemography surveyed in *Between Zeus and the Salmon* established that tapering is reduced but not eliminated when mixed populations of flies or worms are replaced by strains that are (nearly) genetically identical. Experiments that attempt to control environmental heterogeneity have comparable outcomes. Ewbank and Jones (2001), citing and building on work by Yashin, DeBenedictis, and Vaupel, have been able with humans to make the variance in frailty an identifiable parameter within the framework of fixed-frailty models, and they estimate moderate values.

It takes a lot of heterogeneity to make hazard functions visibly bend or drop. When one fits a frailty model to a data set, one can compute how low the lifelong hazards have to be among the least frail few percent of individuals compared to the hazards for the median individual or the most frail few percent of individuals, if simple culling is to explain all the observed tapering. Many authors find these ratios too extreme to be plausible, in comparison with the measured effects of covariates or interventions. Mueller et al. (2003), for instance, obtain a ratio of 1/162 for the 2.5-th percentile compared to the 97.5-th percentile in a recent experimental population of *Drosophila*, excessive, in their view, in comparison to measured effects of dietary change.

Horiuchi, in the crucial endnote 13 of his contribution to this volume, argues the opposite point of view. He likens the hazard ratios for percentiles from frailty fits for a cohort of Swedish males to hazard ratios calculated by multiplying together marginal effects for common covariates estimated from fitting a proportional hazard model to recent US sample data. However, multiplying marginal effects estimated from a model without a full set of interaction terms probably produces overestimates of the ratios among the groups with the largest contrasts. The ratio of 1/32.8 that Horiuchi calculates from the 0.2-th and 99.8-th percentiles of the Swedish frailty fit seems rather to support the implausibility of simple culling as a full explanation of tapering. Further investigation should bring this question to resolution.

Reliability and vitality

Horiuchi observes that fits of frailty models reveal much less heterogeneity on a relative basis for humans and other vertebrates than for insects and other invertebrates. He interprets this contrast, with good reason, as reflecting a greater degree of quality control in the formation of physiological components for organisms whose body plans and lifecourse strategies involve greater investments in individual offspring. Quality control is one aspect of reliability theory, whose importance for biodemography has been emphasized most especially by Gavrilov and Gavrilova (2001). James Vaupel, in his contribution to this volume, joins Horiuchi in turning to reliability theory to make sense of contrasts. The contrasts of interest are relative. Absolute levels of hazard curves tell us only that a day in the life of an insect is something like a year in the life of a human. Horiuchi standardizes across species by expressing time in units of one-hundredth of the modal life span, the age at which the greatest number of adults die. Vaupel standardizes across species by taking ratios of certain percentiles from the tails of the life span distributions. Both approaches bring out clearly the features of low spread and short upper tails that distinguish the life span distributions of vertebrates in general and humans in particular.

Vaupel (p. 266 in this volume) implicates three processes contributing to these patterns:

> The combination of redundancy, repair, and low variability among individuals might be referred to as the "reliability" of a species. Humans are a reliable species in a steady environment; medflies are an unreliable species in an uncertain environment.

Vaupel has a handy way of thinking about redundancy and its impact on hazard functions. Start with any survival curve. Draw two death dates independently at random from the survival curve, and build in redundancy by letting an individual die only at the latter of the two dates. This process generates a new survival curve out of the original one. Vaupel treats only the case of constant hazards, but his main finding is a generic property. This transformation reduces the ratio between the ages when 1 percent and when 10 percent of the population still survive. In other words, on a ratio basis, it shortens the upper tail. In general the effect is not independent of mortality level, as Vaupel shows it to be in the constant hazard case, but renormalizations of level can be shown to preserve the property under reasonable conditions.

What this redundancy transform does, however, is to shift the logarithm of the survival curve upward by an amount that is essentially constant out in the upper tail. It leaves the hazard function, which is the downward slope of this logarithm, unchanged in the tail. It necessarily does so, because any death age found far out in the tail will be, with high probability, the larger of any pair of independently drawn death ages. The same point holds for threefold, fourfold, or higher-fold redundancy transforms. The ratio measure for tail percentiles is reduced, not because of effects at later ages, beyond the direct influence of natural selection, but rather because of reductions in early-age mortality. All three of the reliability processes that Vaupel highlights would seem, on the face of it, to operate primarily on early ages rather than on the upper tails themselves. Theories of senescence do stand in need of something to account for the prolongation of patterns from reproductive and early postreproductive ages out into later postreproductive ages, and considerations of reliability engineering do seem like logical elements for the purpose. But the pieces do not yet fit together.

Gavrilov and Gavrilova (2001) have developed specific formal models grounded in reliability theory to generate hazard curves that are Gompertzian over a range of ages and that taper toward a flat asymptotic plateau at extreme ages. Independent, identically distributed components with constant probabilities of failure over time are arranged together in blocks with built-in redundancy. A block fails only when all its components fail, like electric circuits wired in parallel. The organism dies when any one block fails, like electric circuits wired in series. In the most intriguing of their models, Gavrilov and Gavrilova take the numbers of components in the blocks to be independent Poisson random variables.

Steinsaltz and Evans (2003) correct the derivation of the hazard function in Gavrilov and Gavrilova (2001), give a closed-form formula, and check the implied shapes of hazard curves for a variety of parameter settings. Steinsaltz and Evans are skeptical of the model because it generates a stretch of Gompertzian hazards only for quite specific choices of parameter values. Most choices lead to wholly non-Gompertzian shapes. As with string theory in physics, the lack of a canonical version is a sobering objection to models of this type. However, demographic reasons could be given for preferring certain ranges for the parameters, independent of the shapes to which they lead. The models are more ambitious than frailty models in seeking to explain rather than assume the involvement of exponential functions. The questions surrounding them deserve further review.

The connected component models of Gavrilov and Gavrilova are special cases of a general class of stochastic models for hazard functions that will be called "changing vitality models" in this chapter. The prototype for this class is the Brownian motion model brought to the attention of demographers by Weitz and Fraser (2001) and also put forward by Anderson (2000). An idealized measure of a person's vitality is assumed to change randomly across the person's lifetime like a random walk or like its continuous analogue, a Brownian motion, drifting haphazardly downward. When this random path hits or crosses zero, the person dies. The framework is reminiscent of Strehler and Mildvan (1960), but unlike Strehler and Mildvan's model, an exponential response function is not arbitrarily imposed.

The models of Gavrilov and Gavrilova take on a similar structure when the counts of intact components in each block of the system are treated as a multivariate version of the idealized vitality measure. Steinsaltz and Evans (2003) define a broad class of such models in which vitality changes according to a Markov process. They prove that tapering hazard functions converging to constants at extreme ages are a generic phenomenon common under suitable conditions to all models in this class. The hazard function levels out because the distribution of vitalities among survivors converges to a fixed distribution, for much the same mathematical reason as convergence of age structures in stable population theory. As Steinsaltz and Evans (2003: 3) say, these models help us envision an alternative form of selectivity:

> ...the mortality rate stops increasing, not because we have selected out an exceptional subset of the population, but because the condition of the survivors is reflective of their being survivors, even though they started out the same as everyone else.

These changing vitality models have strong inbuilt tendencies toward producing tapering hazards. The challenge for modelers is to construct special cases with hazard functions that do not taper too early in the age span. The purpose of all these stylized models, at the present stage of inquiry, is to serve as guides to thinking. Although many variations are possible, the underlying logic of changing vitality models has optimistic overtones, as far as opportunities for future extensions in life span are concerned. Improvements in mortality at medium-old ages would be evidence of upward shifts in the vitalities of individuals that would be expected to confer benefits on later ages as well.

Revising theories of senescence

If compositional explanations do not account fully for the observed tapering in hazard functions, then classical evolutionary theories of senescence need to be revised. The theories rest on assumptions about the existence and prominence of genes with certain kinds of actions on survival at older ages, and they also rest on mathematical models for translating specifications of gene effects into predictions of hazard functions. Early evidence about kinds of genes that actually occur will be discussed in the concluding section of this chapter. This section focuses on the interplay between specifications of theory and shapes of hazard curves.

Revising theories of senescence is a tricky matter, if the many successes of the theories are not to be thrown out in the process of repairing the limitations. Most deaths after all occur in the exponentially rising segment of the hazard curve, long treated by the classical theories, not in the tapering segment, where the need for revisions appears.

One strategy for explaining tapering is to argue that mutations with effects exclusively at a range of older ages are rare and aberrant. Pletcher and Curtsinger (1998) give an illuminating discussion. The large number and variety of such genes that are required to drive mutation accumulation theory may be a chimera. Few genes may be expressed for the first time only in later life, and few genes that act on later survival may fail to have some role in development or in somatic maintenance during reproductive adulthood. Debate over the nature of the genes that might figure in mutation accumulation goes all the way back to Haldane, as discussed by Finch (1990: 37), and remains an active subject for research.

It is easy to see that tapering could be explained if mutations with distinctively bad effects on survival at late ages always had at least some bad effects at early ages. Such reinforcing pleiotropic effects would produce tapering, inasmuch as the force of selection would be buoyed up by the early effects and would not fall as rapidly as it otherwise would at older ages. The reverse of antagonistic pleiotropy, reinforcing effects are called "positive pleiotropy" because correlations are positive, but this technical term has misleading connotations when the correlated effects themselves are negative. The nontechnical discussion here will keep reinforcement in the foreground. Reinforcing pleiotropic effects essentially introduce an extra penalty term into calculations of fitness, a plausible ingredient of a wide range of formal models for tapering, as in Wachter (1999).

Charlesworth (2001), aligning himself with the idea of reinforcing pleiotropic effects, regards the idea as a minor revision to mutation accumulation theory. It is, however, major in its demographic consequences. In contrast to classical mutation accumulation, an assumption of possibly small but ubiquitous reinforcing pleiotropy would offer escape from an endless succession of new battles looming against future progress in old-age mortality. Evolution would have cooperated with the goals of medicine, weeding out contributors to late-age failures thanks to their younger-age concomitants.

The more pervasive any patterns of reinforcing pleiotropy, the more strongly interconnected should be the challenges of mortality at medium-old and at older-old ages. A picture of large numbers of largely independent accumulating mutations could be retained if the reinforcing pleiotropic effects were small and diffuse. Stronger and more systematic reinforcing pleiotropy would bolster the case for optimism about life span extension. A few good steps forward might have many good consequences across the age range.

Antagonistic pleiotropy competes with mutation accumulation for the allegiance of experimentalists. The weight of evidence from selection studies such as Rose et al. (2002) swings to and fro. Theories of antagonistic pleiotropy are currently handicapped by the lack of a viable mathematical model to connect the genetic processes with hazard curves of the forms actually observed. The hope would be to represent the hazard curve and an associated age-specific fertility schedule for a population as the limiting equilibrium state of a Markovian stochastic process. The steps or transitions for the process would correspond to mutations going to fixation in a population in the face of genetic drift and natural selection. Mueller and Rose (1996) put forward versions of such a model devised to account for mortality plateaus, that is, for tapering hazard curves. The proof in Wachter (1999) shows that the proposed models fail. The models do not have the intended limiting states.

Mueller et al. (2003: 25) take the position that it is sufficient for their model to produce the intended tapering hazard curves on a purely temporary basis, as transient states. The model is started with a stylized initial state, in their case with a flat adult hazard curve, and after a while it passes through states of the intended form before going on to diverge from the intended form and converge to a limiting state. The problem with this approach is the initial state. For the transient states to be meaningful, evolution would have to have established the specific stylized initial state as the hazard curve at a specific time in the past. An account would therefore be needed of how evolution arranged such a specific starting state in the first place and why the pleiotropic model is meant to apply only after this instant of time and not before. It would also be necessary to explain, for any species to which the model is meant to apply, how we happen to be observing the species at just the right period of time after the establishment of the specific initial state.

Mueller, Rose, and their coauthors contend that their transient states last for millions of generations. The parameters of the model can indeed be set to make the whole process run arbitrarily slowly, slowly arriving at the transient states of interest from other transient states, and slowly departing from them toward other transient states on the way to the limiting states. The parameters can also be set to make the process run quickly. The point of importance is that the transient states of interest are ephemeral relative to the time scale of the process. For Markovian models of pleiotropy, recourse to transient states is not a tenable position.

It is therefore a continuing priority to construct new Markovian models incorporating antagonistic and reinforcing pleiotropy that can successfully generate tapering hazard curves as limiting states. Ideas for the design of such models are set out in Wachter (1999: 10547).

As Mueller et al. (2003: 25) go on to point out, environmental conditions could change over time periods that are shorter than the time scales for the establishment of equilibriums in models of this kind. For arguments based on transient states, this situation would redouble the difficulties associated with the establishment of a specific initial state at a specific initial time. It would also mean that a model incorporating environmental fluctuations would have to be spelled out in order to derive the consequences of pleiotropic processes.

The broader need for ambitious modeling of environmental fluctuations is taken up by Orzack in his contribution to this volume. One set of effects, which he discusses in detail, arises from the way in which symmetric random variability in components of a population projection matrix produces nonsymmetric distributions for rates of growth and associated measures of fitness. More complex effects necessarily arise when temporally correlated fluctuations interact with generational renewal.

From a long-term perspective, the inherent randomness of demographic events in combination with random variations in resources and environments can drive populations to extinction. Orzack advocates extensions of the traditional reckoning of fitness to take account of such impacts on the gene pool. From a short-term perspective, the environments in which organisms find themselves are partly of their own making and choosing. In his contribution to this volume, Marc Mangel presents a remarkable example in which Pacific rockfish adjust their metabolic rates and exposure to oxidative damage by choices of ocean depth across their lifetimes. The subjects of both these contributions take them well beyond the scope of the present chapter, but they represent initiatives likely to figure prominently in the future of biodemographic research.

Mutation accumulation

Charlesworth's (2001) aforementioned formulation of mutation accumulation theory is of particular demographic interest, going beyond the issue of reinforcing pleiotropy and its role in tapering. He spells out a specific explanation for the Gompertzian shape of hazard curves with a wealth of testable consequences.

Charlesworth posits high background levels of extrinsic, age-independent hazards in the wild. The constancy of hazards over age, in a context of near-zero population growth, produces stable age pyramids with sizes of older age groups decreasing exponentially with age at a rate equal to the constant hazard rate. The simple version of the theory makes the assumption (which can be relaxed to some extent) that there is a steady contribution over age for relevant age groups to the production of the next generation either through reproduction or through provisioning, rearing, and protection. For each given age, mutations are assumed that raise hazards only beyond that age, by small amounts, either in a window of ages starting at the given age or at all ages from the given age upward. The density of mutations per unit of time is small and uniform over age.

The mutations tend to add a small perturbation of extra mortality onto the high constant background mortality. If the perturbation is small enough, the age pyramid nearly retains its exponential shape. According to the mathematics of mutation-selection balance, the exponential shape of the age pyramid is impressed onto the shape of the additional term in the hazard function. When populations are taken out of the wild into experimental settings, laboratories, zoos, or, as with humans, into the modern world, the constant background contribution to the hazard function is largely removed, and what remains is the Gompertzian addition.

Mutation accumulation theory in this form has power. It offers to account at a stroke for the ubiquity of Gompertzian hazards across a wide range of species, body plans, life spans, and circumstances. It provides for the exponential function to enter directly from first principles rather than from ad hoc assumptions. It reaffirms the association between the level of extrinsic mortality and the pace of senescent mortality, a widely verified general prediction of the evolutionary theory of senescence.

This volume offers one of the first opportunities to test Charlesworth's proposal by confronting predictions with data. The contribution by Jean-Michel Gaillard and coauthors reports estimates from a dozen wild populations of deer, sheep, moose, and other large herbivorous mammals. Gompertz hazard models have been fitted to data on lengths of life collected from long-term monitoring of individually marked animals. Senescent mortality as measured by the Gompertz slope parameter is substantial. After the onset of adulthood around age 2, the yearly percentage rise in female hazard curves ranges from 11 percent to 40 percent. Initial mortality is low enough to allow many individuals to survive to ages when senescent mortality is dominant. In Gompertz models, the hazard rate at the modal age of death equals the Gompertz slope parameter. In all but one of these populations, at the modal age of death, the background level of mortality comes out to be less than half the overall level of mortality, and in four cases less than one part in six.

The direct evidence reported by Gaillard and his coauthors showing a pervasive role for senescent mortality in the wild among large herbivorous mammals is consistent with circumstantial evidence for birds reviewed by Robert Ricklefs and Alex Scheuerlein (p. 78) in their contribution to this volume. It also accords with the discovery by Bronikowski et al. (2001) of substantial and comparable values of the Gompertz slope parameter for wild (Kenyan) and captive (US) populations of the same species of baboon, set in context by Altmann and Alberts (2003).

The most appealing feature of Charlesworth's proposal, the explanation of the exponential functional form, rests on the claim that mortality in the wild is dominated by extrinsic, age-independent background mortality. This feature is in conflict with these data. The theory further predicts that the background level of extrinsic mortality in the wild over evolutionary time can be read off from the present value of the Gompertz slope parameter in captive or protected populations. For humans one would be talking about values of e_1 on the order of a dozen years. Such a population would immediately go extinct. For the mammals described by Gaillard and his coauthors, in most cases the predicted levels of background mortality would not only be well above the observed levels of background mortality, but above the observed levels of overall mortality under present-day conditions in the wild.

It may be that "the wild" is not what it used to be. Under Charles-worth's proposal, the time periods for which the background level of mortality is relevant are constrained by the appeal to mutation-selection balance. The genetic effects responsible for the Gompertzian shape have to have been continually and (as evolutionary time goes) relatively recently renewed. But this constraint is not a tight one. The analysis so far is restricted to a handful of vertebrate species, and sample sizes are not very large. Although practical difficulties stand in the way of measuring lifetimes for insects or other invertebrates in nature, a broader collection of examples would be invaluable. On the whole, however, the early evidence is not very encouraging for this theoretically appealing version of mutation accumulation theory.

Social support

In his contribution to this volume, Carey (p. 5) lists two clusters of factors favoring extended longevity of species. Resource-based factors are scarcity and uncertainty. Kin-based factors are parental care and sociality. For the evolution of hominid life spans out of primate life spans, social support and exchange must have played large roles. Informally, it has always been recognized that survival into age groups that contribute to the provisioning and nurturing as well as the procreating of the next generation will be favored by natural selection. Recent formal models seek to quantify such components of selective pressure.

In their contribution Hillard Kaplan and coauthors present a model for evolutionary tradeoffs in social species developed by Kaplan and Robson. Mathematical proofs are given in Kaplan and Robson (2002). In this model, transfers of resources occur between age groups within a population. In unpublished work, Ronald Lee has further developed related ideas. In Kaplan and Robson's model, the growth rate for a stable population is chosen, via a support equation, by requiring that the difference between age-specific production and age-specific consumption averaged over all the age groups in the stable population must equal zero at all times. Fertility rates are assumed to adjust in some way to make Lotka's Equation hold. The demands of dependent age groups are covered by excess net production from any other age group. These demands comprise the physiological investments made in creating progeny as well as rearing them, including a cost reckoned at age zero of investment in a brain of a given size.

In this model, the age-specific production function and the initial costs of progeny depend explicitly on a parameter K, and the age-specific survival function depends implicitly on this parameter. Kaplan and coauthors interpret K as an index of investment in embodied capital, including the size of the brain, that governs long-range returns on skills through the balance of production and consumption. Some parametric assumptions and an optimization argument enable them to derive a U-shaped profile for the hazard curve from infancy to old age and to prove that growth is maximized at an optimal intermediate value of K associated with lower than maximal fertility.

Kaplan and Robson's approach has appealing features. Their support equation puts calculations of selective advantage into a context of constrained growth. It sets up a pattern of tradeoffs between juvenile and senior survival that generates relationships between the shapes of hazard functions at both ends, suggesting evolutionary interconnections between them.

Kaplan and Robson primarily seem to envision mutations going to fixation very gradually, fixing K at its optimum and fixing the associated shape of the survival curve and the intrinsic rate of growth. Insofar as K is taken to govern secular changes in brain size and somatic organization, this interpretation is natural. However, in their formulation, K is a free parameter that can be adjusting all sorts of strategic allocations. The tuning of K could easily be partly behavioral, operating rapidly and enabling populations of primates or hominids to adjust their somatic investments and resulting age schedules of survival to prevailing resource constraints. Natural selection, for its part, could be gradually transforming the shape of the dependence on K through mutations that affect, for instance, the costliness of lowered death rates. Response profiles rather than just the optimal values would be evolving. With this interpretation, calculation of selective advantage within the model gives a rationale for certain kinds of pleiotropy.

The Kaplan–Robson framework, through the support equation, is tied to stable populations, and this feature places limits on the range of evolutionary contexts to which it applies. The framework does not help to account for the shapes of hazard curves at extreme ages. Significant provisioning does not go on indefinitely. In its broad outlines, however, the approach does lend support to an optimistic view about the biological potential for future progress against old-age mortality. If evolution equipped social species like our own with the ability to shift across a continuum of survival schedules in response to nutritional and environmental conditions, then gains along a few dimensions could imply mortality reductions over many ages. The picture of endless looming battles tends to fade away. Accounts that emphasize the coevolution of earlier-age and late-age mortality increase the hope that the successes already achieved at earlier ages may be repeatable at later ages too.

Genes for instance

Along with fresh thinking about the mechanisms that must shape the statistical regularities of hazard functions, the bounty of experimental knowledge about the genomes of model organisms with respect to longevity has been growing. In this volume Lawrence Harshman gives a comprehensive guide to findings for *Drosophila melanogaster*, treating developments since Finch's (1990) authoritative survey of the broader field.

The first genes associated with longevity that are coming into view are a little like the first galaxies glimpsed in small telescopes. They serve as instances. There are now instances of genes with effects of certain shapes, as regards hazard functions, just as once there were instances of barred spirals or giant elliptical galaxies. Establishing a morphology of effects will take time, and the early discoveries may be atypical, easier to spot than cases that may prove to predominate. It is too early to try to constrain models for hazard curves on the basis of identified mutations, selection studies, and analyses of quantitative trait loci for life span. However, the roster of instances is likely to expand.

Recently discovered induced mutations in genes in *Drosophila*, discussed in detail by Harshman in this volume (pp. 116–118), make an intriguing contrast. The mutations can increase flies' life spans substantially. The *Indy* gene (Rogina et al. 2000) is involved in metabolic processes, and Harshman sums up by saying that for the *Indy* gene "the mechanism of life span extension could be caloric restriction." On the other hand, the *InR* gene (Tatar et al. 2001) is involved in insulin reception and synthesis of juvenile hormones, and the mutations at issue, while reducing body size and making female flies sterile, produce no measured change in metabolic rate. Enhanced resistance to various kinds of stress is probably part of their advantage.

It seems harder to draw analogies with humans for genes like *Indy* than for genes like *InR*. Whatever the benefits of caloric restriction may be for some people today, over the course of human history large increases in caloric intake have been closely coupled with gains in longevity, as Robert Fogel (e.g., Fogel and Costa 1997) has shown in rich detail. Human bodies appear to have been preadapted to make ready use of diets much richer than could have prevailed over evolutionary time. In itself, that is something of a puzzle. The puzzle would be greater if inbuilt genetic tuning mechanisms were for turning down longevity in conjunction with turning up caloric consumption. On the other hand, it is reasonable to expect genetic tuning mechanisms that turn up longevity in conjunction with turning up capacities for resisting stress and privation and outlasting environmental fluctuations. In a context of homeostatic population regulation, partial reductions in fecundity might have had little cost and even some benefit.

Many questions remain open pending further data. As Harshman writes (p. 106), some experimenters find selection primarily affecting the Gompertz level parameter, others the Gompertz slope parameter. The early quantitative trait loci for longevity along with the selection experiments suggest that there is wide variability keyed to environmental conditions.

An instance of genetic effects particularly pertinent to demography comes from the experiments of Sgro and Partridge (1999), discussed here by Harshman (pp. 107–108), with lines of *Drosophila* artificially selected for longevity. For females, the hazard curve for the selected lines shows both a Gompertzian exponential segment and a tapering late-age segment shifted to the right toward later ages as compared to the hazard curve for a control population. The selected lines also show lower early-age reproduction. When reproduction is suppressed in both selected and control lines, by each of two methods, the shift vanishes. Sgro and Partridge see a delayed direct cost of reproduction likely reflected in the hazard curves. As discussed by Wachter (1999), such push-back mechanisms are the sorts of pleiotropic effects needed in mathematical models for generating hazard curves consistent with observations.

A particular gene in a particular model organism may tell us something general about feasible genetic effects on hazard functions, or it may tell us something specific about the adaptations of that organism. Experimental results have to take their place within our broad knowledge of natural history. In their contribution to this volume, Ricklefs and Scheuerlein (pp. 87–88) give an analysis of variance, partialling out variability in life span by taxonomic levels. The largest entries in the analysis of variance for fitted hazard curves are found at the level of genera, rather than at the broader level of orders and families. The same body plan can accommodate large ranges of variation, suggesting that it is relatively easy for evolutionary change to move along a continuum between shorter and longer lives. That would suggest genetic organization with some high-level control and synchronization of effects on hazard curves.

It is important, as Kaplan has tended to emphasize, not to conflate such genetic capacity for moving along a continuum with abilities of members of a species to adapt less or more flexibly to a range of environments. The mathematical models with which researchers are working probably put excessive emphasis on a picture of hazard curves as outputs of genetic programs, rather than on a picture of individual members of species forging their own hazards through their lives in the rough and tumble of environmental challenges. Although distinct phenomena are involved, both genetic plasticity and behavioral flexibility support prospects for coordinated progress.

Onward

Many of the questions addressed in this volume remain open, and research is in progress to settle them. The next few years should bring more clarity. We shall be learning whether human hazard functions do drop at extreme ages. We shall be building consensus about the plausible efficacy of simple culling and the extent of tapering it may explain. We shall be sorting out the generic implications of reliability theory from the special features of specific models. We shall be seeing how far one can take ideas from mutation accumulation theory and mutation–selection balance while remaining faithful to empirical evidence. We shall be coming to grips with the potential and the limitations of formal models for social support. We shall be giving fuller attention to the evolutionary implications of varying environments, population extinction, and adaptive lifecourse learning. We shall be finding out how to design viable mathematical models for the shaping of hazard functions by processes of antagonistic and reinforcing pleiotropy.

One major need is for gathering more data for more species on aging in the wild. The patient investments in long-term monitoring of populations of primates and some other large mammals are paying off handsomely for aging studies. Stylized "facts" that have long channeled thinking are turning out to be non-facts. But our information is very limited in terms of the range of organisms observed and measured under natural conditions. Such data are very difficult to obtain, but they are becoming indispensable to the interpretation of laboratory experiments and the articulation of evolutionary theory.

Many points of contact exist between the biodemography of longevity and the biodemography of fertility. New developments in that field are summarized in *Offspring: Human Fertility Behavior in Biodemographic Perspective*, edited by Wachter and Bulatao (2003). For humans, extended longevity opens up alternatives for the restructuring of traditional lifecourse stages, as discussed in the contribution to this volume by Ronald Lee and Joshua Goldstein, blending the concerns of biodemography with economic and social demography. Cooperation between specialties should facilitate a fuller integration of lifecourse perspectives.

We may expect the next few years to bring a large collection of genetic studies with model organisms and with humans, including systematic studies of age profiles of gene expression. The picture is likely to be highly complex, and it may be quite a while before generalizations emerge. Among humans, one promising strategy is to concentrate on exceptional longevity. As mortality progressively takes its toll on each human cohort and leaves a rarer and rarer subset of survivors, it may distill away some of the complexity. It may concentrate alleles of certain genes to the point where their roles are easier to identify. It may accentuate dimensions of heterogeneity that can be subject to measurement and observation. The more we learn about the upper tails of the lifetime distribution, the better we can judge whether the oldest-old among us today are truly harbingers of widespread future gains.

Many disciplines and many strands of research come together in the biodemography of longevity. They are well represented in this volume. Broad regularities seen in patterns of human mortality over age and time plausibly reflect structures of biological potential honed by evolution and shared with other species. Many of the new developments in data and theory examined in this chapter appear to have optimistic overtones with respect to the permissiveness of our biological heritage. A wide range of ideas and possibilities, however, remains in play. The work reviewed here suggests but does not yet establish connections between the shapes of hazard curves and the prospects for continued progress, leaving us, as scientists and mortals, in suspense.

References

Altmann, Jeanne, and Susan Alberts. 2003. "Fertility, offspring care, and ontogeny in primates," in K. W. Wachter and R. Bulatao (eds.).

Anderson, James J. 2000. "A vitality-based model relating stressors and environmental properties to organism survival," *Ecological Monographs* 70: 445–470.

Bongaarts, John, and Rodolfo Bulatao. 2000. *Beyond Six Billion: Forecasting the World's Population*. Washington, DC: National Academy Press.

Bronikowski, Ann M., S. C. Alberts, J. Altmann, C. Packer, K. D. Carey, and M. Tatar. 2002. "The aging baboon: Comparative demography in a nonhuman primate," *Proceedings of the National Academy of Sciences* 99: 9591–9595.

Carey, James R., P. Liedo, D. Orozco, and J. W. Vaupel. 1992. "Slowing of mortality rates at older ages in large medfly cohorts," *Science* 258: 457–461.

Charlesworth, Brian. 1994. *Evolution in Age-Structured Populations*. Second Edition. Cambridge: Cambridge University Press.

———. 2001. "Patterns of age-specific means and genetic variances of mortality rates predicted by the mutation-accumulating theory of ageing," *Journal of Theoretical Biology* 210: 47–65.

Curtsinger, James W., H. Fukui, D. Townsend, and J. W. Vaupel. 1992. "Demography of genotypes: Failure of the limited life-span paradigm in *Drosophila melanogaster*," *Science* 258: 461–463.

Ewbank, Douglas and N. Jones. 2001. "Preliminary observations on the variations in the risk of death," paper presented at the Annual Meeting of the Population Association of America, Washington, DC.

Finch, Caleb. 1990. *Longevity, Senescence, and the Genome*. Chicago: University of Chicago Press.

Fogel, Robert and Dora L. Costa. 1997. "A theory of technophysio-evolution, with some implications for forecasting population, health care costs, and pension costs," *Demography* 34: 49–66.

Gavrilov, Leonid and Natalia Gavrilova. 2001. "The reliability theory of aging and longevity," *Journal of Theoretical Biology* 213: 527–545.

Gompertz, Benjamin. 1825. "On the nature of the function expressive of the law of human mortality and on a new mode of determining the value of life contingencies," *Philosophical Transactions of the Royal Society* 115: 513–585.

Kaku, Michio. 1997. *Visions*. New York: Anchor Books.

Kaplan, Hillard and Arthur Robson. 2002. "The emergence of humans: The coevolution of intelligence and longevity with intergenerational transfers," *Proceedings of the National Academy of Sciences, USA* 99: 10221–10226.

Lee, Ronald D. and Larry Carter. 1992. "Modeling and forecasting the time series of U.S. mortality," *Journal of the American Statistical Association* 87: 659–671.

Mueller, Laurence and Michael Rose. 1996. "Evolutionary theory predicts late-life mortality plateaus," *Proceedings of the National Academy of Sciences, USA* 93: 15249–15253.

Mueller, Laurence, Mark Drapeau, Curtis Adams, Christopher Hammerle, Kristy Doyal, Ali

Jazayeri, Tuan Ly, Suhail Beuwala, Avi Mamimdi, and Michael R. Rose. 2003. "Statistical tests of demographic heterogeneity theories," *Experimental Gerontology*, in press.

Oeppen, James and James W. Vaupel. 2002. "Broken limits to life expectancy," *Science* 296: 1029–1031.

Pletcher, Scott and James Curtsinger. 1998. "Mortality plateaus and the evolution of senescence: Why are old-age mortality rates so low?" *Evolution* 52: 454–464.

Robine, Jean-Marie and James W. Vaupel. 2001. "Supercentenarians, slower ageing individuals or senile elderly?" *Experimental Gerontology* 36: 915–930.

Rogina, Blanka, R. Reenan, S. Nilsen, and S. Helfand. 2000. "Extended life-span conferred by cotransporter gene mutations in *Drosophila*," *Science* 290: 2137–2140.

Rose, Michael R. 1991. *The Evolutionary Biology of Aging*. Oxford: Oxford University Press.

Rose, Michael R., Mark Drappeau, Puya Yazdi, Kandarp Shah, Diana Moise, Rena Thakar, Casandra Rauser, and Laurence D. Mueller. 2002. "Evolution of late-life mortality in *Drosophila melanogaster*," *Evolution* 56: 1982–1991.

Sgro, Carla, and Linda Partridge. 1999. "A delayed wave of death from reproduction in Drosophila," *Science* 286: 2521–2524.

Steinsaltz, David and Steve Evans. 2003. "Markov mortality models: Implications of quasistationarity and varying initial distributions," Working Paper, University of California, Berkeley «www.demog.berkeley.edu/~dstein/agingpage.html»

Strehler, B. and A. Mildvan. 1960. "General theory of mortality and aging," *Science* 132: 14–21.

Tatar, Marc, A. Kopelman, D. Epstein, M.-P. Tu, C.-M. Yin, and R. S. Garofalo. 2001. "Mutant *Drosophila* insulin receptor homolog that extends life-span and impairs neuroendocrine function," *Science* 292: 107–110.

Tuljapurkar, Shripad, Nan Li, and Carl Boe. 2000. "A universal pattern of mortality decline in the G-7 countries," *Nature* 405: 789–792.

Vaupel, James W., J. Carey, K. Christensen, T. Johnson, A. Yashin, N. Holm, I. Iachine, A. Khazaeli, P. Liedo, V. Longo, Zeng Yi, K. Manton, and J. Curtsinger. 1998. "Biodemographic trajectories of longevity," *Science* 280: 855–860.

Vaupel, James W., Kenneth Manton and Eric Stallard. 1979. "The impact of heterogeneity in individual frailty on the dynamics of mortality," *Demography* 16: 439–454.

Vaupel, James W. and Anatoli Yashin. 1985. "The deviant dynamics of death in heterogeneous populations," *Sociological Methodology* 15: 179–211.

Wachter, Kenneth W. 1999. "Evolutionary demographic models for mortality plateaus," *Proceedings of the National Academy of Sciences, USA* 96: 10544–10547.

Wachter, Kenneth W. and R. Bulatao (eds.) 2003. *Offspring: Human Fertility Behavior in Biodemographic Perspective*. Washington, DC: National Academy Press.

Wachter, Kenneth W. and Caleb Finch (eds.). 1997. *Between Zeus and the Salmon: The Biodemography of Longevity*. Washington, DC: National Academy Press.

Weitz, Joshua and H. Fraser. 2001. "Explaining mortality rate plateaus," *Proceedings of the National Academy of Sciences, USA* 98: 15383–15386.

AUTHORS

JAMES R. CAREY is Professor, Department of Entomology, University of California at Davis; and Senior Scholar, Center for the Economics and Demography of Aging, University of California, Berkeley.

MARCO FESTA-BIANCHET is Professor, Department of Biology, University of Sherbrooke, Sherbrooke, Québec.

JEAN-MICHEL GAILLARD is Directeur de Recherche, Unité Mixte de Recherche 5558, "Biométrie et Biologie Evolutive," Université Claude Bernard, Lyon1, Villeurbanne, France.

JOSHUA R. GOLDSTEIN is Associate Professor of Sociology and Public Affairs, and Faculty Associate, Office of Population Research, Princeton University.

LAWRENCE G. HARSHMAN is Associate Professor, School of Biological Sciences, University of Nebraska–Lincoln.

SHIRO HORIUCHI is Associate Professor, Rockefeller University, New York.

HILLARD KAPLAN is Professor of Anthropology, Department of Anthropology, University of New Mexico, Albuquerque.

JANE LANCASTER is Professor of Anthropology, Department of Anthropology, University of New Mexico, Albuquerque.

RONALD LEE is Professor of Demography and Economics, University of California, Berkeley.

ANNE LOISON is Chargé de Recherche, Unité Mixte de Recherche 5558, "Biométrie et Biologie Evolutive," Université Claude Bernard, Lyon1, Villeurbanne, France.

MARC MANGEL is Professor, Department of Applied Mathematics and Statistics, Jack Baskin School of Engineering; and Fellow, Stevenson College, University of California, Santa Cruz.

STEVEN HECHT ORZACK is President, Fresh Pond Research Institute, Cambridge, MA.

ROBERT E. RICKLEFS is Curators' Professor of Biology, University of Missouri, St. Louis.

JEAN-MARIE ROBINE is Senior Research Fellow, INSERM, Montpellier, France.

ARTHUR ROBSON is Professor of Economics, Department of Economics, University of Western Ontario, London, Ontario.

YASUHIKO SAITO is Research Associate Professor, Center for Information Networking; and Research Fellow, Population Research Institute, Nihon University, Tokyo.

ALEX SCHEUERLEIN is postdoctoral fellow, Department of Biology, University of Missouri, St. Louis.

ERLING SOLBERG is Senior Scientist, Norwegian Institute for Nature Research, Trondheim, Norway.

SHRIPAD TULJAPURKAR is Morrison Professor of Population Studies and Professor of Biological Sciences, Stanford University.

JAMES W. VAUPEL is Founding Director, Max Planck Institute for Demographic Research, Rostock, Germany.

KENNETH W. WACHTER is Professor of Demography and Statistics, and Chair, Department of Demography, University of California, Berkeley.

JOHN R. WILMOTH is Associate Professor, Department of Demography, University of California, Berkeley.

NIGEL GILLES YOCCOZ is Senior Scientist, Department of Arctic Ecology, Norwegian Institute for Nature Research, Polar Environmental Centre, Troms, Norway.